COMPLEX ANALYSIS

Princeton Lectures in Analysis

I Fourier Analysis: An Introduction

II Complex Analysis

III Real Analysis: Measure Theory,
Integration, and Hilbert Spaces

PRINCETON LECTURES IN ANALYSIS

II

COMPLEX ANALYSIS

Elias M. Stein

&

Rami Shakarchi

PRINCETON UNIVERSITY PRESS
PRINCETON AND OXFORD

Copyright © 2003 by Princeton University Press
Published by Princeton University Press, 41 William Street,
Princeton, New Jersey 08540
In the United Kingdom: Princeton University Press,
3 Market Place, Woodstock, Oxfordshire OX20 1SY

ISBN 0-691-11385-8
Library of Congress Cataloging-in-Publication data has been applied for

British Library Cataloging-in-Publication Data is available

The publisher would like to acknowledge the authors of this volume for
providing the camera-ready copy from which this book was printed

Printed on acid-free paper. ∞
www.pupress.princeton.edu
Printed in the United States of America

19   20   18

ISBN-13: 978-0-691-11385-2

ISBN-10: 0-691-11385-8

To my grandchildren
Carolyn, Alison, Jason

E.M.S.

To my parents
Mohamed & Mireille
and my brother
Karim

R.S.

Foreword

Beginning in the spring of 2000, a series of four one-semester courses were taught at Princeton University whose purpose was to present, in an integrated manner, the core areas of analysis. The objective was to make plain the organic unity that exists between the various parts of the subject, and to illustrate the wide applicability of ideas of analysis to other fields of mathematics and science. The present series of books is an elaboration of the lectures that were given.

While there are a number of excellent texts dealing with individual parts of what we cover, our exposition aims at a different goal: presenting the various sub-areas of analysis not as separate disciplines, but rather as highly interconnected. It is our view that seeing these relations and their resulting synergies will motivate the reader to attain a better understanding of the subject as a whole. With this outcome in mind, we have concentrated on the main ideas and theorems that have shaped the field (sometimes sacrificing a more systematic approach), and we have been sensitive to the historical order in which the logic of the subject developed.

We have organized our exposition into four volumes, each reflecting the material covered in a semester. Their contents may be broadly summarized as follows:

   I. Fourier series and integrals.

  II. Complex analysis.

 III. Measure theory, Lebesgue integration, and Hilbert spaces.

 IV. A selection of further topics, including functional analysis, distributions, and elements of probability theory.

However, this listing does not by itself give a complete picture of the many interconnections that are presented, nor of the applications to other branches that are highlighted. To give a few examples: the elements of (finite) Fourier series studied in Book I, which lead to Dirichlet characters, and from there to the infinitude of primes in an arithmetic progression; the X-ray and Radon transforms, which arise in a number of problems in Book I, and reappear in Book III to play an important role in understanding Besicovitch-like sets in two and three dimensions; Fatou's theorem, which guarantees the existence of boundary values of bounded holomorphic functions in the disc, and whose proof relies on ideas developed in each of the first three books; and the theta function, which first occurs in Book I in the solution of the heat equation, and is then used in Book II to find the number of ways an integer can be represented as the sum of two or four squares, and in the analytic continuation of the zeta function.

A few further words about the books and the courses on which they were based. These courses where given at a rather intensive pace, with 48 lecture-hours a semester. The weekly problem sets played an indispensable part, and as a result exercises and problems have a similarly important role in our books. Each chapter has a series of "Exercises" that are tied directly to the text, and while some are easy, others may require more effort. However, the substantial number of hints that are given should enable the reader to attack most exercises. There are also more involved and challenging "Problems"; the ones that are most difficult, or go beyond the scope of the text, are marked with an asterisk.

Despite the substantial connections that exist between the different volumes, enough overlapping material has been provided so that each of the first three books requires only minimal prerequisites: acquaintance with elementary topics in analysis such as limits, series, differentiable functions, and Riemann integration, together with some exposure to linear algebra. This makes these books accessible to students interested in such diverse disciplines as mathematics, physics, engineering, and finance, at both the undergraduate and graduate level.

It is with great pleasure that we express our appreciation to all who have aided in this enterprise. We are particularly grateful to the students who participated in the four courses. Their continuing interest, enthusiasm, and dedication provided the encouragement that made this project possible. We also wish to thank Adrian Banner and Jose Luis Rodrigo for their special help in running the courses, and their efforts to see that the students got the most from each class. In addition, Adrian Banner also made valuable suggestions that are incorporated in the text.

We wish also to record a note of special thanks for the following individuals: Charles Fefferman, who taught the first week, (successfully launching the whole project!); Paul Hagelstein, who in addition to reading part of the manuscript taught several weeks of one of the courses, and has since taken over the teaching of the second round of the series; Daniel Levine who gave valuable help in proof-reading. Last but not least, our thanks go to Gerree Pecht, for her consummate skill in typesetting and for the time and energy she spent in the preparation of all aspects of the lectures, such as transparencies, notes, and the manuscript.

We are also happy to acknowledge our indebtedness for the support we received from the 250th Anniversary Fund of Princeton University, and the National Science Foundation's VIGRE program.

Elias M. Stein
Rami Shakarchi

Princeton, New Jersey
August 2002

Contents

Introduction

... In effect, if one extends these functions by allowing complex values for the arguments, then there arises a harmony and regularity which without it would remain hidden.

B. Riemann, 1851

When we begin the study of complex analysis we enter a marvelous world, full of wonderful insights. We are tempted to use the adjectives magical, or even miraculous when describing the first theorems we learn; and in pursuing the subject, we continue to be astonished by the elegance and sweep of the results.

The starting point of our study is the idea of extending a function initially given for real values of the argument to one that is defined when the argument is complex. Thus, here the central objects are functions from the complex plane to itself

$$f : \mathbb{C} \to \mathbb{C},$$

or more generally, complex-valued functions defined on open subsets of $\mathbb{C}$. At first, one might object that nothing new is gained from this extension, since any complex number z can be written as $z = x + iy$ where $x, y \in \mathbb{R}$ and z is identified with the point (x, y) in $\mathbb{R}^2$.

However, everything changes drastically if we make a natural, but misleadingly simple-looking assumption on f: that it is differentiable in the complex sense. This condition is called **holomorphicity**, and it shapes most of the theory discussed in this book.

A function $f : \mathbb{C} \to \mathbb{C}$ is holomorphic at the point $z \in \mathbb{C}$ if the limit

$$\lim_{h \to 0} \frac{f(z + h) - f(z)}{h} \quad (h \in \mathbb{C})$$

exists. This is similar to the definition of differentiability in the case of a real argument, except that we allow h to take *complex* values. The reason this assumption is so far-reaching is that, in fact, it encompasses a multiplicity of conditions: so to speak, one for each angle that h can approach zero.

Although one might now be tempted to prove theorems about holo-morphic functions in terms of real variables, the reader will soon discover that complex analysis is a new subject, one which supplies proofs to the theorems that are proper to its own nature. In fact, the proofs of the main properties of holomorphic functions which we discuss in the next chapters are generally very short and quite illuminating.

The study of complex analysis proceeds along two paths that often intersect. In following the first way, we seek to understand the univer-sal characteristics of holomorphic functions, without special regard for specific examples. The second approach is the analysis of some partic-ular functions that have proved to be of great interest in other areas of mathematics. Of course, we cannot go too far along either path without having traveled some way along the other. We shall start our study with some general characteristic properties of holomorphic functions, which are subsumed by three rather miraculous facts:

1. CONTOUR INTEGRATION: If f is holomorphic in Ω, then for appro-priate closed paths in Ω

$$\int_\gamma f(z)dz = 0.$$

2. REGULARITY: If f is holomorphic, then f is indefinitely differen-tiable.

3. ANALYTIC CONTINUATION: If f and g are holomorphic functions in Ω which are equal in an arbitrarily small disc in Ω, then $f = g$ everywhere in Ω.

These three phenomena and other general properties of holomorphic functions are treated in the beginning chapters of this book. Instead of trying to summarize the contents of the rest of this volume, we men-tion briefly several other highlights of the subject.

- The zeta function, which is expressed as an infinite series

$$\zeta(s) = \sum_{n=1}^{\infty} \frac{1}{n^s},$$

and is initially defined and holomorphic in the half-plane $\mathrm{Re}(s) > 1$, where the convergence of the sum is guaranteed. This function and its variants (the L-series) are central in the theory of prime numbers, and have already appeared in Chapter 8 of Book I, where we proved Dirichlet's theorem. Here we will prove that ζ extends to a meromorphic function with a pole at $s = 1$. We shall see that the behavior of $\zeta(s)$ for $\text{Re}(s) = 1$ (and in particular that ζ does not vanish on that line) leads to a proof of the prime number theorem.

- The theta function

$$\Theta(z|\tau) = \sum_{n=-\infty}^{\infty} e^{\pi i n^2 \tau} e^{2\pi i n z},$$

which in fact is a function of the two complex variables z and τ, holomorphic for all z, but only for τ in the half-plane $\text{Im}(\tau) > 0$. On the one hand, when we fix τ, and think of Θ as a function of z, it is closely related to the theory of elliptic (doubly-periodic) functions. On the other hand, when z is fixed, Θ displays features of a modular function in the upper half-plane. The function $\Theta(z|\tau)$ arose in Book I as a fundamental solution of the heat equation on the circle. It will be used again in the study of the zeta function, as well as in the proof of certain results in combinatorics and number theory given in Chapters 6 and 10.

Two additional noteworthy topics that we treat are: the Fourier transform with its elegant connection to complex analysis via contour integration, and the resulting applications of the Poisson summation formula; also conformal mappings, with the mappings of polygons whose inverses are realized by the Schwarz-Christoffel formula, and the particular case of the rectangle, which leads to elliptic integrals and elliptic functions.

1 Preliminaries to Complex Analysis

> The sweeping development of mathematics during the last two centuries is due in large part to the introduction of complex numbers; paradoxically, this is based on the seemingly absurd notion that there are numbers whose squares are negative.
>
> *E. Borel, 1952*

This chapter is devoted to the exposition of basic preliminary material which we use extensively throughout of this book.

We begin with a quick review of the algebraic and analytic properties of complex numbers followed by some topological notions of sets in the complex plane. (See also the exercises at the end of Chapter 1 in Book I.)

Then, we define precisely the key notion of holomorphicity, which is the complex analytic version of differentiability. This allows us to discuss the Cauchy-Riemann equations, and power series.

Finally, we define the notion of a curve and the integral of a function along it. In particular, we shall prove an important result, which we state loosely as follows: if a function f has a primitive, in the sense that there exists a function F that is holomorphic and whose derivative is precisely f, then for any closed curve γ

$$\int_\gamma f(z)\,dz = 0.$$

This is the first step towards Cauchy's theorem, which plays a central role in complex function theory.

1 Complex numbers and the complex plane

Many of the facts covered in this section were already used in Book I.

1.1 Basic properties

A complex number takes the form $z = x + iy$ where x and y are real, and i is an imaginary number that satisfies $i^2 = -1$. We call x and y the

real part and the **imaginary part** of z, respectively, and we write

$$x = \text{Re}(z) \quad \text{and} \quad y = \text{Im}(z).$$

The real numbers are precisely those complex numbers with zero imaginary parts. A complex number with zero real part is said to be **purely imaginary**.

Throughout our presentation, the set of all complex numbers is denoted by $\mathbb{C}$. The complex numbers can be visualized as the usual Euclidean plane by the following simple identification: the complex number $z = x + iy \in \mathbb{C}$ is identified with the point $(x, y) \in \mathbb{R}^2$. For example, 0 corresponds to the origin and i corresponds to $(0, 1)$. Naturally, the x and y axis of $\mathbb{R}^2$ are called the **real axis** and **imaginary axis**, because they correspond to the real and purely imaginary numbers, respectively. (See Figure 1.)

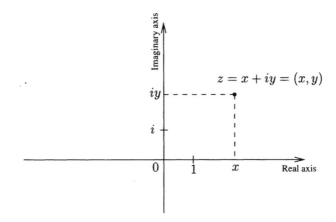

Figure 1. The complex plane

The natural rules for adding and multiplying complex numbers can be obtained simply by treating all numbers as if they were real, and keeping in mind that $i^2 = -1$. If $z_1 = x_1 + iy_1$ and $z_2 = x_2 + iy_2$, then

$$z_1 + z_2 = (x_1 + x_2) + i(y_1 + y_2),$$

and also

$$z_1 z_2 = (x_1 + iy_1)(x_2 + iy_2)$$
$$= x_1 x_2 + ix_1 y_2 + iy_1 x_2 + i^2 y_1 y_2$$
$$= (x_1 x_2 - y_1 y_2) + i(x_1 y_2 + y_1 x_2).$$

If we take the two expressions above as the definitions of addition and multiplication, it is a simple matter to verify the following desirable properties:

- Commutativity: $z_1 + z_2 = z_2 + z_1$ and $z_1 z_2 = z_2 z_1$ for all $z_1, z_2 \in \mathbb{C}$.

- Associativity: $(z_1 + z_2) + z_3 = z_1 + (z_2 + z_3)$; and $(z_1 z_2) z_3 = z_1(z_2 z_3)$ for $z_1, z_2, z_3 \in \mathbb{C}$.

- Distributivity: $z_1(z_2 + z_3) = z_1 z_2 + z_1 z_3$ for all $z_1, z_2, z_3 \in \mathbb{C}$.

Of course, addition of complex numbers corresponds to addition of the corresponding vectors in the plane $\mathbb{R}^2$. Multiplication, however, consists of a rotation composed with a dilation, a fact that will become transparent once we have introduced the polar form of a complex number. At present we observe that multiplication by i corresponds to a rotation by an angle of $\pi/2$.

The notion of length, or absolute value of a complex number is identical to the notion of Euclidean length in $\mathbb{R}^2$. More precisely, we define the **absolute value** of a complex number $z = x + iy$ by

$$|z| = (x^2 + y^2)^{1/2},$$

so that $|z|$ is precisely the distance from the origin to the point (x, y). In particular, the triangle inequality holds:

$$|z + w| \leq |z| + |w| \quad \text{for all } z, w \in \mathbb{C}.$$

We record here other useful inequalities. For all $z \in \mathbb{C}$ we have both $|\text{Re}(z)| \leq |z|$ and $|\text{Im}(z)| \leq |z|$, and for all $z, w \in \mathbb{C}$

$$||z| - |w|| \leq |z - w|.$$

This follows from the triangle inequality since

$$|z| \leq |z - w| + |w| \quad \text{and} \quad |w| \leq |z - w| + |z|.$$

The **complex conjugate** of $z = x + iy$ is defined by

$$\bar{z} = x - iy,$$

and it is obtained by a reflection across the real axis in the plane. In fact a complex number z is real if and only if $z = \bar{z}$, and it is purely imaginary if and only if $z = -\bar{z}$.

The reader should have no difficulty checking that

$$\mathrm{Re}(z) = \frac{z + \bar{z}}{2} \quad \text{and} \quad \mathrm{Im}(z) = \frac{z - \bar{z}}{2i}.$$

Also, one has

$$|z|^2 = z\bar{z} \quad \text{and as a consequence} \quad \frac{1}{z} = \frac{\bar{z}}{|z|^2} \quad \text{whenever } z \neq 0.$$

Any non-zero complex number z can be written in **polar form**

$$z = re^{i\theta},$$

where $r > 0$; also $\theta \in \mathbb{R}$ is called the **argument** of z (defined uniquely up to a multiple of 2π) and is often denoted by $\arg z$, and

$$e^{i\theta} = \cos\theta + i\sin\theta.$$

Since $|e^{i\theta}| = 1$ we observe that $r = |z|$, and θ is simply the angle (with positive counterclockwise orientation) between the positive real axis and the half-line starting at the origin and passing through z. (See Figure 2.)

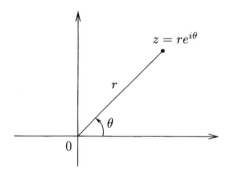

Figure 2. The polar form of a complex number

Finally, note that if $z = re^{i\theta}$ and $w = se^{i\varphi}$, then

$$zw = rse^{i(\theta + \varphi)},$$

so multiplication by a complex number corresponds to a homothety in $\mathbb{R}^2$ (that is, a rotation composed with a dilation).

1.2 Convergence

We make a transition from the arithmetic and geometric properties of complex numbers described above to the key notions of convergence and limits.

A sequence $\{z_1, z_2, \ldots\}$ of complex numbers is said to **converge** to $w \in \mathbb{C}$ if

$$\lim_{n \to \infty} |z_n - w| = 0, \quad \text{and we write} \quad w = \lim_{n \to \infty} z_n.$$

This notion of convergence is not new. Indeed, since absolute values in $\mathbb{C}$ and Euclidean distances in $\mathbb{R}^2$ coincide, we see that z_n converges to w if and only if the corresponding sequence of points in the complex plane converges to the point that corresponds to w.

As an exercise, the reader can check that the sequence $\{z_n\}$ converges to w if and only if the sequence of real and imaginary parts of z_n converge to the real and imaginary parts of w, respectively.

Since it is sometimes not possible to readily identify the limit of a sequence (for example, $\lim_{N \to \infty} \sum_{n=1}^{N} 1/n^3$), it is convenient to have a condition on the sequence itself which is equivalent to its convergence. A sequence $\{z_n\}$ is said to be a **Cauchy sequence** (or simply **Cauchy**) if

$$|z_n - z_m| \to 0 \quad \text{as } n, m \to \infty.$$

In other words, given $\epsilon > 0$ there exists an integer $N > 0$ so that $|z_n - z_m| < \epsilon$ whenever $n, m > N$. An important fact of real analysis is that $\mathbb{R}$ is complete: every Cauchy sequence of real numbers converges to a real number.[1] Since the sequence $\{z_n\}$ is Cauchy if and only if the sequences of real and imaginary parts of z_n are, we conclude that every Cauchy sequence in $\mathbb{C}$ has a limit in $\mathbb{C}$. We have thus the following result.

Theorem 1.1 $\mathbb{C}$, *the complex numbers, is complete.*

We now turn our attention to some simple topological considerations that are necessary in our study of functions. Here again, the reader will note that no new notions are introduced, but rather previous notions are now presented in terms of a new vocabulary.

1.3 Sets in the complex plane

If $z_0 \in \mathbb{C}$ and $r > 0$, we define the **open disc** $D_r(z_0)$ **of radius r centered at** z_0 to be the set of all complex numbers that are at absolute

---

[1] This is sometimes called the Cauchy convergence criterion and is equivalent with the Bolzano-Weierstrass theorem.

value strictly less than r from z_0. In other words,

$$D_r(z_0) = \{z \in \mathbb{C} : |z - z_0| < r\},$$

and this is precisely the usual disc in the plane of radius r centered at z_0. The **closed disc** $\overline{D}_r(z_0)$ **of radius** r **centered at** z_0 is defined by

$$\overline{D}_r(z_0) = \{z \in \mathbb{C} : |z - z_0| \leq r\},$$

and the boundary of either the open or closed disc is the circle

$$C_r(z_0) = \{z \in \mathbb{C} : |z - z_0| = r\}.$$

Since the **unit disc** (that is, the open disc centered at the origin and of radius 1) plays an important role in later chapters, we will often denote it by $\mathbb{D}$,

$$\mathbb{D} = \{z \in \mathbb{C} : |z| < 1\}.$$

Given a set $\Omega \subset \mathbb{C}$, a point z_0 is an **interior point** of Ω if there exists $r > 0$ such that

$$D_r(z_0) \subset \Omega.$$

The **interior** of Ω consists of all its interior points. Finally, a set Ω is **open** if every point in that set is an interior point of Ω. This definition coincides precisely with the definition of an open set in $\mathbb{R}^2$.

A set Ω is **closed** if its complement $\Omega^c = \mathbb{C} - \Omega$ is open. This property can be reformulated in terms of limit points. A point $z \in \mathbb{C}$ is said to be a **limit point** of the set Ω if there exists a sequence of points $z_n \in \Omega$ such that $z_n \neq z$ and $\lim_{n \to \infty} z_n = z$. The reader can now check that a set is closed if and only if it contains all its limit points. The **closure** of any set Ω is the union of Ω and its limit points, and is often denoted by $\overline{\Omega}$.

Finally, the **boundary** of a set Ω is equal to its closure minus its interior, and is often denoted by $\partial\Omega$.

A set Ω is **bounded** if there exists $M > 0$ such that $|z| < M$ whenever $z \in \Omega$. In other words, the set Ω is contained in some large disc. If Ω is bounded, we define its **diameter** by

$$\text{diam}(\Omega) = \sup_{z,w \in \Omega} |z - w|.$$

A set Ω is said to be **compact** if it is closed and bounded. Arguing as in the case of real variables, one can prove the following.

Theorem 1.2 *The set $\Omega \subset \mathbb{C}$ is compact if and only if every sequence $\{z_n\} \subset \Omega$ has a subsequence that converges to a point in Ω.*

An **open covering** of Ω is a family of open sets $\{U_\alpha\}$ (not necessarily countable) such that

$$\Omega \subset \bigcup_\alpha U_\alpha.$$

In analogy with the situation in $\mathbb{R}$, we have the following equivalent formulation of compactness.

Theorem 1.3 *A set Ω is compact if and only if every open covering of Ω has a finite subcovering.*

Another interesting property of compactness is that of **nested sets**. We shall in fact use this result at the very beginning of our study of complex function theory, more precisely in the proof of Goursat's theorem in Chapter 2.

Proposition 1.4 *If $\Omega_1 \supset \Omega_2 \supset \cdots \supset \Omega_n \supset \cdots$ is a sequence of non-empty compact sets in $\mathbb{C}$ with the property that*

$$\mathrm{diam}(\Omega_n) \to 0 \quad \text{as } n \to \infty,$$

then there exists a unique point $w \in \mathbb{C}$ such that $w \in \Omega_n$ for all n.

Proof. Choose a point z_n in each Ω_n. The condition $\mathrm{diam}(\Omega_n) \to 0$ says precisely that $\{z_n\}$ is a Cauchy sequence, therefore this sequence converges to a limit that we call w. Since each set Ω_n is compact we must have $w \in \Omega_n$ for all n. Finally, w is the unique point satisfying this property, for otherwise, if w' satisfied the same property with $w' \neq w$ we would have $|w - w'| > 0$ and the condition $\mathrm{diam}(\Omega_n) \to 0$ would be violated.

The last notion we need is that of connectedness. An open set $\Omega \subset \mathbb{C}$ is said to be **connected** if it is not possible to find two disjoint non-empty open sets Ω_1 and Ω_2 such that

$$\Omega = \Omega_1 \cup \Omega_2.$$

A connected open set in $\mathbb{C}$ will be called a **region**. Similarly, a closed set F is connected if one cannot write $F = F_1 \cup F_2$ where F_1 and F_2 are disjoint non-empty closed sets.

There is an equivalent definition of connectedness for open sets in terms of curves, which is often useful in practice: an open set Ω is connected if and only if any two points in Ω can be joined by a curve γ entirely contained in Ω. See Exercise 5 for more details.

2 Functions on the complex plane

2.1 Continuous functions

Let f be a function defined on a set Ω of complex numbers. We say that f is **continuous** at the point $z_0 \in \Omega$ if for every $\epsilon > 0$ there exists $\delta > 0$ such that whenever $z \in \Omega$ and $|z - z_0| < \delta$ then $|f(z) - f(z_0)| < \epsilon$. An equivalent definition is that for every sequence $\{z_1, z_2, \ldots\} \subset \Omega$ such that $\lim z_n = z_0$, then $\lim f(z_n) = f(z_0)$.

The function f is said to be continuous on Ω if it is continuous at every point of Ω. Sums and products of continuous functions are also continuous.

Since the notions of convergence for complex numbers and points in $\mathbb{R}^2$ are the same, the function f of the complex argument $z = x + iy$ is continuous if and only if it is continuous viewed as a function of the two real variables x and y.

By the triangle inequality, it is immediate that if f is continuous, then the real-valued function defined by $z \mapsto |f(z)|$ is continuous. We say that f attains a **maximum** at the point $z_0 \in \Omega$ if

$$|f(z)| \leq |f(z_0)| \quad \text{for all } z \in \Omega,$$

with the inequality reversed for the definition of a **minimum**.

Theorem 2.1 *A continuous function on a compact set Ω is bounded and attains a maximum and minimum on Ω.*

This is of course analogous to the situation of functions of a real variable, and we shall not repeat the simple proof here.

2.2 Holomorphic functions

We now present a notion that is central to complex analysis, and in distinction to our previous discussion we introduce a definition that is genuinely *complex* in nature.

Let Ω be an open set in $\mathbb{C}$ and f a complex-valued function on Ω. The function f is **holomorphic at the point** $z_0 \in \Omega$ if the quotient

$$(1) \qquad \frac{f(z_0 + h) - f(z_0)}{h}$$

converges to a limit when $h \to 0$. Here $h \in \mathbb{C}$ and $h \neq 0$ with $z_0 + h \in \Omega$, so that the quotient is well defined. The limit of the quotient, when it exists, is denoted by $f'(z_0)$, and is called the **derivative of f at** z_0:

$$f'(z_0) = \lim_{h \to 0} \frac{f(z_0 + h) - f(z_0)}{h}.$$

It should be emphasized that in the above limit, h is a complex number that may approach 0 from any direction.

The function f is said to be **holomorphic on** Ω if f is holomorphic at every point of Ω. If C is a closed subset of $\mathbb{C}$, we say that f is **holomorphic on** C if f is holomorphic in some open set containing C. Finally, if f is holomorphic in all of $\mathbb{C}$ we say that f is **entire**.

Sometimes the terms **regular** or **complex differentiable** are used instead of holomorphic. The latter is natural in view of (1) which mimics the usual definition of the derivative of a function of one real variable. But despite this resemblance, a holomorphic function of one complex variable will satisfy much stronger properties than a differentiable function of one real variable. For example, a holomorphic function will actually be infinitely many times complex differentiable, that is, the existence of the first derivative will guarantee the existence of derivatives of any order. This is in contrast with functions of one real variable, since there are differentiable functions that do not have two derivatives. In fact more is true: every holomorphic function is analytic, in the sense that it has a power series expansion near every point (power series will be discussed in the next section), and for this reason we also use the term **analytic** as a synonym for holomorphic. Again, this is in contrast with the fact that there are indefinitely differentiable functions of one real variable that cannot be expanded in a power series. (See Exercise 23.)

EXAMPLE 1. The function $f(z) = z$ is holomorphic on any open set in $\mathbb{C}$, and $f'(z) = 1$. In fact, any polynomial

$$p(z) = a_0 + a_1 z + \cdots + a_n z^n$$

is holomorphic in the entire complex plane and

$$p'(z) = a_1 + \cdots + n a_n z^{n-1}.$$

This follows from Proposition 2.2 below.

EXAMPLE 2. The function $1/z$ is holomorphic on any open set in $\mathbb{C}$ that does not contain the origin, and $f'(z) = -1/z^2$.

EXAMPLE 3. The function $f(z) = \bar{z}$ is not holomorphic. Indeed, we have

$$\frac{f(z_0 + h) - f(z_0)}{h} = \frac{\bar{h}}{h}$$

which has no limit as $h \to 0$, as one can see by first taking h real and then h purely imaginary.

An important family of examples of holomorphic functions, which we discuss later, are the power series. They contain functions such as e^z, $\sin z$, or $\cos z$, and in fact power series play a crucial role in the theory of holomorphic functions, as we already mentioned in the last paragraph. Some other examples of holomorphic functions that will make their appearance in later chapters were given in the introduction to this book.

It is clear from (1) above that a function f is holomorphic at $z_0 \in \Omega$ if and only if there exists a complex number a such that

$$(2) \qquad f(z_0 + h) - f(z_0) - ah = h\psi(h),$$

where ψ is a function defined for all small h and $\lim_{h \to 0} \psi(h) = 0$. Of course, we have $a = f'(z_0)$. From this formulation, it is clear that f is continuous wherever it is holomorphic. Arguing as in the case of one real variable, using formulation (2) in the case of the chain rule (for example), one proves easily the following desirable properties of holomorphic functions.

Proposition 2.2 *If f and g are holomorphic in Ω, then:*

(i) *$f + g$ is holomorphic in Ω and $(f + g)' = f' + g'$.*

(ii) *fg is holomorphic in Ω and $(fg)' = f'g + fg'$.*

(iii) *If $g(z_0) \neq 0$, then f/g is holomorphic at z_0 and*

$$(f/g)' = \frac{f'g - fg'}{g^2}.$$

Moreover, if $f : \Omega \to U$ and $g : U \to \mathbb{C}$ are holomorphic, the chain rule holds

$$(g \circ f)'(z) = g'(f(z))f'(z) \quad \text{for all } z \in \Omega.$$

Complex-valued functions as mappings

We now clarify the relationship between the complex and real derivatives. In fact, the third example above should convince the reader that the notion of complex differentiability differs significantly from the usual notion of real differentiability of a function of two real variables. Indeed, in terms of real variables, the function $f(z) = \bar{z}$ corresponds to the map $F : (x, y) \mapsto (x, -y)$, which is differentiable in the real sense. Its derivative at a point is the linear map given by its Jacobian, the 2×2 matrix of partial derivatives of the coordinate functions. In fact, F is linear and is therefore equal to its derivative at every point. This implies that F is actually indefinitely differentiable. In particular the existence of the real derivative need not guarantee that f is holomorphic.

This example leads us to associate more generally to each complex-valued function $f = u + iv$ the mapping $F(x, y) = (u(x, y), v(x, y))$ from $\mathbb{R}^2$ to $\mathbb{R}^2$.

Recall that a function $F(x, y) = (u(x, y), v(x, y))$ is said to be differentiable at a point $P_0 = (x_0, y_0)$ if there exists a linear transformation $J : \mathbb{R}^2 \to \mathbb{R}^2$ such that

$$(3) \qquad \frac{|F(P_0 + H) - F(P_0) - J(H)|}{|H|} \to 0 \quad \text{as } |H| \to 0, H \in \mathbb{R}^2.$$

Equivalently, we can write

$$F(P_0 + H) - F(P_0) = J(H) + |H|\Psi(H),$$

with $|\Psi(H)| \to 0$ as $|H| \to 0$. The linear transformation J is unique and is called the derivative of F at P_0. If F is differentiable, the partial derivatives of u and v exist, and the linear transformation J is described in the standard basis of $\mathbb{R}^2$ by the Jacobian matrix of F

$$J = J_F(x, y) = \left(\begin{array}{cc} \partial u/\partial x & \partial u/\partial y \\ \partial v/\partial x & \partial v/\partial y \end{array} \right).$$

In the case of complex differentiation the derivative is a complex number $f'(z_0)$, while in the case of real derivatives, it is a matrix. There is, however, a connection between these two notions, which is given in terms of special relations that are satisfied by the entries of the Jacobian matrix, that is, the partials of u and v. To find these relations, consider the limit in (1) when h is first real, say $h = h_1 + ih_2$ with $h_2 = 0$. Then, if we write $z = x + iy$, $z_0 = x_0 + iy_0$, and $f(z) = f(x, y)$, we find that

$$f'(z_0) = \lim_{h_1 \to 0} \frac{f(x_0 + h_1, y_0) - f(x_0, y_0)}{h_1}$$
$$= \frac{\partial f}{\partial x}(z_0),$$

where $\partial/\partial x$ denotes the usual partial derivative in the x variable. (We fix y_0 and think of f as a complex-valued function of the single real variable x.) Now taking h purely imaginary, say $h = ih_2$, a similar argument yields

$$f'(z_0) = \lim_{h_2 \to 0} \frac{f(x_0, y_0 + h_2) - f(x_0, y_0)}{ih_2}$$
$$= \frac{1}{i} \frac{\partial f}{\partial y}(z_0),$$

where $\partial/\partial y$ is partial differentiation in the y variable. Therefore, if f is holomorphic we have shown that

$$\frac{\partial f}{\partial x} = \frac{1}{i} \frac{\partial f}{\partial y}.$$

Writing $f = u + iv$, we find after separating real and imaginary parts and using $1/i = -i$, that the partials of u and v exist, and they satisfy the following non-trivial relations

$$\frac{\partial u}{\partial x} = \frac{\partial v}{\partial y} \quad \text{and} \quad \frac{\partial u}{\partial y} = -\frac{\partial v}{\partial x}.$$

These are the **Cauchy-Riemann** equations, which link real and complex analysis.

We can clarify the situation further by defining two differential operators

$$\frac{\partial}{\partial z} = \frac{1}{2}\left(\frac{\partial}{\partial x} + \frac{1}{i}\frac{\partial}{\partial y}\right) \quad \text{and} \quad \frac{\partial}{\partial \bar{z}} = \frac{1}{2}\left(\frac{\partial}{\partial x} - \frac{1}{i}\frac{\partial}{\partial y}\right).$$

Proposition 2.3 *If f is holomorphic at z_0, then*

$$\frac{\partial f}{\partial \bar{z}}(z_0) = 0 \quad \text{and} \quad f'(z_0) = \frac{\partial f}{\partial z}(z_0) = 2\frac{\partial u}{\partial z}(z_0).$$

Also, if we write $F(x, y) = f(z)$, then F is differentiable in the sense of real variables, and

$$\det J_F(x_0, y_0) = |f'(z_0)|^2.$$

Proof. Taking real and imaginary parts, it is easy to see that the Cauchy-Riemann equations are equivalent to $\partial f/\partial \bar{z} = 0$. Moreover, by our earlier observation

$$f'(z_0) = \frac{1}{2}\left(\frac{\partial f}{\partial x}(z_0) + \frac{1}{i}\frac{\partial f}{\partial y}(z_0)\right)$$
$$= \frac{\partial f}{\partial z}(z_0),$$

and the Cauchy-Riemann equations give $\partial f/\partial z = 2\partial u/\partial z$. To prove that F is differentiable it suffices to observe that if $H = (h_1, h_2)$ and $h = h_1 + ih_2$, then the Cauchy-Riemann equations imply

$$J_F(x_0, y_0)(H) = \left(\frac{\partial u}{\partial x} - i\frac{\partial u}{\partial y} \right)(h_1 + ih_2) = f'(z_0)h \,,$$

where we have identified a complex number with the pair of real and imaginary parts. After a final application of the Cauchy-Riemann equations, the above results imply that

(4)
$$\det J_F(x_0, y_0) = \frac{\partial u}{\partial x}\frac{\partial v}{\partial y} - \frac{\partial v}{\partial x}\frac{\partial u}{\partial y} = \left(\frac{\partial u}{\partial x}\right)^2 + \left(\frac{\partial u}{\partial y}\right)^2 = \left|2\frac{\partial u}{\partial z}\right|^2 = |f'(z_0)|^2.$$

So far, we have assumed that f is holomorphic and deduced relations satisfied by its real and imaginary parts. The next theorem contains an important converse, which completes the circle of ideas presented here.

Theorem 2.4 *Suppose $f = u + iv$ is a complex-valued function defined on an open set Ω. If u and v are continuously differentiable and satisfy the Cauchy-Riemann equations on Ω, then f is holomorphic on Ω and $f'(z) = \partial f/\partial z$.*

Proof. Write

$$u(x + h_1, y + h_2) - u(x, y) = \frac{\partial u}{\partial x}h_1 + \frac{\partial u}{\partial y}h_2 + |h|\psi_1(h)$$

and

$$v(x + h_1, y + h_2) - v(x, y) = \frac{\partial v}{\partial x}h_1 + \frac{\partial v}{\partial y}h_2 + |h|\psi_2(h),$$

where $\psi_j(h) \to 0$ (for $j = 1, 2$) as $|h|$ tends to 0, and $h = h_1 + ih_2$. Using the Cauchy-Riemann equations we find that

$$f(z + h) - f(z) = \left(\frac{\partial u}{\partial x} - i\frac{\partial u}{\partial y} \right)(h_1 + ih_2) + |h|\psi(h),$$

where $\psi(h) = \psi_1(h) + \psi_2(h) \to 0$, as $|h| \to 0$. Therefore f is holomorphic and

$$f'(z) = 2\frac{\partial u}{\partial z} = \frac{\partial f}{\partial z}.$$

2.3 Power series

The prime example of a power series is the complex **exponential** function, which is defined for $z \in \mathbb{C}$ by

$$e^z = \sum_{n=0}^{\infty} \frac{z^n}{n!}.$$

When z is real, this definition coincides with the usual exponential function, and in fact, the series above converges absolutely for every $z \in \mathbb{C}$. To see this, note that

$$\left| \frac{z^n}{n!} \right| = \frac{|z|^n}{n!},$$

so $|e^z|$ can be compared to the series $\sum |z|^n / n! = e^{|z|} < \infty$. In fact, this estimate shows that the series defining e^z is uniformly convergent in every disc in $\mathbb{C}$.

In this section we will prove that e^z is holomorphic in all of $\mathbb{C}$ (it is entire), and that its derivative can be found by differentiating the series term by term. Hence

$$(e^z)' = \sum_{n=0}^{\infty} n \frac{z^{n-1}}{n!} = \sum_{m=0}^{\infty} \frac{z^m}{m!} = e^z,$$

and therefore e^z is its own derivative.

In contrast, the geometric series

$$\sum_{n=0}^{\infty} z^n$$

converges absolutely only in the disc $|z| < 1$, and its sum there is the function $1/(1-z)$, which is holomorphic in the open set $\mathbb{C} - \{1\}$. This identity is proved exactly as when z is real: we first observe

$$\sum_{n=0}^{N} z^n = \frac{1 - z^{N+1}}{1 - z},$$

and then note that if $|z| < 1$ we must have $\lim_{N \to \infty} z^{N+1} = 0$.

In general, a **power series** is an expansion of the form

$$(5) \qquad \sum_{n=0}^{\infty} a_n z^n,$$

where $a_n \in \mathbb{C}$. To test for absolute convergence of this series, we must investigate

$$\sum_{n=0}^{\infty} |a_n| \, |z|^n \, ,$$

and we observe that if the series (5) converges absolutely for some z_0, then it will also converge for all z in the disc $|z| \leq |z_0|$. We now prove that there always exists an open disc (possibly empty) on which the power series converges absolutely.

Theorem 2.5 *Given a power series $\sum_{n=0}^{\infty} a_n z^n$, there exists $0 \leq R \leq \infty$ such that:*

(i) *If $|z| < R$ the series converges absolutely.*

(ii) *If $|z| > R$ the series diverges.*

Moreover, if we use the convention that $1/0 = \infty$ and $1/\infty = 0$, then R is given by Hadamard's formula

$$1/R = \limsup |a_n|^{1/n}.$$

The number R is called the **radius of convergence** of the power series, and the region $|z| < R$ the **disc of convergence**. In particular, we have $R = \infty$ in the case of the exponential function, and $R = 1$ for the geometric series.

Proof. Let $L = 1/R$ where R is defined by the formula in the statement of the theorem, and suppose that $L \neq 0, \infty$. (These two easy cases are left as an exercise.) If $|z| < R$, choose $\epsilon > 0$ so small that

$$(L + \epsilon)|z| = r < 1.$$

By the definition L, we have $|a_n|^{1/n} \leq L + \epsilon$ for all large n, therefore

$$|a_n| \, |z|^n \leq \{(L + \epsilon)|z|\}^n = r^n.$$

Comparison with the geometric series $\sum r^n$ shows that $\sum a_n z^n$ converges.

If $|z| > R$, then a similar argument proves that there exists a sequence of terms in the series whose absolute value goes to infinity, hence the series diverges.

Remark. On the boundary of the disc of convergence, $|z| = R$, the situation is more delicate as one can have either convergence or divergence. (See Exercise 19.)

Further examples of power series that converge in the whole complex plane are given by the standard **trigonometric functions**; these are defined by

$$\cos z = \sum_{n=0}^{\infty}(-1)^n \frac{z^{2n}}{(2n)!}, \quad \text{and} \quad \sin z = \sum_{n=0}^{\infty}(-1)^n \frac{z^{2n+1}}{(2n+1)!},$$

and they agree with the usual cosine and sine of a real argument whenever $z \in \mathbb{R}$. A simple calculation exhibits a connection between these two functions and the complex exponential, namely,

$$\cos z = \frac{e^{iz}+e^{-iz}}{2} \quad \text{and} \quad \sin z = \frac{e^{iz}-e^{-iz}}{2i}.$$

These are called the **Euler formulas** for the cosine and sine functions.

Power series provide a very important class of analytic functions that are particularly simple to manipulate.

Theorem 2.6 *The power series $f(z) = \sum_{n=0}^{\infty} a_n z^n$ defines a holomorphic function in its disc of convergence. The derivative of f is also a power series obtained by differentiating term by term the series for f, that is,*

$$f'(z) = \sum_{n=0}^{\infty} n a_n z^{n-1}.$$

Moreover, f' has the same radius of convergence as f.

Proof. The assertion about the radius of convergence of f' follows from Hadamard's formula. Indeed, $\lim_{n\to\infty} n^{1/n} = 1$, and therefore

$$\limsup |a_n|^{1/n} = \limsup |na_n|^{1/n},$$

so that $\sum a_n z^n$ and $\sum n a_n z^n$ have the same radius of convergence, and hence so do $\sum a_n z^n$ and $\sum n a_n z^{n-1}$.

To prove the first assertion, we must show that the series

$$g(z) = \sum_{n=0}^{\infty} n a_n z^{n-1}$$

gives the derivative of f. For that, let R denote the radius of convergence of f, and suppose $|z_0| < r < R$. Write

$$f(z) = S_N(z) + E_N(z),$$

where

$$S_N(z) = \sum_{n=0}^{N} a_n z^n \quad \text{and} \quad E_N(z) = \sum_{n=N+1}^{\infty} a_n z^n.$$

Then, if h is chosen so that $|z_0 + h| < r$ we have

$$\frac{f(z_0 + h) - f(z_0)}{h} - g(z_0) = \left(\frac{S_N(z_0 + h) - S_N(z_0)}{h} - S_N'(z_0) \right)$$
$$+ (S_N'(z_0) - g(z_0)) + \left(\frac{E_N(z_0 + h) - E_N(z_0)}{h} \right).$$

Since $a^n - b^n = (a - b)(a^{n-1} + a^{n-2}b + \cdots + ab^{n-2} + b^{n-1})$, we see that

$$\left| \frac{E_N(z_0 + h) - E_N(z_0)}{h} \right| \leq \sum_{n=N+1}^{\infty} |a_n| \left| \frac{(z_0 + h)^n - z_0^n}{h} \right| \leq \sum_{n=N+1}^{\infty} |a_n| n r^{n-1},$$

where we have used the fact that $|z_0| < r$ and $|z_0 + h| < r$. The expression on the right is the tail end of a convergent series, since g converges absolutely on $|z| < R$. Therefore, given $\epsilon > 0$ we can find N_1 so that $N > N_1$ implies

$$\left| \frac{E_N(z_0 + h) - E_N(z_0)}{h} \right| < \epsilon.$$

Also, since $\lim_{N \to \infty} S_N'(z_0) = g(z_0)$, we can find N_2 so that $N > N_2$ implies

$$|S_N'(z_0) - g(z_0)| < \epsilon.$$

If we fix N so that both $N > N_1$ and $N > N_2$ hold, then we can find $\delta > 0$ so that $|h| < \delta$ implies

$$\left| \frac{S_N(z_0 + h) - S_N(z_0)}{h} - S_N'(z_0) \right| < \epsilon,$$

simply because the derivative of a polynomial is obtained by differentiating it term by term. Therefore,

$$\left| \frac{f(z_0 + h) - f(z_0)}{h} - g(z_0) \right| < 3\epsilon$$

whenever $|h| < \delta$, thereby concluding the proof of the theorem.

Successive applications of this theorem yield the following.

Corollary 2.7 *A power series is infinitely complex differentiable in its disc of convergence, and the higher derivatives are also power series obtained by termwise differentiation.*

We have so far dealt only with power series centered at the origin. More generally, a power series centered at $z_0 \in \mathbb{C}$ is an expression of the form

$$f(z) = \sum_{n=0}^{\infty} a_n(z - z_0)^n.$$

The disc of convergence of f is now centered at z_0 and its radius is still given by Hadamard's formula. In fact, if

$$g(z) = \sum_{n=0}^{\infty} a_n z^n,$$

then f is simply obtained by translating g, namely $f(z) = g(w)$ where $w = z - z_0$. As a consequence everything about g also holds for f after we make the appropriate translation. In particular, by the chain rule,

$$f'(z) = g'(w) = \sum_{n=0}^{\infty} na_n(z - z_0)^{n-1}.$$

A function f defined on an open set Ω is said to be **analytic** (or have a **power series expansion**) at a point $z_0 \in \Omega$ if there exists a power series $\sum a_n(z - z_0)^n$ centered at z_0, with positive radius of convergence, such that

$$f(z) = \sum_{n=0}^{\infty} a_n(z - z_0)^n \quad \text{for all } z \text{ in a neighborhood of } z_0.$$

If f has a power series expansion at every point in Ω, we say that f is **analytic on** Ω.

By Theorem 2.6, an analytic function on Ω is also holomorphic there. A deep theorem which we prove in the next chapter says that the converse is true: every holomorphic function is analytic. For that reason, we use the terms holomorphic and analytic interchangeably.

3 Integration along curves

In the definition of a curve, we distinguish between the one-dimensional geometric object in the plane (endowed with an orientation), and its parametrization, which is a mapping from a closed interval to $\mathbb{C}$, that is not uniquely determined.

A **parametrized curve** is a function $z(t)$ which maps a closed interval $[a, b] \subset \mathbb{R}$ to the complex plane. We shall impose regularity conditions on the parametrization which are always verified in the situations that occur in this book. We say that the parametrized curve is **smooth** if $z'(t)$ exists and is continuous on $[a, b]$, and $z'(t) \neq 0$ for $t \in [a, b]$. At the points $t = a$ and $t = b$, the quantities $z'(a)$ and $z'(b)$ are interpreted as the one-sided limits

$$z'(a) = \lim_{\substack{h \to 0 \\ h > 0}} \frac{z(a + h) - z(a)}{h} \quad \text{and} \quad z'(b) = \lim_{\substack{h \to 0 \\ h < 0}} \frac{z(b + h) - z(b)}{h}.$$

In general, these quantities are called the right-hand derivative of $z(t)$ at a, and the left-hand derivative of $z(t)$ at b, respectively.

Similarly we say that the parametrized curve is **piecewise-smooth** if z is continuous on $[a, b]$ and if there exist points

$$a = a_0 < a_1 < \cdots < a_n = b,$$

where $z(t)$ is smooth in the intervals $[a_k, a_{k+1}]$. In particular, the right-hand derivative at a_k may differ from the left-hand derivative at a_k for $k = 1, \ldots, n - 1$.

Two parametrizations,

$$z : [a, b] \to \mathbb{C} \quad \text{and} \quad \tilde{z} : [c, d] \to \mathbb{C},$$

are **equivalent** if there exists a continuously differentiable bijection $s \mapsto t(s)$ from $[c, d]$ to $[a, b]$ so that $t'(s) > 0$ and

$$\tilde{z}(s) = z(t(s)).$$

The condition $t'(s) > 0$ says precisely that the orientation is preserved: as s travels from c to d, then $t(s)$ travels from a to b. The family of all parametrizations that are equivalent to $z(t)$ determines a **smooth curve** $\gamma \subset \mathbb{C}$, namely the image of $[a, b]$ under z with the orientation given by z as t travels from a to b. We can define a curve γ^- obtained from the curve γ by reversing the orientation (so that γ and γ^- consist of the same points in the plane). As a particular parametrization for γ^- we can take $z^- : [a, b] \to \mathbb{R}^2$ defined by

$$z^-(t) = z(b + a - t).$$

It is also clear how to define a **piecewise-smooth curve**. The points $z(a)$ and $z(b)$ are called the **end-points** of the curve and are independent on the parametrization. Since γ carries an orientation, it is natural to say that γ begins at $z(a)$ and ends at $z(b)$.

A smooth or piecewise-smooth curve is **closed** if $z(a) = z(b)$ for any of its parametrizations. Finally, a smooth or piecewise-smooth curve is **simple** if it is not self-intersecting, that is, $z(t) \neq z(s)$ unless $s = t$. Of course, if the curve is closed to begin with, then we say that it is simple whenever $z(t) \neq z(s)$ unless $s = t$, or $s = a$ and $t = b$.

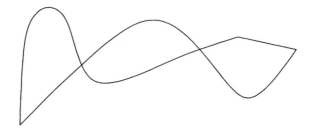

Figure 3. A closed piecewise-smooth curve

For brevity, we shall call any piecewise-smooth curve a **curve**, since these will be the objects we shall be primarily concerned with.

A basic example consists of a circle. Consider the circle $C_r(z_0)$ centered at z_0 and of radius r, which by definition is the set

$$C_r(z_0) = \{z \in \mathbb{C} : |z - z_0| = r\}.$$

The **positive orientation** (counterclockwise) is the one that is given by the standard parametrization

$$z(t) = z_0 + re^{it}, \quad \text{where } t \in [0, 2\pi],$$

while the **negative orientation** (clockwise) is given by

$$z(t) = z_0 + re^{-it}, \quad \text{where } t \in [0, 2\pi].$$

In the following chapters, we shall denote by C a general *positively* oriented circle.

An important tool in the study of holomorphic functions is integration of functions along curves. Loosely speaking, a key theorem in complex analysis says that if a function is holomorphic in the interior of a closed curve γ, then

$$\int_\gamma f(z)\, dz = 0,$$

and we shall turn our attention to a version of this theorem (called Cauchy's theorem) in the next chapter. Here we content ourselves with the necessary definitions and properties of the integral.

Given a smooth curve γ in $\mathbb{C}$ parametrized by $z : [a, b] \to \mathbb{C}$, and f a continuous function on γ, we define the **integral of f along γ** by

$$\int_\gamma f(z)\, dz = \int_a^b f(z(t)) z'(t)\, dt.$$

In order for this definition to be meaningful, we must show that the right-hand integral is independent of the parametrization chosen for γ. Say that $\tilde{z}$ is an equivalent parametrization as above. Then the change of variables formula and the chain rule imply that

$$\int_a^b f(z(t)) z'(t)\, dt = \int_c^d f(z(t(s))) z'(t(s)) t'(s)\, ds = \int_c^d f(\tilde{z}(s)) \tilde{z}'(s)\, ds.$$

This proves that the integral of f over γ is well defined.

If γ is piecewise smooth, then the integral of f over γ is simply the sum of the integrals of f over the smooth parts of γ, so if $z(t)$ is a piecewise-smooth parametrization as before, then

$$\int_\gamma f(z)\, dz = \sum_{k=0}^{n-1} \int_{a_k}^{a_{k+1}} f(z(t)) z'(t)\, dt.$$

By definition, the **length** of the smooth curve γ is

$$\text{length}(\gamma) = \int_a^b |z'(t)|\, dt.$$

Arguing as we just did, it is clear that this definition is also independent of the parametrization. Also, if γ is only piecewise-smooth, then its length is the sum of the lengths of its smooth parts.

Proposition 3.1 *Integration of continuous functions over curves satisfies the following properties:*

(i) *It is linear, that is, if $\alpha, \beta \in \mathbb{C}$, then*

$$\int_\gamma (\alpha f(z) + \beta g(z)) \, dz = \alpha \int_\gamma f(z) \, dz + \beta \int_\gamma g(z) \, dz.$$

(ii) *If γ^- is γ with the reverse orientation, then*

$$\int_\gamma f(z) \, dz = - \int_{\gamma^-} f(z) \, dz.$$

(iii) *One has the inequality*

$$\left| \int_\gamma f(z) \, dz \right| \leq \sup_{z \in \gamma} |f(z)| \cdot \text{length}(\gamma).$$

Proof. The first property follows from the definition and the linearity of the Riemann integral. The second property is left as an exercise. For the third, note that

$$\left| \int_\gamma f(z) \, dz \right| \leq \sup_{t \in [a,b]} |f(z(t))| \int_a^b |z'(t)| \, dt \leq \sup_{z \in \gamma} |f(z)| \cdot \text{length}(\gamma)$$

as was to be shown.

As we have said, Cauchy's theorem states that for appropriate closed curves γ in an open set Ω on which f is holomorphic, then

$$\int_\gamma f(z) \, dz = 0.$$

The existence of primitives gives a first manifestation of this phenomenon. Suppose f is a function on the open set Ω. A **primitive** for f on Ω is a function F that is holomorphic on Ω and such that $F'(z) = f(z)$ for all $z \in \Omega$.

Theorem 3.2 *If a continuous function f has a primitive F in Ω, and γ is a curve in Ω that begins at w_1 and ends at w_2, then*

$$\int_\gamma f(z) \, dz = F(w_2) - F(w_1).$$

Proof. If γ is smooth, the proof is a simple application of the chain rule and the fundamental theorem of calculus. Indeed, if $z(t) : [a, b] \to \mathbb{C}$ is a parametrization for γ, then $z(a) = w_1$ and $z(b) = w_2$, and we have

$$
\begin{aligned}
\int_\gamma f(z)\, dz &= \int_a^b f(z(t))z'(t)\, dt \\
&= \int_a^b F'(z(t))z'(t)\, dt \\
&= \int_a^b \frac{d}{dt} F(z(t))\, dt \\
&= F(z(b)) - F(z(a)).
\end{aligned}
$$

If γ is only piecewise-smooth, then arguing as we just did, we obtain a telescopic sum, and we have

$$
\begin{aligned}
\int_\gamma f(z)\, dz &= \sum_{k=0}^{n-1} F(z(a_{k+1})) - F(z(a_k)) \\
&= F(z(a_n)) - F(z(a_0)) \\
&= F(z(b)) - F(z(a)).
\end{aligned}
$$

Corollary 3.3 *If γ is a closed curve in an open set Ω, and f is continuous and has a primitive in Ω, then*

$$
\int_\gamma f(z)\, dz = 0.
$$

This is immediate since the end-points of a closed curve coincide.

For example, the function $f(z) = 1/z$ does not have a primitive in the open set $\mathbb{C} - \{0\}$, since if C is the unit circle parametrized by $z(t) = e^{it}$, $0 \le t \le 2\pi$, we have

$$
\int_C f(z)\, dz = \int_0^{2\pi} \frac{ie^{it}}{e^{it}}\, dt = 2\pi i \ne 0.
$$

In subsequent chapters, we shall see that this innocent calculation, which provides an example of a function f and closed curve γ for which $\int_\gamma f(z)\, dz \ne 0$, lies at the heart of the theory.

Corollary 3.4 *If f is holomorphic in a region Ω and $f' = 0$, then f is constant.*

Proof. Fix a point $w_0 \in \Omega$. It suffices to show that $f(w) = f(w_0)$ for all $w \in \Omega$.

Since Ω is connected, for any $w \in \Omega$, there exists a curve γ which joins w_0 to w. Since f is clearly a primitive for f', we have

$$\int_\gamma f'(z)\, dz = f(w) - f(w_0).$$

By assumption, $f' = 0$ so the integral on the left is 0, and we conclude that $f(w) = f(w_0)$ as desired.

Remark on notation. When convenient, we follow the practice of using the notation $f(z) = O(g(z))$ to mean that there is a constant $C > 0$ such that $|f(z)| \leq C|g(z)|$ for z in a neighborhood of the point in question. In addition, we say $f(z) = o(g(z))$ when $|f(z)/g(z)| \to 0$. We also write $f(z) \sim g(z)$ to mean that $f(z)/g(z) \to 1$.

4 Exercises

1. Describe geometrically the sets of points z in the complex plane defined by the following relations:

    (a) $|z - z_1| = |z - z_2|$ where $z_1, z_2 \in \mathbb{C}$.

    (b) $1/z = \bar{z}$.

    (c) $\mathrm{Re}(z) = 3$.

    (d) $\mathrm{Re}(z) > c$, (resp., $\geq c$) where $c \in \mathbb{R}$.

    (e) $\mathrm{Re}(az + b) > 0$ where $a, b \in \mathbb{C}$.

    (f) $|z| = \mathrm{Re}(z) + 1$.

    (g) $\mathrm{Im}(z) = c$ with $c \in \mathbb{R}$.

2. Let $\langle \cdot, \cdot \rangle$ denote the usual inner product in $\mathbb{R}^2$. In other words, if $Z = (x_1, y_1)$ and $W = (x_2, y_2)$, then

$$\langle Z, W \rangle = x_1 x_2 + y_1 y_2.$$

Similarly, we may define a Hermitian inner product $(\cdot, \cdot)$ in $\mathbb{C}$ by

$$(z, w) = z\bar{w}.$$

The term Hermitian is used to describe the fact that $(\cdot, \cdot)$ is not symmetric, but rather satisfies the relation

$$(z, w) = \overline{(w, z)} \qquad \text{for all } z, w \in \mathbb{C}.$$

Show that

$$\langle z, w \rangle = \frac{1}{2}[(z, w) + (w, z)] = \text{Re}(z, w),$$

where we use the usual identification $z = x + iy \in \mathbb{C}$ with $(x, y) \in \mathbb{R}^2$.

3. With $\omega = se^{i\varphi}$, where $s \geq 0$ and $\varphi \in \mathbb{R}$, solve the equation $z^n = \omega$ in $\mathbb{C}$ where n is a natural number. How many solutions are there?

4. Show that it is impossible to define a total ordering on $\mathbb{C}$. In other words, one cannot find a relation $\succ$ between complex numbers so that:

(i) For any two complex numbers z, w, one and only one of the following is true:
   $z \succ w$, $w \succ z$ or $z = w$.

(ii) For all $z_1, z_2, z_3 \in \mathbb{C}$ the relation $z_1 \succ z_2$ implies $z_1 + z_3 \succ z_2 + z_3$.

(iii) Moreover, for all $z_1, z_2, z_3 \in \mathbb{C}$ with $z_3 \succ 0$, then $z_1 \succ z_2$ implies $z_1 z_3 \succ z_2 z_3$.

[Hint: First check if $i \succ 0$ is possible.]

5. A set Ω is said to be **pathwise connected** if any two points in Ω can be joined by a (piecewise-smooth) curve entirely contained in Ω. The purpose of this exercise is to prove that an *open* set Ω is pathwise connected if and only if Ω is connected.

(a) Suppose first that Ω is open and pathwise connected, and that it can be written as $\Omega = \Omega_1 \cup \Omega_2$ where Ω_1 and Ω_2 are disjoint non-empty open sets. Choose two points $w_1 \in \Omega_1$ and $w_2 \in \Omega_2$ and let γ denote a curve in Ω joining w_1 to w_2. Consider a parametrization $z : [0, 1] \to \Omega$ of this curve with $z(0) = w_1$ and $z(1) = w_2$, and let

$$t^* = \sup_{0 \leq t \leq 1} \{t : z(s) \in \Omega_1 \text{ for all } 0 \leq s < t\}.$$

Arrive at a contradiction by considering the point $z(t^*)$.

(b) Conversely, suppose that Ω is open and connected. Fix a point $w \in \Omega$ and let $\Omega_1 \subset \Omega$ denote the set of all points that can be joined to w by a curve contained in Ω. Also, let $\Omega_2 \subset \Omega$ denote the set of all points that cannot be joined to w by a curve in Ω. Prove that both Ω_1 and Ω_2 are open, disjoint and their union is Ω. Finally, since Ω_1 is non-empty (why?) conclude that $\Omega = \Omega_1$ as desired.

The proof actually shows that the regularity and type of curves we used to define pathwise connectedness can be relaxed without changing the equivalence between the two definitions when Ω is open. For instance, we may take all curves to be continuous, or simply polygonal lines.[2]

6. Let Ω be an open set in $\mathbb{C}$ and $z \in \Omega$. The **connected component** (or simply the **component**) of Ω containing z is the set C_z of all points w in Ω that can be joined to z by a curve entirely contained in Ω.

(a) Check first that C_z is open and connected. Then, show that $w \in C_z$ defines an equivalence relation, that is: (i) $z \in C_z$, (ii) $w \in C_z$ implies $z \in C_w$, and (iii) if $w \in C_z$ and $z \in C_\zeta$, then $w \in C_\zeta$.

Thus Ω is the union of all its connected components, and two components are either disjoint or coincide.

(b) Show that Ω can have only countably many distinct connected components.

(c) Prove that if Ω is the complement of a compact set, then Ω has only one unbounded component.

[Hint: For (b), one would otherwise obtain an uncountable number of disjoint open balls. Now, each ball contains a point with rational coordinates. For (c), note that the complement of a large disc containing the compact set is connected.]

7. The family of mappings introduced here plays an important role in complex analysis. These mappings, sometimes called **Blaschke factors**, will reappear in various applications in later chapters.

(a) Let z, w be two complex numbers such that $\bar{z}w \neq 1$. Prove that

$$\left| \frac{w - z}{1 - \bar{w}z} \right| < 1 \quad \text{if } |z| < 1 \text{ and } |w| < 1,$$

and also that

$$\left| \frac{w - z}{1 - \bar{w}z} \right| = 1 \quad \text{if } |z| = 1 \text{ or } |w| = 1.$$

[Hint: Why can one assume that z is real? It then suffices to prove that

$$(r - w)(r - \bar{w}) \leq (1 - rw)(1 - r\bar{w})$$

with equality for appropriate r and $|w|$.]

(b) Prove that for a fixed w in the unit disc $\mathbb{D}$, the mapping

$$F : z \mapsto \frac{w - z}{1 - \bar{w}z}$$

satisfies the following conditions:

---

[2] A polygonal line is a piecewise-smooth curve which consists of finitely many straight line segments.

(i) F maps the unit disc to itself (that is, $F : \mathbb{D} \to \mathbb{D}$), and is holomorphic.

(ii) F interchanges 0 and w, namely $F(0) = w$ and $F(w) = 0$.

(iii) $|F(z)| = 1$ if $|z| = 1$.

(iv) $F : \mathbb{D} \to \mathbb{D}$ is bijective. [Hint: Calculate $F \circ F$.]

8. Suppose U and V are open sets in the complex plane. Prove that if $f : U \to V$ and $g : V \to \mathbb{C}$ are two functions that are differentiable (in the real sense, that is, as functions of the two real variables x and y), and $h = g \circ f$, then

$$\frac{\partial h}{\partial z} = \frac{\partial g}{\partial z}\frac{\partial f}{\partial z} + \frac{\partial g}{\partial \bar{z}}\frac{\partial \bar{f}}{\partial z}$$

and

$$\frac{\partial h}{\partial \bar{z}} = \frac{\partial g}{\partial z}\frac{\partial f}{\partial \bar{z}} + \frac{\partial g}{\partial \bar{z}}\frac{\partial \bar{f}}{\partial \bar{z}}.$$

This is the complex version of the chain rule.

9. Show that in polar coordinates, the Cauchy-Riemann equations take the form

$$\frac{\partial u}{\partial r} = \frac{1}{r}\frac{\partial v}{\partial \theta} \quad \text{and} \quad \frac{1}{r}\frac{\partial u}{\partial \theta} = -\frac{\partial v}{\partial r}.$$

Use these equations to show that the logarithm function defined by

$$\log z = \log r + i\theta \quad \text{where } z = re^{i\theta} \text{ with } -\pi < \theta < \pi$$

is holomorphic in the region $r > 0$ and $-\pi < \theta < \pi$.

10. Show that

$$4\frac{\partial}{\partial z}\frac{\partial}{\partial \bar{z}} = 4\frac{\partial}{\partial \bar{z}}\frac{\partial}{\partial z} = \triangle,$$

where $\triangle$ is the **Laplacian**

$$\triangle = \frac{\partial^2}{\partial x^2} + \frac{\partial^2}{\partial y^2}.$$

11. Use Exercise 10 to prove that if f is holomorphic in the open set Ω, then the real and imaginary parts of f are **harmonic**; that is, their Laplacian is zero.

12. Consider the function defined by

$$f(x + iy) = \sqrt{|x||y|}, \quad \text{whenever } x, y \in \mathbb{R}.$$

Show that f satisfies the Cauchy-Riemann equations at the origin, yet f is not holomorphic at 0.

13. Suppose that f is holomorphic in a region Ω. Prove that in any one of the following cases:

  (a) $\text{Re}(f)$ is constant;

  (b) $\text{Im}(f)$ is constant;

  (c) $|f|$ is constant;

one can conclude that f is constant.

14. Suppose $\{a_n\}_{n=1}^N$ and $\{b_n\}_{n=1}^N$ are two finite sequences of complex numbers. Let $B_k = \sum_{n=1}^k b_n$ denote the partial sums of the series $\sum b_n$ with the convention $B_0 = 0$. Prove the **summation by parts** formula

$$\sum_{n=M}^N a_n b_n = a_N B_N - a_M B_{M-1} - \sum_{n=M}^{N-1} (a_{n+1} - a_n) B_n.$$

15. Abel's theorem. Suppose $\sum_{n=1}^\infty a_n$ converges. Prove that

$$\lim_{r \to 1,\ r<1} \sum_{n=1}^\infty r^n a_n = \sum_{n=1}^\infty a_n.$$

[Hint: Sum by parts.] In other words, if a series converges, then it is Abel summable with the same limit. For the precise definition of these terms, and more information on summability methods, we refer the reader to Book I, Chapter 2.

16. Determine the radius of convergence of the series $\sum_{n=1}^\infty a_n z^n$ when:

  (a) $a_n = (\log n)^2$

  (b) $a_n = n!$

  (c) $a_n = \frac{n^2}{4^n + 3n}$

  (d) $a_n = (n!)^3/(3n)!$ [Hint: Use Stirling's formula, which says that $n! \sim cn^{n+\frac12} e^{-n}$ for some $c > 0$..]

  (e) Find the radius of convergence of the **hypergeometric series**

$$F(\alpha, \beta, \gamma; z) = 1 + \sum_{n=1}^\infty \frac{\alpha(\alpha+1)\cdots(\alpha+n-1)\beta(\beta+1)\cdots(\beta+n-1)}{n!\gamma(\gamma+1)\cdots(\gamma+n-1)} z^n.$$

Here $\alpha, \beta \in \mathbb{C}$ and $\gamma \neq 0, -1, -2, \ldots$.

(f) Find the radius of convergence of the Bessel function of order r:

$$J_r(z) = \left(\frac{z}{2}\right)^r \sum_{n=0}^{\infty} \frac{(-1)^n}{n!(n+r)!} \left(\frac{z}{2}\right)^{2n},$$

where r is a positive integer.

17. Show that if $\{a_n\}_{n=0}^{\infty}$ is a sequence of non-zero complex numbers such that

$$\lim_{n \to \infty} \frac{|a_{n+1}|}{|a_n|} = L,$$

then

$$\lim_{n \to \infty} |a_n|^{1/n} = L.$$

In particular, this exercise shows that when applicable, the ratio test can be used to calculate the radius of convergence of a power series.

18. Let f be a power series centered at the origin. Prove that f has a power series expansion around any point in its disc of convergence.
[Hint: Write $z = z_0 + (z - z_0)$ and use the binomial expansion for z^n.]

19. Prove the following:

(a) The power series $\sum nz^n$ does not converge on any point of the unit circle.

(b) The power series $\sum z^n/n^2$ converges at every point of the unit circle.

(c) The power series $\sum z^n/n$ converges at every point of the unit circle except $z = 1$. [Hint: Sum by parts.]

20. Expand $(1 - z)^{-m}$ in powers of z. Here m is a fixed positive integer. Also, show that if

$$(1 - z)^{-m} = \sum_{n=0}^{\infty} a_n z^n,$$

then one obtains the following asymptotic relation for the coefficients:

$$a_n \sim \frac{1}{(m-1)!} n^{m-1} \qquad \text{as } n \to \infty.$$

21. Show that for $|z| < 1$, one has

$$\frac{z}{1-z^2} + \frac{z^2}{1-z^4} + \cdots + \frac{z^{2^n}}{1-z^{2^{n+1}}} + \cdots = \frac{z}{1-z},$$

and

$$\frac{z}{1+z} + \frac{2z^2}{1+z^2} + \cdots + \frac{2^k z^{2^k}}{1+z^{2^k}} + \cdots = \frac{z}{1-z}.$$

Justify any change in the order of summation.

[Hint: Use the dyadic expansion of an integer and the fact that $2^{k+1} - 1 = 1 + 2 + 2^2 + \cdots + 2^k$.]

22. Let $\mathbb{N} = \{1, 2, 3, \ldots\}$ denote the set of positive integers. A subset $S \subset \mathbb{N}$ is said to be in arithmetic progression if

$$S = \{a, a + d, a + 2d, a + 3d, \ldots\}$$

where $a, d \in \mathbb{N}$. Here d is called the step of S.

Show that $\mathbb{N}$ cannot be partitioned into a finite number of subsets that are in arithmetic progression with distinct steps (except for the trivial case $a = d = 1$).

[Hint: Write $\sum_{n \in \mathbb{N}} z^n$ as a sum of terms of the type $\frac{z^a}{1-z^d}$.]

23. Consider the function f defined on $\mathbb{R}$ by

$$f(x) = \begin{cases} 0 & \text{if } x \leq 0, \\ e^{-1/x^2} & \text{if } x > 0. \end{cases}$$

Prove that f is indefinitely differentiable on $\mathbb{R}$, and that $f^{(n)}(0) = 0$ for all $n \geq 1$. Conclude that f does not have a converging power series expansion $\sum_{n=0}^{\infty} a_n x^n$ for x near the origin.

24. Let γ be a smooth curve in $\mathbb{C}$ parametrized by $z(t) : [a, b] \to \mathbb{C}$. Let γ^- denote the curve with the same image as γ but with the reverse orientation. Prove that for any continuous function f on γ

$$\int_{\gamma} f(z) \, dz = - \int_{\gamma^-} f(z) \, dz.$$

25. The next three calculations provide some insight into Cauchy's theorem, which we treat in the next chapter.

(a) Evaluate the integrals

$$\int_{\gamma} z^n \, dz$$

for all integers n. Here γ is any circle centered at the origin with the positive (counterclockwise) orientation.

(b) Same question as before, but with γ any circle not enclosing the origin.

(c) Show that if $|a| < r < |b|$, then

$$\int_\gamma \frac{1}{(z-a)(z-b)}\, dz = \frac{2\pi i}{a-b},$$

where γ denotes the circle centered at the origin, of radius r, with the positive orientation.

26. Suppose f is continuous in a region Ω. Prove that any two primitives of f (if they exist) differ by a constant.

2 Cauchy's Theorem and Its Applications

> The solution of a large number of problems can be reduced, in the last analysis, to the evaluation of definite integrals; thus mathematicians have been much occupied with this task... However, among many results obtained, a number were initially discovered by the aid of a type of induction based on the passage from real to imaginary. Often passage of this kind led directly to remarkable results. Nevertheless this part of the theory, as has been observed by Laplace, is subject to various difficulties...
>
> After having reflected on this subject and brought together various results mentioned above, I hope to establish the passage from the real to the imaginary based on a direct and rigorous analysis; my researches have thus led me to the method which is the object of this memoir...
>
> *A. L. Cauchy, 1827*

In the previous chapter, we discussed several preliminary ideas in complex analysis: open sets in $\mathbb{C}$, holomorphic functions, and integration along curves. The first remarkable result of the theory exhibits a deep connection between these notions. Loosely stated, Cauchy's theorem says that if f is holomorphic in an open set Ω and $\gamma \subset \Omega$ is a closed curve whose interior is also contained in Ω then

$$(1) \qquad \int_\gamma f(z)\, dz = 0.$$

Many results that follow, and in particular the calculus of residues, are related in one way or another to this fact.

A precise and general formulation of Cauchy's theorem requires defining unambiguously the "interior" of a curve, and this is not always an easy task. At this early stage of our study, we shall make use of the device of limiting ourselves to regions whose boundaries are curves that are "toy contours." As the name suggests, these are closed curves whose visualization is so simple that the notion of their interior will be unambiguous, and the proof of Cauchy's theorem in this setting will be quite direct. For many applications, it will suffice to restrict ourselves to these types of curves. At a later stage, we take up the questions related to more general curves, their interiors, and the extended form of Cauchy's theorem.

Our initial version of Cauchy's theorem begins with the observation that it suffices that f have a primitive in Ω, by Corollary 3.3 in Chapter 1. The existence of such a primitive for toy contours will follow from a theorem of Goursat (which is itself a simple special case)[1] that asserts that if f is holomorphic in an open set that contains a triangle T and its interior, then

$$\int_T f(z)\, dz = 0.$$

It is noteworthy that this simple case of Cauchy's theorem suffices to prove some of its more complicated versions. From there, we can prove the existence of primitives in the interior of some simple regions, and therefore prove Cauchy's theorem in that setting. As a first application of this viewpoint, we evaluate several real integrals by using appropriate toy contours.

The above ideas also lead us to a central result of this chapter, the Cauchy integral formula; this states that if f is holomorphic in an open set containing a circle C and its interior, then for all z inside C,

$$f(z) = \frac{1}{2\pi i} \int_C \frac{f(\zeta)}{\zeta - z}\, d\zeta.$$

Differentiation of this identity yields other integral formulas, and in particular we obtain the regularity of holomorphic functions. This is remarkable, since holomorphicity assumed only the existence of the first derivative, and yet we obtain as a consequence the existence of derivatives of all orders. (An analogous statement is decisively false in the case of real variables!)

The theory developed up to that point already has a number of noteworthy consequences:

- The property at the base of "analytic continuation," namely that a holomorphic function is determined by its restriction to any open subset of its domain of definition. This is a consequence of the fact that holomorphic functions have power series expansions.

---

[1] Goursat's result came after Cauchy's theorem, and its interest is the technical fact that its proof requires only the existence of the complex derivative at each point, and not its continuity. For the earlier proof, see Exercise 5.

- Liouville's theorem, which yields a quick proof of the fundamental theorem of algebra.

- Morera's theorem, which gives a simple integral characterization of holomorphic functions, and shows that these functions are preserved under uniform limits.

1 Goursat's theorem

Corollary 3.3 in the previous chapter says that if f has a primitive in an open set Ω, then

$$\int_\gamma f(z)\, dz = 0$$

for any closed curve γ in Ω. Conversely, if we can show that the above relation holds for some types of curves γ, then a primitive will exist. Our starting point is Goursat's theorem, from which in effect we shall deduce most of the other results in this chapter.

Theorem 1.1 *If Ω is an open set in $\mathbb{C}$, and $T \subset \Omega$ a triangle whose interior is also contained in Ω, then*

$$\int_T f(z)\, dz = 0$$

whenever f is holomorphic in Ω.

Proof.   We call $T^{(0)}$ our original triangle (with a fixed orientation which we choose to be positive), and let $d^{(0)}$ and $p^{(0)}$ denote the diameter and perimeter of $T^{(0)}$, respectively. The first step in our construction consists of bisecting each side of the triangle and connecting the midpoints. This creates four new smaller triangles, denoted $T_1^{(1)}, T_2^{(1)}, T_3^{(1)}$, and $T_4^{(1)}$ that are similar to the original triangle. The construction and orientation of each triangle are illustrated in Figure 1. The orientation is chosen to be consistent with that of the original triangle, and so after cancellations arising from integrating over the same side in two opposite directions, we have

(2)
$$\int_{T^{(0)}} f(z)\, dz = \int_{T_1^{(1)}} f(z)\, dz + \int_{T_2^{(1)}} f(z)\, dz + \int_{T_3^{(1)}} f(z)\, dz + \int_{T_4^{(1)}} f(z)\, dz.$$

For some j we must have

$$\left| \int_{T^{(0)}} f(z)\, dz \right| \leq 4 \left| \int_{T_j^{(1)}} f(z)\, dz \right|,$$

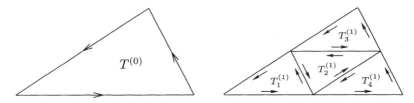

Figure 1. Bisection of $T^{(0)}$

for otherwise (2) would be contradicted. We choose a triangle that satisfies this inequality, and rename it $T^{(1)}$. Observe that if $d^{(1)}$ and $p^{(1)}$ denote the diameter and perimeter of $T^{(1)}$, respectively, then $d^{(1)} = (1/2)d^{(0)}$ and $p^{(1)} = (1/2)p^{(0)}$. We now repeat this process for the triangle $T^{(1)}$, bisecting it into four smaller triangles. Continuing this process, we obtain a sequence of triangles

$$T^{(0)}, T^{(1)}, \ldots, T^{(n)}, \ldots$$

with the properties that

$$\left| \int_{T^{(0)}} f(z)\, dz \right| \leq 4^n \left| \int_{T^{(n)}} f(z)\, dz \right|$$

and

$$d^{(n)} = 2^{-n} d^{(0)}, \qquad p^{(n)} = 2^{-n} p^{(0)}$$

where $d^{(n)}$ and $p^{(n)}$ denote the diameter and perimeter of $T^{(n)}$, respectively. We also denote by $\mathcal{T}^{(n)}$ the *solid* closed triangle with boundary $T^{(n)}$, and observe that our construction yields a sequence of nested compact sets

$$\mathcal{T}^{(0)} \supset \mathcal{T}^{(1)} \supset \cdots \supset \mathcal{T}^{(n)} \supset \cdots$$

whose diameter goes to 0. By Proposition 1.4 in Chapter 1, there exists a unique point z_0 that belongs to all the solid triangles $\mathcal{T}^{(n)}$. Since f is holomorphic at z_0 we can write

$$f(z) = f(z_0) + f'(z_0)(z - z_0) + \psi(z)(z - z_0),$$

where $\psi(z) \to 0$ as $z \to z_0$. Since the constant $f(z_0)$ and the linear function $f'(z_0)(z - z_0)$ have primitives, we can integrate the above equality using Corollary 3.3 in the previous chapter, and obtain

(3) $$\int_{T^{(n)}} f(z)\, dz = \int_{T^{(n)}} \psi(z)(z - z_0)\, dz.$$

Now z_0 belongs to the closure of the solid triangle $T^{(n)}$ and z to its boundary, so we must have $|z - z_0| \leq d^{(n)}$, and using (3) we get, by (iii) in Proposition 3.1 of the previous chapter, the estimate

$$\left| \int_{T^{(n)}} f(z)\, dz \right| \leq \epsilon_n d^{(n)} p^{(n)},$$

where $\epsilon_n = \sup_{z \in T^{(n)}} |\psi(z)| \to 0$ as $n \to \infty$. Therefore

$$\left| \int_{T^{(n)}} f(z)\, dz \right| \leq \epsilon_n 4^{-n} d^{(0)} p^{(0)},$$

which yields our final estimate

$$\left| \int_{T^{(0)}} f(z)\, dz \right| \leq 4^n \left| \int_{T^{(n)}} f(z)\, dz \right| \leq \epsilon_n d^{(0)} p^{(0)}.$$

Letting $n \to \infty$ concludes the proof since $\epsilon_n \to 0$.

Corollary 1.2 *If f is holomorphic in an open set Ω that contains a rectangle R and its interior, then*

$$\int_R f(z)\, dz = 0.$$

This is immediate since we first choose an orientation as in Figure 2 and note that

$$\int_R f(z)\, dz = \int_{T_1} f(z)\, dz + \int_{T_2} f(z)\, dz.$$

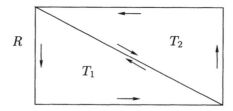

Figure 2. A rectangle as the union of two triangles

2 Local existence of primitives and Cauchy's theorem in a disc

We first prove the existence of primitives in a disc as a consequence of Goursat's theorem.

Theorem 2.1 *A holomorphic function in an open disc has a primitive in that disc.*

Proof. After a translation, we may assume without loss of generality that the disc, say D, is centered at the origin. Given a point $z \in D$, consider the piecewise-smooth curve that joins 0 to z first by moving in the horizontal direction from 0 to $\tilde{z}$ where $\tilde{z} = \mathrm{Re}(z)$, and then in the vertical direction from $\tilde{z}$ to z. We choose the orientation from 0 to z, and denote this polygonal line (which consists of at most two segments) by γ_z, as shown on Figure 3.

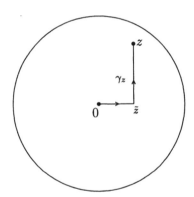

Figure 3. The polygonal line γ_z

Define

$$F(z) = \int_{\gamma_z} f(w)\,dw.$$

The choice of γ_z gives an unambiguous definition of the function $F(z)$. We contend that F is holomorphic in D and $F'(z) = f(z)$. To prove this, fix $z \in D$ and let $h \in \mathbb{C}$ be so small that $z + h$ also belongs to the disc. Now consider the difference

$$F(z + h) - F(z) = \int_{\gamma_{z+h}} f(w)\,dw - \int_{\gamma_z} f(w)\,dw.$$

The function f is first integrated along γ_{z+h} with the original orientation, and then along γ_z with the reverse orientation (because of the minus sign in front of the second integral). This corresponds to (a) in Figure 4. Since we integrate f over the line segment starting at the origin in two opposite directions, it cancels, leaving us with the contour in (b). Then, we complete the square and triangle as shown in (c), so that after an application of Goursat's theorem for triangles and rectangles we are left with the line segment from z to $z + h$ as given in (d).

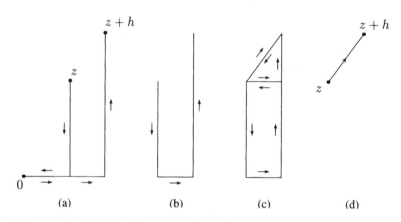

Figure 4. Relation between the polygonal lines γ_z and γ_{z+h}

Hence the above cancellations yield

$$F(z + h) - F(z) = \int_\eta f(w)\, dw$$

where η is the straight line segment from z to $z + h$. Since f is continuous at z we can write

$$f(w) = f(z) + \psi(w)$$

where $\psi(w) \to 0$ as $w \to z$. Therefore

(4)
$$F(z + h) - F(z) = \int_\eta f(z)\, dw + \int_\eta \psi(w)\, dw = f(z) \int_\eta dw + \int_\eta \psi(w)\, dw.$$

On the one hand, the constant 1 has w as a primitive, so the first integral is simply h by an application of Theorem 3.2 in Chapter 1. On the other hand, we have the following estimate:

$$\left| \int_\eta \psi(w)\, dw \right| \leq \sup_{w \in \eta} |\psi(w)|\, |h|.$$

Since the supremum above goes to 0 as h tends to 0, we conclude from equation (4) that

$$\lim_{h \to 0} \frac{F(z+h) - F(z)}{h} = f(z),$$

thereby proving that F is a primitive for f on the disc.

This theorem says that locally, every holomorphic function has a primitive. It is crucial to realize, however, that the theorem is true not only for arbitrary discs, but also for other sets as well. We shall return to this point shortly in our discussion of "toy contours."

Theorem 2.2 (Cauchy's theorem for a disc) *If f is holomorphic in a disc, then*

$$\int_\gamma f(z)\, dz = 0$$

for any closed curve γ in that disc.

Proof. Since f has a primitive, we can apply Corollary 3.3 of Chapter 1.

Corollary 2.3 *Suppose f is holomorphic in an open set containing the circle C and its interior. Then*

$$\int_C f(z)\, dz = 0.$$

Proof. Let D be the disc with boundary circle C. Then there exists a slightly larger disc D' which contains D and so that f is holomorphic on D'. We may now apply Cauchy's theorem in D' to conclude that $\int_C f(z)\, dz = 0$.

In fact, the proofs of the theorem and its corollary apply whenever we can define without ambiguity the "interior" of a contour, and construct appropriate polygonal paths in an open neighborhood of that contour and its interior. In the case of the circle, whose interior is the disc, there was no problem since the geometry of the disc made it simple to travel horizontally and vertically inside it.

The following definition is loosely stated, although its applications will be clear and unambiguous. We call a **toy contour** any closed curve where the notion of interior is obvious, and a construction similar to that in Theorem 2.1 is possible in a neighborhood of the curve and its interior. Its positive orientation is that for which the interior is to the left as we travel along the toy contour. This is consistent with the definition of the positive orientation of a circle. For example, circles, triangles, and rectangles are toy contours, since in each case we can modify (and actually copy) the argument given previously.

Another important example of a toy contour is the "keyhole" Γ (illustrated in Figure 5), which we shall put to use in the proof of the Cauchy integral formula. It consists of two almost complete circles, one large

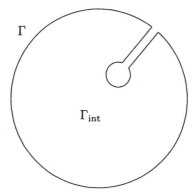

Figure 5. The keyhole contour

and one small, connected by a narrow corridor. The interior of Γ, which we denote by Γ_{int}, is clearly that region enclosed by the curve, and can be given precise meaning with enough work. We fix a point z_0 in that interior. If f is holomorphic in a neighborhood of Γ and its interior, then it is holomorphic in the inside of a slightly larger keyhole, say Λ, whose interior Λ_{int} contains $\Gamma \cup \Gamma_{\text{int}}$. If $z \in \Lambda_{\text{int}}$, let γ_z denote any curve contained inside Λ_{int} connecting z_0 to z, and which consists of finitely many horizontal or vertical segments (as in Figure 6). If η_z is any other such curve, the rectangle version of Goursat's theorem (Corollary 1.2) makes it possible to see that

$$\int_{\gamma_z} f(w)\,dw = \int_{\eta_z} f(w)\,dw\,,$$

and we may therefore define F unambiguously in Λ_{int}.

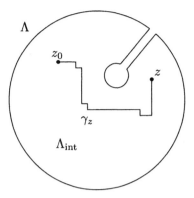

Figure 6. A curve γ_z

Arguing as above allows us to show that F is a primitive of f in Λ_{int} and therefore $\int_\Gamma f(z)\,dz = 0$.

The important point is that for a toy contour γ we easily have that

$$\int_\gamma f(z)\,dz = 0\,,$$

whenever f is holomorphic in an open set that contains the contour γ and its interior.

We shall understand this better when we see in Corollary 5.3 of the next chapter that a similar conclusion holds if γ is a closed curve in a "simply connected" region in which f is holomorphic. Related to this is the fact that the interior of a simple closed and piecewise-smooth curve (in particular, a toy contour) is simply connected, as shown in Theorem 2.2 of Appendix B.

Other examples of toy contours which we shall encounter in applications and for which Cauchy's theorem and its corollary also hold are given in Figure 7.

3 Evaluation of some integrals

Here we take up the idea that originally motivated Cauchy. We shall show by several examples how some integrals may be evaluated by the use of his theorem. A more systematic approach, in terms of the calculus of residues, may be found in the next chapter.

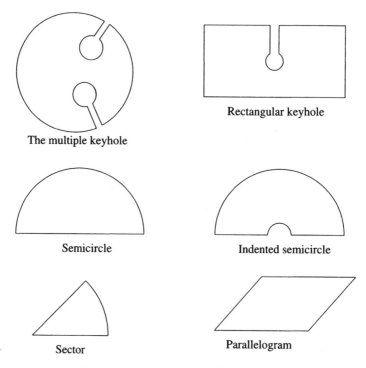

The multiple keyhole

Rectangular keyhole

Semicircle

Indented semicircle

Sector

Parallelogram

Figure 7. Examples of toy contours

EXAMPLE 1. We show that if $\xi \in \mathbb{R}$, then

$$(5) \qquad e^{-\pi\xi^2} = \int_{-\infty}^{\infty} e^{-\pi x^2} e^{-2\pi i x \xi} \, dx.$$

This gives a new proof of the fact that $e^{-\pi x^2}$ is its own Fourier transform, a fact we proved in Theorem 1.4 of Chapter 5 in Book I.

If $\xi = 0$, the formula is precisely the known integral[2]

$$1 = \int_{-\infty}^{\infty} e^{-\pi x^2} \, dx.$$

Now suppose that $\xi > 0$, and consider the function $f(z) = e^{-\pi z^2}$, which is entire, and in particular holomorphic in the interior of the toy contour γ_R depicted in Figure 8.

---

[2]An alternate derivation follows from the fact that $\Gamma(1/2) = \sqrt{\pi}$, where Γ is the gamma function in Chapter 6.

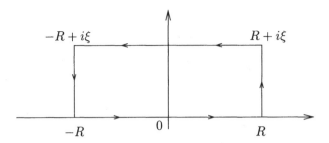

Figure 8. The contour γ_R in Example 1

The contour γ_R consists of a rectangle with vertices $R, R + i\xi, -R + i\xi, -R$ and the positive counterclockwise orientation. By Cauchy's theorem,

$$(6) \qquad \int_{\gamma_R} f(z)\, dz = 0.$$

The integral over the real segment is simply

$$\int_{-R}^{R} e^{-\pi x^2}\, dx\,,$$

which converges to 1 as $R \to \infty$. The integral on the vertical side on the right is

$$I(R) = \int_0^\xi f(R + iy) i\, dy = \int_0^\xi e^{-\pi(R^2 + 2iRy - y^2)} i\, dy.$$

This integral goes to 0 as $R \to \infty$ since ξ is fixed and we may estimate it by

$$|I(R)| \le C e^{-\pi R^2}.$$

Similarly, the integral over the vertical segment on the left also goes to 0 as $R \to \infty$ for the same reasons. Finally, the integral over the horizontal segment on top is

$$\int_R^{-R} e^{-\pi(x+i\xi)^2}\, dx = -e^{\pi \xi^2} \int_{-R}^{R} e^{-\pi x^2} e^{-2\pi i x\xi}\, dx.$$

Therefore, we find in the limit as $R \to \infty$ that (6) gives

$$0 = 1 - e^{\pi \xi^2} \int_{-\infty}^{\infty} e^{-\pi x^2} e^{-2\pi i x\xi}\, dx,$$

and our desired formula is established. In the case $\xi < 0$, we then consider
the symmetric rectangle, in the lower half-plane.

The technique of shifting the contour of integration, which was used
in the previous example, has many other applications. Note that the
original integral (5) is taken over the real line, which by an application
of Cauchy's theorem is then shifted upwards or downwards (depending
on the sign of ξ) in the complex plane.

EXAMPLE 2. Another classical example is

$$\int_0^\infty \frac{1 - \cos x}{x^2}\, dx = \frac{\pi}{2}.$$

Here we consider the function $f(z) = (1 - e^{iz})/z^2$, and we integrate over
the indented semicircle in the upper half-plane positioned on the x-axis,
as shown in Figure 9.

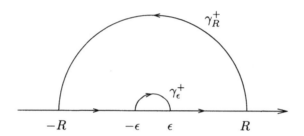

Figure 9. The indented semicircle of Example 2

If we denote by γ_ϵ^+ and γ_R^+ the semicircles of radii ϵ and R with negative
and positive orientations respectively, Cauchy's theorem gives

$$\int_{-R}^{-\epsilon} \frac{1 - e^{ix}}{x^2}\, dx + \int_{\gamma_\epsilon^+} \frac{1 - e^{iz}}{z^2}\, dz + \int_{\epsilon}^{R} \frac{1 - e^{ix}}{x^2}\, dx + \int_{\gamma_R^+} \frac{1 - e^{iz}}{z^2}\, dz = 0.$$

First we let $R \to \infty$ and observe that

$$\left| \frac{1 - e^{iz}}{z^2} \right| \le \frac{2}{|z|^2},$$

so the integral over γ_R^+ goes to zero. Therefore

$$\int_{|x|\ge\epsilon} \frac{1 - e^{ix}}{x^2}\, dx = -\int_{\gamma_\epsilon^+} \frac{1 - e^{iz}}{z^2}\, dz.$$

Next, note that

$$f(z) = \frac{-iz}{z^2} + E(z)$$

where $E(z)$ is bounded as $z \to 0$, while on γ_ϵ^+ we have $z = \epsilon e^{i\theta}$ and $dz = i\epsilon e^{i\theta} d\theta$. Thus

$$\int_{\gamma_\epsilon^+} \frac{1 - e^{iz}}{z^2} dz \to \int_\pi^0 (-ii)\, d\theta = -\pi \quad \text{as } \epsilon \to 0.$$

Taking real parts then yields

$$\int_{-\infty}^{\infty} \frac{1 - \cos x}{x^2} dx = \pi.$$

Since the integrand is even, the desired formula is proved.

4 Cauchy's integral formulas

Representation formulas, and in particular integral representation formulas, play an important role in mathematics, since they allow us to recover a function on a large set from its behavior on a smaller set. For example, we saw in Book I that a solution of the steady-state heat equation in the disc was completely determined by its boundary values on the circle via a convolution with the Poisson kernel

$$(7) \qquad u(r,\theta) = \frac{1}{2\pi} \int_0^{2\pi} P_r(\theta - \varphi) u(1, \varphi)\, d\varphi.$$

In the case of holomorphic functions, the situation is analogous, which is not surprising since the real and imaginary parts of a holomorphic function are harmonic.[3] Here, we will prove an integral representation formula in a manner that is independent of the theory of harmonic functions. In fact, it is also possible to recover the Poisson integral formula (7) as a consequence of the next theorem (see Exercises 11 and 12).

Theorem 4.1 *Suppose f is holomorphic in an open set that contains the closure of a disc D. If C denotes the boundary circle of this disc with the positive orientation, then*

$$f(z) = \frac{1}{2\pi i} \int_C \frac{f(\zeta)}{\zeta - z} d\zeta \quad \text{for any point } z \in D.$$

---

[3]This fact is an immediate consequence of the Cauchy-Riemann equations. We refer the reader to Exercise 11 in Chapter 1.

Proof.   Fix $z \in D$ and consider the "keyhole" $\Gamma_{\delta,\epsilon}$ which omits the point z as shown in Figure 10.

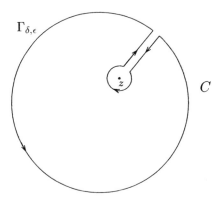

Figure 10. The keyhole $\Gamma_{\delta,\epsilon}$

Here δ is the width of the corridor, and ϵ the radius of the small circle centered at z. Since the function $F(\zeta) = f(\zeta)/(\zeta - z)$ is holomorphic away from the point $\zeta = z$, we have

$$\int_{\Gamma_{\delta,\epsilon}} F(\zeta)\, d\zeta = 0$$

by Cauchy's theorem for the chosen toy contour. Now we make the corridor narrower by letting δ tend to 0, and use the continuity of F to see that in the limit, the integrals over the two sides of the corridor cancel out. The remaining part consists of two curves, the large boundary circle C with the positive orientation, and a small circle C_ϵ centered at z of radius ϵ and oriented negatively, that is, clockwise. To see what happens to the integral over the small circle we write

(8) $$F(\zeta) = \frac{f(\zeta) - f(z)}{\zeta - z} + \frac{f(z)}{\zeta - z}$$

and note that since f is holomorphic the first term on the right-hand side of (8) is bounded so that its integral over C_ϵ goes to 0 as $\epsilon \to 0$. To conclude the proof, it suffices to observe that

$$\int_{C_\epsilon} \frac{f(z)}{\zeta - z}\, d\zeta = f(z) \int_{C_\epsilon} \frac{d\zeta}{\zeta - z}$$

$$= -f(z) \int_0^{2\pi} \frac{\epsilon i e^{-it}}{\epsilon e^{-it}}\, dt$$

$$= -f(z) 2\pi i\,,$$

so that in the limit we find

$$0 = \int_C \frac{f(\zeta)}{\zeta - z}\, d\zeta - 2\pi i f(z)\,,$$

as was to be shown.

Remarks. Our earlier discussion of toy contours provides simple extensions of the Cauchy integral formula; for instance, if f is holomorphic in an open set that contains a (positively oriented) rectangle R and its interior, then

$$f(z) = \frac{1}{2\pi i} \int_R \frac{f(\zeta)}{\zeta - z}\, d\zeta\,,$$

whenever z belongs to the interior of R. To establish this result, it suffices to repeat the proof of Theorem 4.1 replacing the "circular" keyhole by a "rectangular" keyhole.

It should also be noted that the above integral vanishes when z is outside R, since in this case $F(\zeta) = f(\zeta)/(\zeta - z)$ is holomorphic inside R. Of course, a similar result also holds for the circle or any other toy contour.

As a corollary to the Cauchy integral formula, we arrive at a second remarkable fact about holomorphic functions, namely their regularity. We also obtain further integral formulas expressing the derivatives of f inside the disc in terms of the values of f on the boundary.

Corollary 4.2 *If f is holomorphic in an open set Ω, then f has infinitely many complex derivatives in Ω. Moreover, if $C \subset \Omega$ is a circle whose interior is also contained in Ω, then*

$$f^{(n)}(z) = \frac{n!}{2\pi i} \int_C \frac{f(\zeta)}{(\zeta - z)^{n+1}}\, d\zeta$$

for all z in the interior of C.

We recall that, as in the above theorem, we take the circle C to have positive orientation.

Proof. The proof is by induction on n, the case $n = 0$ being simply the Cauchy integral formula. Suppose that f has up to $n - 1$ complex derivatives and that

$$f^{(n-1)}(z) = \frac{(n-1)!}{2\pi i} \int_C \frac{f(\zeta)}{(\zeta - z)^n} \, d\zeta.$$

Now for h small, the difference quotient for $f^{(n-1)}$ takes the form

(9)
$$\frac{f^{(n-1)}(z+h) - f^{(n-1)}(z)}{h} =$$

$$\frac{(n-1)!}{2\pi i} \int_C f(\zeta) \frac{1}{h} \left[\frac{1}{(\zeta - z - h)^n} - \frac{1}{(\zeta - z)^n} \right] d\zeta.$$

We now recall that

$$A^n - B^n = (A - B)[A^{n-1} + A^{n-2}B + \cdots + AB^{n-2} + B^{n-1}].$$

With $A = 1/(\zeta - z - h)$ and $B = 1/(\zeta - z)$, we see that the term in brackets in equation (9) is equal to

$$\frac{h}{(\zeta - z - h)(\zeta - z)}[A^{n-1} + A^{n-2}B + \cdots + AB^{n-2} + B^{n-1}].$$

But observe that if h is small, then $z + h$ and z stay at a finite distance from the boundary circle C, so in the limit as h tends to 0, we find that the quotient converges to

$$\frac{(n-1)!}{2\pi i} \int_C f(\zeta) \left[\frac{1}{(\zeta - z)^2} \right] \left[\frac{n}{(\zeta - z)^{n-1}} \right] d\zeta = \frac{n!}{2\pi i} \int_C \frac{f(\zeta)}{(\zeta - z)^{n+1}} \, d\zeta,$$

which completes the induction argument and proves the theorem.

From now on, we call the formulas of Theorem 4.1 and Corollary 4.2 the **Cauchy integral formulas**.

Corollary 4.3 (Cauchy inequalities) *If f is holomorphic in an open set that contains the closure of a disc D centered at z_0 and of radius R, then*

$$|f^{(n)}(z_0)| \leq \frac{n! \|f\|_C}{R^n},$$

where $\|f\|_C = \sup_{z \in C} |f(z)|$ denotes the supremum of $|f|$ on the boundary circle C.

Proof. Applying the Cauchy integral formula for $f^{(n)}(z_0)$, we obtain

$$
\begin{aligned}
|f^{(n)}(z_0)| &= \left| \frac{n!}{2\pi i} \int_C \frac{f(\zeta)}{(\zeta - z_0)^{n+1}} \, d\zeta \right| \\
&= \frac{n!}{2\pi} \left| \int_0^{2\pi} \frac{f(z_0 + Re^{i\theta})}{(Re^{i\theta})^{n+1}} Rie^{i\theta} \, d\theta \right| \\
&\leq \frac{n!}{2\pi} \frac{\|f\|_C}{R^n} 2\pi.
\end{aligned}
$$

Another striking consequence of the Cauchy integral formula is its connection with power series. In Chapter 1, we proved that a power series is holomorphic in the interior of its disc of convergence, and promised a proof of a converse, which is the content of the next theorem.

Theorem 4.4 *Suppose f is holomorphic in an open set Ω. If D is a disc centered at z_0 and whose closure is contained in Ω, then f has a power series expansion at z_0*

$$
f(z) = \sum_{n=0}^{\infty} a_n (z - z_0)^n
$$

for all $z \in D$, and the coefficients are given by

$$
a_n = \frac{f^{(n)}(z_0)}{n!} \quad \text{for all } n \geq 0.
$$

Proof. Fix $z \in D$. By the Cauchy integral formula, we have

$$
\tag{10} f(z) = \frac{1}{2\pi i} \int_C \frac{f(\zeta)}{\zeta - z} \, d\zeta,
$$

where C denotes the boundary of the disc and $z \in D$. The idea is to write

$$
\tag{11} \frac{1}{\zeta - z} = \frac{1}{\zeta - z_0 - (z - z_0)} = \frac{1}{\zeta - z_0} \frac{1}{1 - \left(\frac{z - z_0}{\zeta - z_0} \right)},
$$

and use the geometric series expansion. Since $\zeta \in C$ and $z \in D$ is fixed, there exists $0 < r < 1$ such that

$$
\left| \frac{z - z_0}{\zeta - z_0} \right| < r,
$$

therefore

(12)
$$\frac{1}{1 - \left(\frac{z - z_0}{\zeta - z_0}\right)} = \sum_{n=0}^{\infty} \left(\frac{z - z_0}{\zeta - z_0}\right)^n,$$

where the series converges uniformly for $\zeta \in C$. This allows us to interchange the infinite sum with the integral when we combine (10), (11), and (12), thereby obtaining

$$f(z) = \sum_{n=0}^{\infty} \left(\frac{1}{2\pi i} \int_C \frac{f(\zeta)}{(\zeta - z_0)^{n+1}} \, d\zeta\right) \cdot (z - z_0)^n.$$

This proves the power series expansion; further the use of the Cauchy integral formulas for the derivatives (or simply differentiation of the series) proves the formula for a_n.

Observe that since power series define indefinitely (complex) differentiable functions, the theorem gives another proof that a holomorphic function is automatically indefinitely differentiable.

Another important observation is that the power series expansion of f centered at z_0 converges in any disc, no matter how large, as long as its closure is contained in Ω. In particular, if f is entire (that is, holomorphic on all of $\mathbb{C}$), the theorem implies that f has a power series expansion around 0, say $f(z) = \sum_{n=0}^{\infty} a_n z^n$, that converges in all of $\mathbb{C}$.

Corollary 4.5 (Liouville's theorem) *If f is entire and bounded, then f is constant.*

Proof. It suffices to prove that $f' = 0$, since $\mathbb{C}$ is connected, and we may then apply Corollary 3.4 in Chapter 1.

For each $z_0 \in \mathbb{C}$ and all $R > 0$, the Cauchy inequalities yield

$$|f'(z_0)| \leq \frac{B}{R}$$

where B is a bound for f. Letting $R \to \infty$ gives the desired result.

As an application of our work so far, we can give an elegant proof of the fundamental theorem of algebra.

Corollary 4.6 *Every non-constant polynomial $P(z) = a_n z^n + \cdots + a_0$ with complex coefficients has a root in $\mathbb{C}$.*

Proof. If P has no roots, then $1/P(z)$ is a bounded holomorphic function. To see this, we can of course assume that $a_n \neq 0$, and write

$$\frac{P(z)}{z^n} = a_n + \left(\frac{a_{n-1}}{z} + \cdots + \frac{a_0}{z^n}\right)$$

whenever $z \neq 0$. Since each term in the parentheses goes to 0 as $|z| \to \infty$ we conclude that there exists $R > 0$ so that if $c = |a_n|/2$, then

$$|P(z)| \geq c|z|^n \quad \text{whenever } |z| > R.$$

In particular, P is bounded from below when $|z| > R$. Since P is continuous and has no roots in the disc $|z| \leq R$, it is bounded from below in that disc as well, thereby proving our claim.

By Liouville's theorem we then conclude that $1/P$ is constant. This contradicts our assumption that P is non-constant and proves the corollary.

Corollary 4.7 *Every polynomial* $P(z) = a_n z^n + \cdots + a_0$ *of degree* $n \geq 1$ *has precisely* n *roots in* $\mathbb{C}$. *If these roots are denoted by* $w_1, \ldots, w_n$, *then* P *can be factored as*

$$P(z) = a_n(z - w_1)(z - w_2) \cdots (z - w_n).$$

Proof. By the previous result P has a root, say w_1. Then, writing $z = (z - w_1) + w_1$, inserting this expression for z in P, and using the binomial formula we get

$$P(z) = b_n(z - w_1)^n + \cdots + b_1(z - w_1) + b_0,$$

where $b_0, \ldots, b_{n-1}$ are new coefficients, and $b_n = a_n$. Since $P(w_1) = 0$, we find that $b_0 = 0$, therefore

$$P(z) = (z - w_1)\left[b_n(z - w_1)^{n-1} + \cdots + b_1\right] = (z - w_1)Q(z),$$

where Q is a polynomial of degree $n - 1$. By induction on the degree of the polynomial, we conclude that $P(z)$ has n roots and can be expressed as

$$P(z) = c(z - w_1)(z - w_2) \cdots (z - w_n)$$

for some $c \in \mathbb{C}$. Expanding the right-hand side, we realize that the coefficient of z^n is c and therefore $c = a_n$ as claimed.

Finally, we end this section with a discussion of analytic continuation (the third of the "miracles" we mentioned in the introduction). It states that the "genetic code" of a holomorphic function is determined (that is, the function is fixed) if we know its values on appropriate arbitrarily small subsets. Note that in the theorem below, Ω is assumed connected.

Theorem 4.8 *Suppose f is a holomorphic function in a region Ω that vanishes on a sequence of distinct points with a limit point in Ω. Then f is identically 0.*

In other words, if the zeros of a holomorphic function f in the connected open set Ω accumulate *in Ω*, then $f = 0$.

Proof. Suppose that $z_0 \in \Omega$ is a limit point for the sequence $\{w_k\}_{k=1}^{\infty}$ and that $f(w_k) = 0$. First, we show that f is identically zero in a small disc containing z_0. For that, we choose a disc D centered at z_0 and contained in Ω, and consider the power series expansion of f in that disc

$$f(z) = \sum_{n=0}^{\infty} a_n (z - z_0)^n.$$

If f is not identically zero, there exists a smallest integer m such that $a_m \neq 0$. But then we can write

$$f(z) = a_m (z - z_0)^m (1 + g(z - z_0)),$$

where $g(z - z_0)$ converges to 0 as $z \to z_0$. Taking $z = w_k \neq z_0$ for a sequence of points converging to z_0, we get a contradiction since $a_m(w_k - z_0)^m \neq 0$ and $1 + g(w_k - z_0) \neq 0$, but $f(w_k) = 0$.

We conclude the proof using the fact that Ω is connected. Let U denote the interior of the set of points where $f(z) = 0$. Then U is open by definition and non-empty by the argument just given. The set U is also closed since if $z_n \in U$ and $z_n \to z$, then $f(z) = 0$ by continuity, and f vanishes in a neighborhood of z by the argument above. Hence $z \in U$. Now if we let V denote the complement of U in Ω, we conclude that U and V are both open, disjoint, and

$$\Omega = U \cup V.$$

Since Ω is connected we conclude that either U or V is empty. (Here we use one of the two equivalent definitions of connectedness discussed in Chapter 1.) Since $z_0 \in U$, we find that $U = \Omega$ and the proof is complete.

An immediate consequence of the theorem is the following.

Corollary 4.9 *Suppose f and g are holomorphic in a region Ω and $f(z) = g(z)$ for all z in some non-empty open subset of Ω (or more generally for z in some sequence of distinct points with limit point in Ω). Then $f(z) = g(z)$ throughout Ω.*

Suppose we are given a pair of functions f and F analytic in regions Ω and Ω', respectively, with $\Omega \subset \Omega'$. If the two functions agree on the smaller set Ω, we say that F is an **analytic continuation** of f into the region Ω'. The corollary then guarantees that there can be only one such analytic continuation, since F is uniquely determined by f.

5 Further applications

We gather in this section various consequences of the results proved so far.

5.1 Morera's theorem

A direct application of what was proved here is a converse of Cauchy's theorem.

Theorem 5.1 *Suppose f is a continuous function in the open disc D such that for any triangle T contained in D*

$$\int_T f(z)\, dz = 0,$$

then f is holomorphic.

Proof. By the proof of Theorem 2.1 the function f has a primitive F in D that satisfies $F' = f$. By the regularity theorem, we know that F is indefinitely (and hence twice) complex differentiable, and therefore f is holomorphic.

5.2 Sequences of holomorphic functions

Theorem 5.2 *If $\{f_n\}_{n=1}^{\infty}$ is a sequence of holomorphic functions that converges uniformly to a function f in every compact subset of Ω, then f is holomorphic in Ω.*

Proof. Let D be any disc whose closure is contained in Ω and T any triangle in that disc. Then, since each f_n is holomorphic, Goursat's theorem implies

$$\int_T f_n(z)\, dz = 0 \quad \text{for all } n.$$

By assumption $f_n \to f$ uniformly in the closure of D, so f is continuous and

$$\int_T f_n(z)\, dz \to \int_T f(z)\, dz.$$

As a result, we find $\int_T f(z)\,dz = 0$, and by Morera's theorem, we conclude that f is holomorphic in D. Since this conclusion is true for every D whose closure is contained in Ω, we find that f is holomorphic in all of Ω.

This is a striking result that is obviously not true in the case of real variables: the uniform limit of continuously differentiable functions need not be differentiable. For example, we know that every continuous function on $[0,1]$ can be approximated uniformly by polynomials, by Weierstrass's theorem (see Chapter 5, Book I), yet not every continuous function is differentiable.

We can go one step further and deduce convergence theorems for the sequence of derivatives. Recall that if f is a power series with radius of convergence R, then f' can be obtained by differentiating term by term the series for f, and moreover f' has radius of convergence R. (See Theorem 2.6 in Chapter 1.) This implies in particular that if S_n are the partial sums of f, then S'_n converges uniformly to f' on every compact subset of the disc of convergence of f. The next theorem generalizes this fact.

Theorem 5.3 *Under the hypotheses of the previous theorem, the sequence of derivatives $\{f'_n\}_{n=1}^\infty$ converges uniformly to f' on every compact subset of Ω.*

Proof. We may assume without loss of generality that the sequence of functions in the theorem converges uniformly on all of Ω. Given $\delta > 0$, let Ω_δ denote the subset of Ω defined by

$$\Omega_\delta = \{z \in \Omega : \overline{D_\delta}(z) \subset \Omega\}.$$

In other words, Ω_δ consists of all points in Ω which are at distance $> \delta$ from its boundary. To prove the theorem, it suffices to show that $\{f'_n\}$ converges uniformly to f' on Ω_δ for each δ. This is achieved by proving the following inequality:

$$(13) \qquad \sup_{z \in \Omega_\delta} |F'(z)| \leq \frac{1}{\delta} \sup_{\zeta \in \Omega} |F(\zeta)|$$

whenever F is holomorphic in Ω, since it can then be applied to $F = f_n - f$ to prove the desired fact. The inequality (13) follows at once from the Cauchy integral formula and the definition of Ω_δ, since for every $z \in \Omega_\delta$ the closure of $D_\delta(z)$ is contained in Ω and

$$F'(z) = \frac{1}{2\pi i} \int_{C_\delta(z)} \frac{F(\zeta)}{(\zeta - z)^2}\,d\zeta.$$

Hence,

$$
\begin{aligned}
|F'(z)| &\le \frac{1}{2\pi} \int_{C_\delta(z)} \frac{|F(\zeta)|}{|\zeta - z|^2} \, |d\zeta| \\
&\le \frac{1}{2\pi} \sup_{\zeta \in \Omega} |F(\zeta)| \frac{1}{\delta^2} \, 2\pi\delta \\
&= \frac{1}{\delta} \sup_{\zeta \in \Omega} |F(\zeta)|,
\end{aligned}
$$

as was to be shown.

Of course, there is nothing special about the first derivative, and in fact under the hypotheses of the last theorem, we may conclude (arguing as above) that for every $k \ge 0$ the sequence of kth derivatives $\{f_n^{(k)}\}_{n=1}^\infty$ converges uniformly to $f^{(k)}$ on every compact subset of Ω.

In practice, one often uses Theorem 5.2 to construct holomorphic functions (say, with a prescribed property) as a series

(14) $$F(z) = \sum_{n=1}^\infty f_n(z).$$

Indeed, if each f_n is holomorphic in a given region Ω of the complex plane, and the series converges uniformly in compact subsets of Ω, then Theorem 5.2 guarantees that F is also holomorphic in Ω. For instance, various special functions are often expressed in terms of a converging series like (14). A specific example is the Riemann zeta function discussed in Chapter 6.

We now turn to a variant of this idea, which consists of functions defined in terms of integrals.

5.3 Holomorphic functions defined in terms of integrals

As we shall see later in this book, a number of other special functions are defined in terms of integrals of the type

$$f(z) = \int_a^b F(z, s) \, ds,$$

or as limits of such integrals. Here, the function F is holomorphic in the first argument, and continuous in the second. The integral is taken in the sense of Riemann integration over the bounded interval $[a, b]$. The problem then is to establish that f is holomorphic.

In the next theorem, we impose a sufficient condition on F, often satisfied in practice, that easily implies that f is holomorphic.

After a simple linear change of variables, we may assume that $a = 0$ and $b = 1$.

Theorem 5.4 *Let $F(z, s)$ be defined for $(z, s) \in \Omega \times [0, 1]$ where Ω is an open set in $\mathbb{C}$. Suppose F satisfies the following properties:*

(i)   *$F(z, s)$ is holomorphic in z for each s.*

(ii) *F is continuous on $\Omega \times [0, 1]$.*

Then the function f defined on Ω by

$$f(z) = \int_0^1 F(z, s)\, ds$$

is holomorphic.

The second condition says that F is jointly continuous in both arguments.

To prove this result, it suffices to prove that f is holomorphic in any disc D contained in Ω, and by Morera's theorem this could be achieved by showing that for any triangle T contained in D we have

$$\int_T \int_0^1 F(z, s)\, ds\, dz = 0.$$

Interchanging the order of integration, and using property (i) would then yield the desired result. We can, however, get around the issue of justifying the change in the order of integration by arguing differently. The idea is to interpret the integral as a "uniform" limit of Riemann sums, and then apply the results of the previous section.

Proof. For each $n \geq 1$, we consider the Riemann sum

$$f_n(z) = (1/n) \sum_{k=1}^{n} F(z, k/n).$$

Then f_n is holomorphic in all of Ω by property (i), and we claim that on any disc D whose closure is contained in Ω, the sequence $\{f_n\}_{n=1}^{\infty}$ converges uniformly to f. To see this, we recall that a continuous function on a compact set is uniformly continuous, so if $\epsilon > 0$ there exists $\delta > 0$ such that

$$\sup_{z \in D} |F(z, s_1) - F(z, s_2)| < \epsilon \quad \text{whenever } |s_1 - s_2| < \delta.$$

Then, if $n > 1/\delta$, and $z \in D$ we have

$$
\begin{aligned}
|f_n(z) - f(z)| &= \left| \sum_{k=1}^{n} \int_{(k-1)/n}^{k/n} F(z, k/n) - F(z, s) \, ds \right| \\
&\leq \sum_{k=1}^{n} \int_{(k-1)/n}^{k/n} |F(z, k/n) - F(z, s)| \, ds \\
&< \sum_{k=1}^{n} \frac{\epsilon}{n} \\
&< \epsilon.
\end{aligned}
$$

This proves the claim, and by Theorem 5.2 we conclude that f is holomorphic in D. As a consequence, f is holomorphic in Ω, as was to be shown.

5.4 Schwarz reflection principle

In real analysis, there are various situations where one wishes to extend a function from a given set to a larger one. Several techniques exist that provide extensions for continuous functions, and more generally for functions with varying degrees of smoothness. Of course, the difficulty of the technique increases as we impose more conditions on the extension.

The situation is very different for holomorphic functions. Not only are these functions indefinitely differentiable in their domain of definition, but they also have additional characteristically rigid properties, which make them difficult to mold. For example, there exist holomorphic functions in a disc which are continuous on the closure of the disc, but which cannot be continued (analytically) into any region larger than the disc. (This phenomenon is discussed in Problem 1.) Another fact we have seen above is that holomorphic functions must be identically zero if they vanish on small open sets (or even, for example, a non-zero line segment).

It turns out that the theory developed in this chapter provides a simple extension phenomenon that is very useful in applications: the Schwarz reflection principle. The proof consists of two parts. First we define the extension, and then check that the resulting function is still holomorphic. We begin with this second point.

Let Ω be an open subset of $\mathbb{C}$ that is symmetric with respect to the real line, that is

$$
z \in \Omega \quad \text{if and only if} \quad \bar{z} \in \Omega.
$$

Let Ω^+ denote the part of Ω that lies in the upper half-plane and Ω^- that part that lies in the lower half-plane.

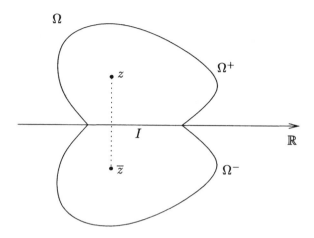

Figure 11. An open set symmetric across the real axis

Also, let $I = \Omega \cap \mathbb{R}$ so that I denotes the interior of that part of the boundary of Ω^+ and Ω^- that lies on the real axis. Then we have

$$\Omega^+ \cup I \cup \Omega^- = \Omega$$

and the only interesting case of the next theorem occurs, of course, when I is non-empty.

Theorem 5.5 (Symmetry principle) *If f^+ and f^- are holomorphic functions in Ω^+ and Ω^- respectively, that extend continuously to I and*

$$f^+(x) = f^-(x) \quad \text{for all } x \in I,$$

then the function f defined on Ω by

$$f(z) = \begin{cases} f^+(z) & \text{if } z \in \Omega^+, \\ f^+(z) = f^-(z) & \text{if } z \in I, \\ f^-(z) & \text{if } z \in \Omega^- \end{cases}$$

is holomorphic on all of Ω.

Proof. One notes first that f is continuous throughout Ω. The only difficulty is to prove that f is holomorphic at points of I. Suppose D is a disc centered at a point on I and entirely contained in Ω. We prove that f is holomorphic in D by Morera's theorem. Suppose T is a triangle in D. If T does not intersect I, then

$$\int_T f(z)\,dz = 0$$

since f is holomorphic in the upper and lower half-discs. Suppose now that one side or vertex of T is contained in I, and the rest of T is in, say, the upper half-disc. If T_ϵ is the triangle obtained from T by slightly raising the edge or vertex which lies on I, we have $\int_{T_\epsilon} f = 0$ since T_ϵ is entirely contained in the upper half-disc (an illustration of the case when an edge lies on I is given in Figure 12(a)). We then let $\epsilon \to 0$, and by continuity we conclude that

$$\int_T f(z)\,dz = 0.$$

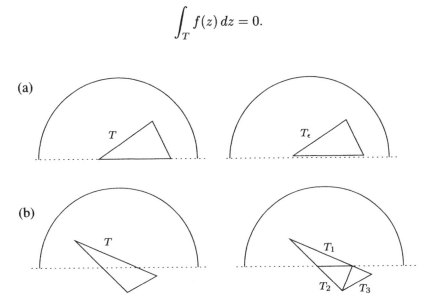

Figure 12. (a) Raising a vertex; (b) splitting a triangle

If the interior of T intersects I, we can reduce the situation to the previous one by writing T as the union of triangles each of which has an edge or vertex on I as shown in Figure 12(b). By Morera's theorem we conclude that f is holomorphic in D, as was to be shown.

We can now state the extension principle, where we use the above notation.

Theorem 5.6 (Schwarz reflection principle) *Suppose that f is a holomorphic function in Ω^+ that extends continuously to I and such that f is real-valued on I. Then there exists a function F holomorphic in all of Ω such that $F = f$ on Ω^+.*

Proof. The idea is simply to define $F(z)$ for $z \in \Omega^-$ by

$$F(z) = \overline{f(\bar{z})}.$$

To prove that F is holomorphic in Ω^- we note that if $z, z_0 \in \Omega^-$, then $\bar{z}, \overline{z_0} \in \Omega^+$ and hence, the power series expansion of f near $\overline{z_0}$ gives

$$f(\bar{z}) = \sum a_n (\bar{z} - \overline{z_0})^n.$$

As a consequence we see that

$$F(z) = \sum \overline{a_n} (z - z_0)^n$$

and F is holomorphic in Ω^-. Since f is real valued on I we have $\overline{f(x)} = f(x)$ whenever $x \in I$ and hence F extends continuously up to I. The proof is complete once we invoke the symmetry principle.

5.5 Runge's approximation theorem

We know by Weierstrass's theorem that any continuous function on a compact interval can be approximated uniformly by polynomials.[4] With this result in mind, one may inquire about similar approximations in complex analysis. More precisely, we ask the following question: what conditions on a compact set $K \subset \mathbb{C}$ guarantee that any function holomorphic in a neighborhood of this set can be approximated uniformly by polynomials on K?

An example of this is provided by power series expansions. We recall that if f is a holomorphic function in a disc D, then it has a power series expansion $f(z) = \sum_{n=0}^{\infty} a_n z^n$ that converges uniformly on every compact set $K \subset D$. By taking partial sums of this series, we conclude that f can be approximated uniformly by polynomials on any compact subset of D.

In general, however, some condition on K must be imposed, as we see by considering the function $f(z) = 1/z$ on the unit circle $K = C$. Indeed, recall that $\int_C f(z)\,dz = 2\pi i$, and if p is any polynomial, then Cauchy's theorem implies $\int_C p(z)\,dz = 0$, and this quickly leads to a contradiction.

---

[4]A proof may be found in Section 1.8, Chapter 5, of Book I.

A restriction on K that guarantees the approximation pertains to the topology of its complement: K^c must be connected. In fact, a slight modification of the above example when $f(z) = 1/z$ proves that this condition on K is also necessary; see Problem 4.

Conversely, uniform approximations exist when K^c is connected, and this result follows from a theorem of Runge which states that for *any* K a uniform approximation exists by *rational functions* with "singularities" in the complement of K.[5] This result is remarkable since rational functions are globally defined, while f is given only in a neighborhood of K. In particular, f could be defined independently on different components of K, making the conclusion of the theorem even more striking.

Theorem 5.7 *Any function holomorphic in a neighborhood of a compact set K can be approximated uniformly on K by rational functions whose singularities are in K^c.*

If K^c is connected, any function holomorphic in a neighborhood of K can be approximated uniformly on K by polynomials.

We shall see how the second part of the theorem follows from the first: when K^c is connected, one can "push" the singularities to infinity thereby transforming the rational functions into polynomials.

The key to the theorem lies in an integral representation formula that is a simple consequence of the Cauchy integral formula applied to a square.

Lemma 5.8 *Suppose f is holomorphic in an open set Ω, and $K \subset \Omega$ is compact. Then, there exists finitely many segments $\gamma_1, \ldots, \gamma_N$ in $\Omega - K$ such that*

$$(15) \qquad f(z) = \sum_{n=1}^{N} \frac{1}{2\pi i} \int_{\gamma_n} \frac{f(\zeta)}{\zeta - z} \, d\zeta \qquad \text{for all } z \in K.$$

Proof. Let $d = c \cdot d(K, \Omega^c)$, where c is any constant $< 1/\sqrt{2}$, and consider a grid formed by (solid) squares with sides parallel to the axis and of length d.

We let $\mathcal{Q} = \{Q_1, \ldots, Q_M\}$ denote the finite collection of squares in this grid that intersect K, with the boundary of each square given the positive orientation. (We denote by ∂Q_m the boundary of the square Q_m.) Finally, we let $\gamma_1, \ldots, \gamma_N$ denote the sides of squares in $\mathcal{Q}$ that do not belong to two adjacent squares in $\mathcal{Q}$. (See Figure 13.) The choice of d guarantees that for each n, $\gamma_n \subset \Omega$, and γ_n does not intersect K; for if it did, then it would belong to two adjacent squares in $\mathcal{Q}$, contradicting our choice of γ_n.

---

[5] These singularities are points where the function is not holomorphic, and are "poles", as defined in the next chapter.

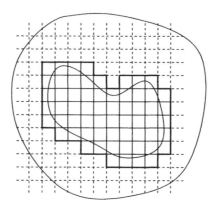

Figure 13. The union of the γ_n's is in bold-face

Since for any $z \in K$ that is not on the boundary of a square in $\mathcal{Q}$ there exists j so that $z \in Q_j$, Cauchy's theorem implies

$$\frac{1}{2\pi i} \int_{\partial Q_m} \frac{f(\zeta)}{\zeta - z} \, d\zeta = \begin{cases} f(z) & \text{if } m = j, \\ 0 & \text{if } m \neq j. \end{cases}$$

Thus, for all such z we have

$$f(z) = \sum_{m=1}^{M} \frac{1}{2\pi i} \int_{\partial Q_m} \frac{f(\zeta)}{\zeta - z} \, d\zeta.$$

However, if Q_m and $Q_{m'}$ are adjacent, the integral over their common side is taken once in each direction, and these cancel. This establishes (15) when z is in K and not on the boundary of a square in $\mathcal{Q}$. Since $\gamma_n \subset K^c$, continuity guarantees that this relation continues to hold for all $z \in K$, as was to be shown.

The first part of Theorem 5.7 is therefore a consequence of the next lemma.

Lemma 5.9 *For any line segment γ entirely contained in $\Omega - K$, there exists a sequence of rational functions with singularities on γ that approximate the integral $\int_\gamma f(\zeta)/(\zeta - z) \, d\zeta$ uniformly on K.*

Proof. If $\gamma(t) : [0, 1] \to \mathbb{C}$ is a parametrization for γ, then

$$\int_\gamma \frac{f(\zeta)}{\zeta - z} \, d\zeta = \int_0^1 \frac{f(\gamma(t))}{\gamma(t) - z} \gamma'(t) \, dt.$$

Since γ does not intersect K, the integrand $F(z,t)$ in this last integral is jointly continuous on $K \times [0,1]$, and since K is compact, given $\epsilon > 0$, there exists $\delta > 0$ such that

$$\sup_{z \in K} |F(z,t_1) - F(z,t_2)| < \epsilon \quad \text{whenever } |t_1 - t_2| < \delta.$$

Arguing as in the proof of Theorem 5.4, we see that the Riemann sums of the integral $\int_0^1 F(z,t)\,dt$ approximate it uniformly on K. Since each of these Riemann sums is a rational function with singularities on γ, the lemma is proved.

Finally, the process of pushing the poles to infinity is accomplished by using the fact that K^c is connected. Since any rational function whose only singularity is at the point z_0 is a polynomial in $1/(z - z_0)$, it suffices to establish the next lemma to complete the proof of Theorem 5.7.

Lemma 5.10 *If K^c is connected and $z_0 \notin K$, then the function $1/(z - z_0)$ can be approximated uniformly on K by polynomials.*

Proof. First, we choose a point z_1 that is outside a large open disc D centered at the origin and which contains K. Then

$$\frac{1}{z - z_1} = -\frac{1}{z_1} \frac{1}{1 - z/z_1} = \sum_{n=1}^{\infty} -\frac{z^n}{z_1^{n+1}},$$

where the series converges uniformly for $z \in K$. The partial sums of this series are polynomials that provide a uniform approximation to $1/(z - z_1)$ on K. In particular, this implies that any power $1/(z - z_1)^k$ can also be approximated uniformly on K by polynomials.

It now suffices to prove that $1/(z - z_0)$ can be approximated uniformly on K by polynomials in $1/(z - z_1)$. To do so, we use the fact that K^c is connected to travel from z_0 to the point z_1. Let γ be a curve in K^c that is parametrized by $\gamma(t)$ on $[0,1]$, and such that $\gamma(0) = z_0$ and $\gamma(1) = z_1$. If we let $\rho = \frac{1}{2}d(K,\gamma)$, then $\rho > 0$ since γ and K are compact. We then choose a sequence of points $\{w_1, \ldots, w_\ell\}$ on γ such that $w_0 = z_0$, $w_\ell = z_1$, and $|w_j - w_{j+1}| < \rho$ for all $0 \le j < \ell$.

We claim that if w is a point on γ, and w' any other point with $|w - w'| < \rho$, then $1/(z - w)$ can be approximated uniformly on K by polynomials in $1/(z - w')$. To see this, note that

$$\frac{1}{z - w} = \frac{1}{z - w'} \frac{1}{1 - \frac{w - w'}{z - w'}}$$

$$= \sum_{n=0}^{\infty} \frac{(w - w')^n}{(z - w')^{n+1}},$$

and since the sum converges uniformly for $z \in K$, the approximation by partial sums proves our claim.

This result allows us to travel from z_0 to z_1 through the finite sequence $\{w_j\}$ to find that $1/(z - z_0)$ can be approximated uniformly on K by polynomials in $1/(z - z_1)$. This concludes the proof of the lemma, and also that of the theorem.

6 Exercises

1. Prove that

$$\int_0^\infty \sin(x^2)\, dx = \int_0^\infty \cos(x^2)\, dx = \frac{\sqrt{2\pi}}{4}.$$

These are the **Fresnel integrals**. Here, $\int_0^\infty$ is interpreted as $\lim_{R \to \infty} \int_0^R$.

[Hint: Integrate the function e^{-z^2} over the path in Figure 14. Recall that $\int_{-\infty}^\infty e^{-x^2}\, dx = \sqrt{\pi}$.]

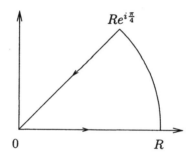

Figure 14. The contour in Exercise 1

2. Show that $\displaystyle\int_0^\infty \frac{\sin x}{x}\, dx = \frac{\pi}{2}$.

[Hint: The integral equals $\frac{1}{2i} \int_{-\infty}^\infty \frac{e^{ix}-1}{x}\, dx$. Use the indented semicircle.]

3. Evaluate the integrals

$$\int_0^\infty e^{-ax} \cos bx\, dx \quad \text{and} \quad \int_0^\infty e^{-ax} \sin bx\, dx, \quad a > 0$$

by integrating e^{-Az}, $A = \sqrt{a^2 + b^2}$, over an appropriate sector with angle ω, with $\cos\omega = a/A$.

4. Prove that for all $\xi \in \mathbb{C}$ we have $e^{-\pi\xi^2} = \int_{-\infty}^{\infty} e^{-\pi x^2} e^{2\pi i x \xi} \, dx$.

5. Suppose f is continuously *complex* differentiable on Ω, and $T \subset \Omega$ is a triangle whose interior is also contained in Ω. Apply Green's theorem to show that

$$\int_T f(z) \, dz = 0.$$

This provides a proof of Goursat's theorem under the additional assumption that f' is continuous.

[Hint: Green's theorem says that if (F, G) is a continuously differentiable vector field, then

$$\int_T F \, dx + G \, dy = \int_{\text{Interior of } T} \left(\frac{\partial G}{\partial x} - \frac{\partial F}{\partial y} \right) \, dx dy.$$

For appropriate F and G, one can then use the Cauchy-Riemann equations.]

6. Let Ω be an open subset of $\mathbb{C}$ and let $T \subset \Omega$ be a triangle whose interior is also contained in Ω. Suppose that f is a function holomorphic in Ω except possibly at a point w inside T. Prove that if f is bounded near w, then

$$\int_T f(z) \, dz = 0.$$

7. Suppose $f : \mathbb{D} \to \mathbb{C}$ is holomorphic. Show that the diameter $d = \sup_{z, w \in \mathbb{D}} |f(z) - f(w)|$ of the image of f satisfies

$$2|f'(0)| \leq d.$$

Moreover, it can be shown that equality holds precisely when f is linear, $f(z) = a_0 + a_1 z$.

Note. In connection with this result, see the relationship between the diameter of a curve and Fourier series described in Problem 1, Chapter 4, Book I.

[Hint: $2f'(0) = \frac{1}{2\pi i} \int_{|\zeta|=r} \frac{f(\zeta) - f(-\zeta)}{\zeta^2} \, d\zeta$ whenever $0 < r < 1$.]

8. If f is a holomorphic function on the strip $-1 < y < 1$, $x \in \mathbb{R}$ with

$$|f(z)| \leq A(1 + |z|)^\eta, \qquad \eta \text{ a fixed real number}$$

for all z in that strip, show that for each integer $n \geq 0$ there exists $A_n \geq 0$ so that

$$|f^{(n)}(x)| \leq A_n (1 + |x|)^\eta, \qquad \text{for all } x \in \mathbb{R}.$$

[Hint: Use the Cauchy inequalities.]

9. Let Ω be a bounded open subset of $\mathbb{C}$, and $\varphi : \Omega \to \Omega$ a holomorphic function. Prove that if there exists a point $z_0 \in \Omega$ such that

$$\varphi(z_0) = z_0 \quad \text{and} \quad \varphi'(z_0) = 1$$

then φ is linear.

[Hint: Why can one assume that $z_0 = 0$? Write $\varphi(z) = z + a_n z^n + O(z^{n+1})$ near 0, and prove that if $\varphi_k = \varphi \circ \cdots \circ \varphi$ (where φ appears k times), then $\varphi_k(z) = z + k a_n z^n + O(z^{n+1})$. Apply the Cauchy inequalities and let $k \to \infty$ to conclude the proof. Here we use the standard O notation, where $f(z) = O(g(z))$ as $z \to 0$ means that $|f(z)| \le C|g(z)|$ for some constant C as $|z| \to 0$.]

10. Weierstrass's theorem states that a continuous function on $[0, 1]$ can be uniformly approximated by polynomials. Can every continuous function on the closed unit disc be approximated uniformly by polynomials in the variable z?

11. Let f be a holomorphic function on the disc D_{R_0} centered at the origin and of radius R_0.

(a) Prove that whenever $0 < R < R_0$ and $|z| < R$, then

$$f(z) = \frac{1}{2\pi} \int_0^{2\pi} f(Re^{i\varphi}) \text{Re} \left(\frac{Re^{i\varphi} + z}{Re^{i\varphi} - z} \right) d\varphi.$$

(b) Show that

$$\text{Re} \left(\frac{Re^{i\gamma} + r}{Re^{i\gamma} - r} \right) = \frac{R^2 - r^2}{R^2 - 2Rr\cos\gamma + r^2}.$$

[Hint: For the first part, note that if $w = R^2/\bar{z}$, then the integral of $f(\zeta)/(\zeta - w)$ around the circle of radius R centered at the origin is zero. Use this, together with the usual Cauchy integral formula, to deduce the desired identity.]

12. Let u be a real-valued function defined on the unit disc $\mathbb{D}$. Suppose that u is twice continuously differentiable and harmonic, that is,

$$\triangle u(x, y) = 0$$

for all $(x, y) \in \mathbb{D}$.

(a) Prove that there exists a holomorphic function f on the unit disc such that

$$\text{Re}(f) = u.$$

Also show that the imaginary part of f is uniquely defined up to an additive (real) constant. [Hint: From the previous chapter we would have $f'(z) = 2\partial u/\partial z$. Therefore, let $g(z) = 2\partial u/\partial z$ and prove that g is holomorphic. Why can one find F with $F' = g$? Prove that $\text{Re}(F)$ differs from u by a real constant.]

(b) Deduce from this result, and from Exercise 11, the Poisson integral representation formula from the Cauchy integral formula: If u is harmonic in the unit disc and continuous on its closure, then if $z = re^{i\theta}$ one has

$$u(z) = \frac{1}{2\pi} \int_0^{2\pi} P_r(\theta - \varphi) u(e^{i\varphi}) \, d\varphi$$

where $P_r(\gamma)$ is the Poisson kernel for the unit disc given by

$$P_r(\gamma) = \frac{1 - r^2}{1 - 2r \cos \gamma + r^2}.$$

13. Suppose f is an analytic function defined everywhere in $\mathbb{C}$ and such that for each $z_0 \in \mathbb{C}$ at least one coefficient in the expansion

$$f(z) = \sum_{n=0}^{\infty} c_n(z - z_0)^n$$

is equal to 0. Prove that f is a polynomial.

[Hint: Use the fact that $c_n n! = f^{(n)}(z_0)$ and use a countability argument.]

14. Suppose that f is holomorphic in an open set containing the closed unit disc, except for a pole at z_0 on the unit circle. Show that if

$$\sum_{n=0}^{\infty} a_n z^n$$

denotes the power series expansion of f in the open unit disc, then

$$\lim_{n \to \infty} \frac{a_n}{a_{n+1}} = z_0.$$

15. Suppose f is a non-vanishing continuous function on $\overline{\mathbb{D}}$ that is holomorphic in $\mathbb{D}$. Prove that if

$$|f(z)| = 1 \quad \text{whenever } |z| = 1,$$

then f is constant.

[Hint: Extend f to all of $\mathbb{C}$ by $f(z) = 1/\overline{f(1/\overline{z})}$ whenever $|z| > 1$, and argue as in the Schwarz reflection principle.]

7 Problems

1. Here are some examples of analytic functions on the unit disc that cannot be extended analytically past the unit circle. The following definition is needed. Let

f be a function defined in the unit disc $\mathbb{D}$, with boundary circle C. A point w on C is said to be *regular* for f if there is an open neighborhood U of w and an analytic function g on U, so that $f = g$ on $\mathbb{D} \cap U$. A function f defined on $\mathbb{D}$ cannot be continued analytically past the unit circle if no point of C is regular for f.

(a) Let

$$f(z) = \sum_{n=0}^{\infty} z^{2^n} \quad \text{for } |z| < 1.$$

Notice that the radius of convergence of the above series is 1. Show that f cannot be continued analytically past the unit disc. [Hint: Suppose $\theta = 2\pi p/2^k$, where p and k are positive integers. Let $z = re^{i\theta}$; then $|f(re^{i\theta})| \to \infty$ as $r \to 1$.]

(b) * Fix $0 < \alpha < \infty$. Show that the analytic function f defined by

$$f(z) = \sum_{n=0}^{\infty} 2^{-n\alpha} z^{2^n} \quad \text{for } |z| < 1$$

extends continuously to the unit circle, but cannot be analytically continued past the unit circle. [Hint: There is a nowhere differentiable function lurking in the background. See Chapter 4 in Book I.]

2. * Let

$$F(z) = \sum_{n=1}^{\infty} d(n)z^n \quad \text{for } |z| < 1$$

where $d(n)$ denotes the number of divisors of n. Observe that the radius of convergence of this series is 1. Verify the identity

$$\sum_{n=1}^{\infty} d(n)z^n = \sum_{n=1}^{\infty} \frac{z^n}{1 - z^n}.$$

Using this identity, show that if $z = r$ with $0 < r < 1$, then

$$|F(r)| \geq c\frac{1}{1-r} \log(1/(1-r))$$

as $r \to 1$. Similarly, if $\theta = 2\pi p/q$ where p and q are positive integers and $z = re^{i\theta}$, then

$$|F(re^{i\theta})| \geq c_{p/q}\frac{1}{1-r} \log(1/(1-r))$$

as $r \to 1$. Conclude that F cannot be continued analytically past the unit disc.

3. Morera's theorem states that if f is continuous in $\mathbb{C}$, and $\int_T f(z)\,dz = 0$ for all triangles T, then f is holomorphic in $\mathbb{C}$. Naturally, we may ask if the conclusion still holds if we replace triangles by other sets.

(a) Suppose that f is continuous on $\mathbb{C}$, and

(16)
$$\int_C f(z)\, dz = 0$$

for every circle C. Prove that f is holomorphic.

(b) More generally, let Γ be any toy contour, and $\mathcal{F}$ the collection of all translates and dilates of Γ. Show that if f is continuous on $\mathbb{C}$, and

$$\int_\gamma f(z)\, dz = 0 \qquad \text{for all } \gamma \in \mathcal{F}$$

then f is holomorphic. In particular, Morera's theorem holds under the weaker assumption that $\int_T f(z)\, dz = 0$ for all equilateral triangles.

[Hint: As a first step, assume that f is twice real differentiable, and write $f(z) = f(z_0) + a(z - z_0) + b(\overline{z - z_0}) + O(|z - z_0|^2)$ for z near z_0. Integrating this expansion over small circles around z_0 yields $\partial f/\partial \bar{z} = b = 0$ at z_0. Alternatively, suppose only that f is differentiable and apply Green's theorem to conclude that the real and imaginary parts of f satisfy the Cauchy-Riemann equations.

In general, let $\varphi(w) = \varphi(x, y)$ (when $w = x + iy$) denote a smooth function with $0 \le \varphi(w) \le 1$, and $\int_{\mathbb{R}^2} \varphi(w)\, dV(w) = 1$, where $dV(w) = dx\, dy$, and $\int$ denotes the usual integral of a function of two variables in $\mathbb{R}^2$. For each $\epsilon > 0$, let $\varphi_\epsilon(z) = \epsilon^{-2} \varphi(\epsilon^{-1} z)$, as well as

$$f_\epsilon(z) = \int_{\mathbb{R}^2} f(z - w)\varphi_\epsilon(w)\, dV(w),$$

where the integral denotes the usual integral of functions of two variables, with $dV(w)$ the area element of $\mathbb{R}^2$. Then f_ϵ is smooth, satisfies condition (16), and $f_\epsilon \to f$ uniformly on any compact subset of $\mathbb{C}$.]

4. Prove the converse to Runge's theorem: if K is a compact set whose complement if not connected, then there exists a function f holomorphic in a neighborhood of K which cannot be approximated uniformly by polynomial on K.

[Hint: Pick a point z_0 in a bounded component of K^c, and let $f(z) = 1/(z - z_0)$. If f can be approximated uniformly by polynomials on K, show that there exists a polynomial p such that $|(z - z_0)p(z) - 1| < 1$. Use the maximum modulus principle (Chapter 3) to show that this inequality continues to hold for all z in the component of K^c that contains z_0.]

5.[*] There exists an entire function F with the following "universal" property: given any entire function h, there is an increasing sequence $\{N_k\}_{k=1}^\infty$ of positive integers, so that

$$\lim_{k \to \infty} F(z + N_k) = h(z)$$

uniformly on every compact subset of $\mathbb{C}$.

(a) Let $p_1, p_2, \ldots$ denote an enumeration of the collection of polynomials whose coefficients have rational real and imaginary parts. Show that it suffices to find an entire function F and an increasing sequence $\{M_n\}$ of positive integers, such that

(17) $|F(z) - p_n(z - M_n)| < \dfrac{1}{n}$ whenever $z \in D_n$,

where D_n denotes the disc centered at M_n and of radius n. [Hint: Given h entire, there exists a sequence $\{n_k\}$ such that $\lim_{k \to \infty} p_{n_k}(z) = h(z)$ uniformly on every compact subset of $\mathbb{C}$.]

(b) Construct F satisfying (17) as an infinite series

$$F(z) = \sum_{n=1}^{\infty} u_n(z)$$

where $u_n(z) = p_n(z - M_n)e^{-c_n(z - M_n)^2}$, and the quantities $c_n > 0$ and $M_n > 0$ are chosen appropriately with $c_n \to 0$ and $M_n \to \infty$. [Hint: The function e^{-z^2} vanishes rapidly as $|z| \to \infty$ in the sectors $\{|\arg z| < \pi/4 - \delta\}$ and $\{|\pi - \arg z| < \pi/4 - \delta\}$.]

In the same spirit, there exists an alternate "universal" entire function G with the following property: given any entire function h, there is an increasing sequence $\{N_k\}_{k=1}^{\infty}$ of positive integers, so that

$$\lim_{k \to \infty} D^{N_k} G(z) = h(z)$$

uniformly on every compact subset of $\mathbb{C}$. Here $D^j G$ denotes the jth (complex) derivative of G.

3 Meromorphic Functions and the Logarithm

> One knows that the differential calculus, which has contributed so much to the progress of analysis, is founded on the consideration of differential coefficients, that is derivatives of functions. When one attributes an infinitesimal increase ϵ to the variable x, the function $f(x)$ of this variable undergoes in general an infinitesimal increase of which the first term is proportional to ϵ, and the finite coefficient of ϵ of this increase is what is called its differential coefficient... If considering the values of x where $f(x)$ becomes infinite, we add to one of these values designated by x_1, the infinitesimal ϵ, and then develop $f(x_1 + \epsilon)$ in increasing power of the same quantity, the first terms of this development contain negative powers of ϵ; one of these will be the product of $1/\epsilon$ with a finite coefficient, which we will call the residue of the function $f(x)$, relative to the particular value x_1 of the variable x. Residues of this kind present themselves naturally in several branches of algebraic and infinitesimal analysis. Their consideration furnish methods that can be simply used, that apply to a large number of diverse questions, and that give new formulae that would seem to be of interest to mathematicians...
>
> *A. L. Cauchy, 1826*

There is a general principle in the theory, already implicit in Riemann's work, which states that analytic functions are in an essential way characterized by their singularities. That is to say, globally analytic functions are "effectively" determined by their zeros, and meromorphic functions by their zeros and poles. While these assertions cannot be formulated as precise general theorems, there are nevertheless significant instances where this principle applies.

We begin this chapter by considering singularities, in particular the different kind of point singularities ("isolated" singularities) that a holomorphic function can have. In order of increasing severity, these are:

- removable singularities

- poles

- essential singularities.

The first type is harmless since a function can actually be extended to be holomorphic at its removable singularities (hence the name). Near the third type, the function oscillates and may grow faster than any power, and a complete understanding of its behavior is not easy. For the second type the analysis is more straight forward and is connected with the calculus of residues, which arises as follows.

Recall that by Cauchy's theorem a holomorphic function f in an open set which contains a closed curve γ and its interior satisfies

$$\int_\gamma f(z)\,dz = 0.$$

The question that occurs is: what happens if f has a pole in the interior of the curve? To try to answer this question consider the example $f(z) = 1/z$, and recall that if C is a (positively oriented) circle centered at 0, then

$$\int_C \frac{dz}{z} = 2\pi i.$$

This turns out to be the key ingredient in the calculus of residues.

A new aspect appears when we consider indefinite integrals of holomorphic functions that have singularities. As the basic example $f(z) = 1/z$ shows, the resulting "function" (in this case the logarithm) may not be single-valued, and understanding this phenomenon is of importance for a number of subjects. Exploiting this multi-valuedness leads in effect to the "argument principle." We can use this principle to count the number of zeros of a holomorphic function inside a suitable curve. As a simple consequence of this result, we obtain a significant geometric property of holomorphic functions: they are open mappings. From this, the maximum principle, another important feature of holomorphic functions, is an easy step.

In order to turn to the logarithm itself, and come to grips with the precise nature of its multi-valuedness, we introduce the notions of homotopy of curves and simply connected domains. It is on the latter type of open sets that single-valued branches of the logarithm can be defined.

1 Zeros and poles

By definition, a **point singularity** of a function f is a complex number z_0 such that f is defined in a neighborhood of z_0 but not at the point

z_0 itself. We shall also call such points **isolated singularities**. For example, if the function f is defined only on the punctured plane by $f(z) = z$, then the origin is a point singularity. Of course, in that case, the function f can actually be defined at 0 by setting $f(0) = 0$, so that the resulting extension is continuous and in fact entire. (Such points are then called removable singularities.) More interesting is the case of the function $g(z) = 1/z$ defined in the punctured plane. It is clear now that g cannot be defined as a continuous function, much less as a holomorphic function, at the point 0. In fact, $g(z)$ grows to infinity as z approaches 0, and we shall say that the origin is a pole singularity. Finally, the case of the function $h(z) = e^{1/z}$ on the punctured plane shows that removable singularities and poles do not tell the whole story. Indeed, the function $h(z)$ grows indefinitely as z approaches 0 on the positive real line, while h approaches 0 as z goes to 0 on the negative real axis. Finally h oscillates rapidly, yet remains bounded, as z approaches the origin on the imaginary axis.

Since singularities often appear because the denominator of a fraction vanishes, we begin with a local study of the zeros of a holomorphic function.

A complex number z_0 is a **zero** for the holomorphic function f if $f(z_0) = 0$. In particular, analytic continuation shows that the zeros of a non-trivial holomorphic function are isolated. In other words, if f is holomorphic in Ω and $f(z_0) = 0$ for some $z_0 \in \Omega$, then there exists an open neighborhood U of z_0 such that $f(z) \neq 0$ for all $z \in U - \{z_0\}$, unless f is identically zero. We start with a local description of a holomorphic function near a zero.

Theorem 1.1 *Suppose that f is holomorphic in a connected open set Ω, has a zero at a point $z_0 \in \Omega$, and does not vanish identically in Ω. Then there exists a neighborhood $U \subset \Omega$ of z_0, a non-vanishing holomorphic function g on U, and a unique positive integer n such that*

$$f(z) = (z - z_0)^n g(z) \quad \text{for all } z \in U.$$

Proof. Since Ω is connected and f is not identically zero, we conclude that f is not identically zero in a neighborhood of z_0. In a small disc centered at z_0 the function f has a power series expansion

$$f(z) = \sum_{k=0}^{\infty} a_k (z - z_0)^k.$$

Since f is not identically zero near z_0, there exists a smallest integer n

such that $a_n \neq 0$. Then, we can write

$$f(z) = (z - z_0)^n [a_n + a_{n+1}(z - z_0) + \cdots] = (z - z_0)^n g(z),$$

where g is defined by the series in brackets, and hence is holomorphic, and is nowhere vanishing for all z close to z_0 (since $a_n \neq 0$). To prove the uniqueness of the integer n, suppose that we can also write

$$f(z) = (z - z_0)^n g(z) = (z - z_0)^m h(z)$$

where $h(z_0) \neq 0$. If $m > n$, then we may divide by $(z - z_0)^n$ to see that

$$g(z) = (z - z_0)^{m-n} h(z)$$

and letting $z \rightarrow z_0$ yields $g(z_0) = 0$, a contradiction. If $m < n$ a similar argument gives $h(z_0) = 0$, which is also a contradiction. We conclude that $m = n$, thus $h = g$, and the theorem is proved.

In the case of the above theorem, we say that f has a **zero of order** n (or **multiplicity** n) at z_0. If a zero is of order 1, we say that it is **simple**. We observe that, quantitatively, the order describes the rate at which the function vanishes.

The importance of the previous theorem comes from the fact that we can now describe precisely the type of singularity possessed by the function $1/f$ at z_0.

For this purpose, it is now convenient to define a **deleted neighborhood** of z_0 to be an open disc centered at z_0, minus the point z_0, that is, the set

$$\{z : 0 < |z - z_0| < r\}$$

for some $r > 0$. Then, we say that a function f defined in a deleted neighborhood of z_0 has a **pole** at z_0, if the function $1/f$, defined to be zero at z_0, is holomorphic in a full neighborhood of z_0.

Theorem 1.2 *If f has a pole at $z_0 \in \Omega$, then in a neighborhood of that point there exist a non-vanishing holomorphic function h and a unique positive integer n such that*

$$f(z) = (z - z_0)^{-n} h(z).$$

Proof.   By the previous theorem we have $1/f(z) = (z - z_0)^n g(z)$, where g is holomorphic and non-vanishing in a neighborhood of z_0, so the result follows with $h(z) = 1/g(z)$.

The integer n is called the **order** (or **multiplicity**) of the pole, and describes the rate at which the function grows near z_0. If the pole is of order 1, we say that it is **simple**.

The next theorem should be reminiscent of power series expansion, except that now we allow terms of negative order, to account for the presence of a pole.

Theorem 1.3 *If f has a pole of order n at z_0, then*

$$\text{(1)} \qquad f(z) = \frac{a_{-n}}{(z - z_0)^n} + \frac{a_{-n+1}}{(z - z_0)^{n-1}} + \cdots + \frac{a_{-1}}{(z - z_0)} + G(z),$$

where G is a holomorphic function in a neighborhood of z_0.

Proof. The proof follows from the multiplicative statement in the previous theorem. Indeed, the function h has a power series expansion

$$h(z) = A_0 + A_1(z - z_0) + \cdots$$

so that

$$f(z) = (z - z_0)^{-n}(A_0 + A_1(z - z_0) + \cdots)$$
$$= \frac{a_{-n}}{(z - z_0)^n} + \frac{a_{-n+1}}{(z - z_0)^{n-1}} + \cdots + \frac{a_{-1}}{(z - z_0)} + G(z).$$

The sum

$$\frac{a_{-n}}{(z - z_0)^n} + \frac{a_{-n+1}}{(z - z_0)^{n-1}} + \cdots + \frac{a_{-1}}{(z - z_0)}$$

is called the **principal part** of f at the pole z_0, and the coefficient a_{-1} is the **residue** of f at that pole. We write $\operatorname{res}_{z_0} f = a_{-1}$. The importance of the residue comes from the fact that all the other terms in the principal part, that is, those of order strictly greater than 1, have primitives in a deleted neighborhood of z_0. Therefore, if $P(z)$ denotes the principal part above and C is any circle centered at z_0, we get

$$\frac{1}{2\pi i} \int_C P(z)\, dz = a_{-1}.$$

We shall return to this important point in the section on the residue formula.

As we shall see, in many cases, the evaluation of integrals reduces to the calculation of residues. In the case when f has a simple pole at z_0, it is clear that

$$\operatorname{res}_{z_0} f = \lim_{z \to z_0} (z - z_0) f(z).$$

If the pole is of higher order, a similar formula holds, one that involves differentiation as well as taking a limit.

Theorem 1.4 *If f has a pole of order n at z_0, then*

$$\operatorname{res}_{z_0} f = \lim_{z \to z_0} \frac{1}{(n-1)!} \left(\frac{d}{dz}\right)^{n-1} (z - z_0)^n f(z).$$

The theorem is an immediate consequence of formula (1), which implies

$$(z - z_0)^n f(z) = a_{-n} + a_{-n+1}(z - z_0) + \cdots + a_{-1}(z - z_0)^{n-1} +$$

$$+ G(z)(z - z_0)^n.$$

2 The residue formula

We now discuss the celebrated residue formula. Our approach follows the discussion of Cauchy's theorem in the last chapter: we first consider the case of the circle and its interior the disc, and then explain generalizations to toy contours and their interiors.

Theorem 2.1 *Suppose that f is holomorphic in an open set containing a circle C and its interior, except for a pole at z_0 inside C. Then*

$$\int_C f(z)\, dz = 2\pi i \operatorname{res}_{z_0} f.$$

Proof. Once again, we may choose a keyhole contour that avoids the pole, and let the width of the corridor go to zero to see that

$$\int_C f(z)\, dz = \int_{C_\epsilon} f(z)\, dz$$

where C_ϵ is the small circle centered at the pole z_0 and of radius ϵ.

Now we observe that

$$\frac{1}{2\pi i} \int_{C_\epsilon} \frac{a_{-1}}{z - z_0}\, dz = a_{-1}$$

is an immediate consequence of the Cauchy integral formula (Theorem 4.1 of the previous chapter), applied to the constant function $f = a_{-1}$. Similarly,

$$\frac{1}{2\pi i} \int_{C_\epsilon} \frac{a_{-k}}{(z - z_0)^k}\, dz = 0$$

when $k > 1$, by using the corresponding formulae for the derivatives (Corollary 4.2 also in the previous chapter). But we know that in a neighborhood of z_0 we can write

$$f(z) = \frac{a_{-n}}{(z - z_0)^n} + \frac{a_{-n+1}}{(z - z_0)^{n-1}} + \cdots + \frac{a_{-1}}{z - z_0} + G(z),$$

where G is holomorphic. By Cauchy's theorem, we also know that $\int_{C_\epsilon} G(z)\,dz = 0$, hence $\int_{C_\epsilon} f(z)\,dz = a_{-1}$. This implies the desired result.

This theorem can be generalized to the case of finitely many poles in the circle, as well as to the case of toy contours.

Corollary 2.2 *Suppose that f is holomorphic in an open set containing a circle C and its interior, except for poles at the points $z_1, \ldots, z_N$ inside C. Then*

$$\int_C f(z)\,dz = 2\pi i \sum_{k=1}^{N} \operatorname{res}_{z_k} f.$$

For the proof, consider a multiple keyhole which has a loop avoiding each one of the poles. Let the width of the corridors go to zero. In the limit, the integral over the large circle equals a sum of integrals over small circles to which Theorem 2.1 applies.

Corollary 2.3 *Suppose that f is holomorphic in an open set containing a toy contour γ and its interior, except for poles at the points $z_1, \ldots, z_N$ inside γ. Then*

$$\int_\gamma f(z)\,dz = 2\pi i \sum_{k=1}^{N} \operatorname{res}_{z_k} f.$$

In the above, we take γ to have positive orientation.

The proof consists of choosing a keyhole appropriate for the given toy contour, so that, as we have seen previously, we can reduce the situation to integrating over small circles around the poles where Theorem 2.1 applies.

The identity $\int_\gamma f(z)\,dz = 2\pi i \sum_{k=1}^{N} \operatorname{res}_{z_k} f$ is referred to as the **residue formula**.

2.1 Examples

The calculus of residues provides a powerful technique to compute a wide range of integrals. In the examples we give next, we evaluate three improper Riemann integrals of the form

$$\int_{-\infty}^{\infty} f(x)\, dx.$$

The main idea is to extend f to the complex plane, and then choose a family γ_R of toy contours so that

$$\lim_{R \to \infty} \int_{\gamma_R} f(z)\, dz = \int_{-\infty}^{\infty} f(x)\, dx.$$

By computing the residues of f at its poles, we easily obtain $\int_{\gamma_R} f(z)\, dz$. The challenging part is to choose the contours γ_R, so that the above limit holds. Often, this choice is motivated by the decay behavior of f.

EXAMPLE 1. First, we prove that

$$(2) \qquad\qquad \int_{-\infty}^{\infty} \frac{dx}{1+x^2} = \pi$$

by using contour integration. Note that if we make the change of variables $x \mapsto x/y$, this yields

$$\frac{1}{\pi} \int_{-\infty}^{\infty} \frac{y\, dx}{y^2 + x^2} = \int_{-\infty}^{\infty} \mathcal{P}_y(x)\, dx.$$

In other words, formula (2) says that the integral of the Poisson kernel $\mathcal{P}_y(x)$ is equal to 1 for each $y > 0$. This was proved quite easily in Lemma 2.5 of Chapter 5 in Book I, since $1/(1+x^2)$ is the derivative of $\arctan x$. Here we provide a residue calculation that leads to another proof of (2).

Consider the function

$$f(z) = \frac{1}{1+z^2},$$

which is holomorphic in the complex plane except for simple poles at the points i and $-i$. Also, we choose the contour γ_R shown in Figure 1. The contour consists of the segment $[-R, R]$ on the real axis and of a large half-circle centered at the origin in the upper half-plane. Since we may write

$$f(z) = \frac{1}{(z-i)(z+i)}$$

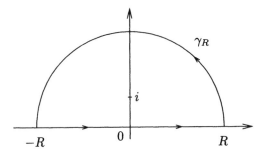

Figure 1. The contour γ_R in Example 1

we see that the residue of f at i is simply $1/2i$. Therefore, if R is large enough, we have

$$\int_{\gamma_R} f(z)\,dz = \frac{2\pi i}{2i} = \pi.$$

If we denote by C_R^+ the large half-circle of radius R, we see that

$$\left| \int_{C_R^+} f(z)\,dz \right| \leq \pi R \frac{B}{R^2} \leq \frac{M}{R},$$

where we have used the fact that $|f(z)| \leq B/|z|^2$ when $z \in C_R^+$ and R is large. So this integral goes to 0 as $R \to \infty$. Therefore, in the limit we find that

$$\int_{-\infty}^{\infty} f(x)\,dx = \pi,$$

as desired. We remark that in this example, there is nothing special about our choice of the semicircle in the upper half-plane. One gets the same conclusion if one uses the semicircle in the lower half-plane, with the other pole and the appropriate residue.

EXAMPLE 2. An integral that will play an important role in Chapter 6 is

$$\int_{-\infty}^{\infty} \frac{e^{ax}}{1+e^x}\,dx = \frac{\pi}{\sin \pi a}, \qquad 0 < a < 1.$$

To prove this formula, let $f(z) = e^{az}/(1+e^z)$, and consider the contour consisting of a rectangle in the upper half-plane with a side lying

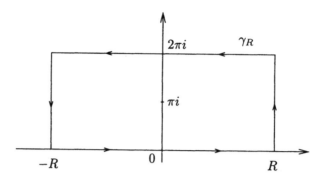

Figure 2. The contour γ_R in Example 2

on the real axis, and a parallel side on the line $\text{Im}(z) = 2\pi$, as shown in Figure 2.

The only point in the rectangle γ_R where the denominator of f vanishes is $z = \pi i$. To compute the residue of f at that point, we argue as follows: First, note

$$(z - \pi i)f(z) = e^{az}\frac{z - \pi i}{1 + e^z} = e^{az}\frac{z - \pi i}{e^z - e^{\pi i}}.$$

We recognize on the right the inverse of a difference quotient, and in fact

$$\lim_{z \to \pi i} \frac{e^z - e^{\pi i}}{z - \pi i} = e^{\pi i} = -1$$

since e^z is its own derivative. Therefore, the function f has a simple pole at πi with residue

$$\text{res}_{\pi i}f = -e^{a\pi i}.$$

As a consequence, the residue formula says that

(3)
$$\int_{\gamma_R} f = -2\pi i e^{a\pi i}.$$

We now investigate the integrals of f over each side of the rectangle. Let I_R denote

$$\int_{-R}^{R} f(x)\,dx$$

and I the integral we wish to compute, so that $I_R \to I$ as $R \to \infty$. Then, it is clear that the integral of f over the top side of the rectangle (with the orientation from right to left) is

$$-e^{2\pi i a} I_R.$$

Finally, if $A_R = \{R + it : 0 \le t \le 2\pi\}$ denotes the vertical side on the right, then

$$\left| \int_{A_R} f \right| \le \int_0^{2\pi} \left| \frac{e^{a(R+it)}}{1 + e^{R+it}} \right| dt \le C e^{(a-1)R},$$

and since $a < 1$, this integral tends to 0 as $R \to \infty$. Similarly, the integral over the vertical segment on the left goes to 0, since it can be bounded by Ce^{-aR} and $a > 0$. Therefore, in the limit as R tends to infinity, the identity (3) yields

$$I - e^{2\pi i a} I = -2\pi i e^{a\pi i},$$

from which we deduce

$$\begin{aligned}
I &= -2\pi i \frac{e^{a\pi i}}{1 - e^{2\pi i a}} \\
&= \frac{2\pi i}{e^{\pi i a} - e^{-\pi i a}} \\
&= \frac{\pi}{\sin \pi a},
\end{aligned}$$

and the computation is complete.

EXAMPLE 3. Now we calculate another Fourier transform, namely

$$\int_{-\infty}^{\infty} \frac{e^{-2\pi i x \xi}}{\cosh \pi x} dx = \frac{1}{\cosh \pi \xi}$$

where

$$\cosh z = \frac{e^z + e^{-z}}{2}.$$

In other words, the function $1/\cosh \pi x$ is its own Fourier transform, a property also shared by $e^{-\pi x^2}$ (see Example 1, Chapter 2). To see this, we use a rectangle γ_R as shown on Figure 3 whose width goes to infinity, but whose height is fixed.

For a fixed $\xi \in \mathbb{R}$, let

$$f(z) = \frac{e^{-2\pi i z \xi}}{\cosh \pi z}.$$

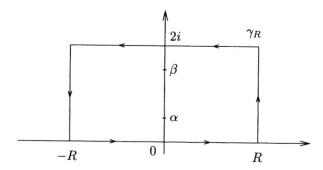

Figure 3. The contour γ_R in Example 3

and note that the denominator of f vanishes precisely when $e^{\pi z} = -e^{-\pi z}$, that is, when $e^{2\pi z} = -1$. In other words, the only poles of f inside the rectangle are at the points $\alpha = i/2$ and $\beta = 3i/2$. To find the residue of f at α, we note that

$$(z - \alpha)f(z) = e^{-2\pi i z \xi}\frac{2(z - \alpha)}{e^{\pi z} + e^{-\pi z}}$$

$$= 2e^{-2\pi i z \xi}e^{\pi z}\frac{(z - \alpha)}{e^{2\pi z} - e^{2\pi \alpha}}.$$

We recognize on the right the reciprocal of the difference quotient for the function $e^{2\pi z}$ at $z = \alpha$. Therefore

$$\lim_{z \to \alpha}(z - \alpha)f(z) = 2e^{-2\pi i \alpha \xi}e^{\pi \alpha}\frac{1}{2\pi e^{2\pi \alpha}} = \frac{e^{\pi \xi}}{\pi i},$$

which shows that f has a simple pole at α with residue $e^{\pi \xi}/(\pi i)$. Similarly, we find that f has a simple pole at β with residue $-e^{3\pi \xi}/(\pi i)$.

We dispense with the integrals of f on the vertical sides by showing that they go to zero as R tends to infinity. Indeed, if $z = R + iy$ with $0 \leq y \leq 2$, then

$$|e^{-2\pi i z \xi}| \leq e^{4\pi|\xi|},$$

and

$$|\cosh \pi z| = \left| \frac{e^{\pi z} + e^{-\pi z}}{2} \right|$$

$$\geq \frac{1}{2} \left| |e^{\pi z}| - |e^{-\pi z}| \right|$$

$$\geq \frac{1}{2}(e^{\pi R} - e^{-\pi R})$$

$$\rightarrow \infty \quad \text{as } R \rightarrow \infty,$$

which shows that the integral over the vertical segment on the right goes to 0 as $R \rightarrow \infty$. A similar argument shows that the integral of f over the vertical segment on the left also goes to 0 as $R \rightarrow \infty$. Finally, we see that if I denotes the integral we wish to calculate, then the integral of f over the top side of the rectangle (with the orientation from right to left) is simply $-e^{4\pi\xi}I$ where we have used the fact that $\cosh \pi\zeta$ is periodic with period $2i$. In the limit as R tends to infinity, the residue formula gives

$$I - e^{4\pi\xi}I = 2\pi i \left(\frac{e^{\pi\xi}}{\pi i} - \frac{e^{3\pi\xi}}{\pi i} \right)$$

$$= -2e^{2\pi\xi}(e^{\pi\xi} - e^{-\pi\xi}),$$

and since $1 - e^{4\pi\xi} = -e^{2\pi\xi}(e^{2\pi\xi} - e^{-2\pi\xi})$, we find that

$$I = 2\frac{e^{\pi\xi} - e^{-\pi\xi}}{e^{2\pi\xi} - e^{-2\pi\xi}} = 2\frac{e^{\pi\xi} - e^{-\pi\xi}}{(e^{\pi\xi} - e^{-\pi\xi})(e^{\pi\xi} + e^{-\pi\xi})} = \frac{2}{e^{\pi\xi} + e^{-\pi\xi}} = \frac{1}{\cosh \pi\xi}$$

as claimed.

A similar argument actually establishes the following formula:

$$\int_{-\infty}^{\infty} e^{-2\pi i x \xi} \frac{\sin \pi a}{\cosh \pi x + \cos \pi a} \, dx = \frac{2 \sinh 2\pi a \xi}{\sinh 2\pi \xi}$$

whenever $0 < a < 1$, and where $\sinh z = (e^z - e^{-z})/2$. We have proved above the particular case $a = 1/2$. This identity can be used to determine an explicit formula for the Poisson kernel for the strip (see Problem 3 in Chapter 5 of Book I), or to prove the sum of two squares theorem, as we shall see in Chapter 10.

3 Singularities and meromorphic functions

Returning to Section 1, we see that we have described the analytical character of a function near a pole. We now turn our attention to the other types of isolated singularities.

Let f be a function holomorphic in an open set Ω except possibly at one point z_0 in Ω. If we can define f at z_0 in such a way that f becomes holomorphic in all of Ω, we say that z_0 is a **removable** singularity for f.

Theorem 3.1 (Riemann's theorem on removable singularities)
Suppose that f is holomorphic in an open set Ω except possibly at a point z_0 in Ω. If f is bounded on $\Omega - \{z_0\}$, then z_0 is a removable singularity.

Proof.   Since the problem is local we may consider a small disc D centered at z_0 and whose closure is contained in Ω. Let C denote the boundary circle of that disc with the usual positive orientation. We shall prove that if $z \in D$ and $z \neq z_0$, then under the assumptions of the theorem we have

$$(4) \qquad\qquad f(z) = \frac{1}{2\pi i} \int_C \frac{f(\zeta)}{\zeta - z}\, d\zeta.$$

Since an application of Theorem 5.4 in the previous chapter proves that the right-hand side of equation (4) defines a holomorphic function on *all* of D that agrees with $f(z)$ when $z \neq z_0$, this give us the desired extension.

To prove formula (4) we fix $z \in D$ with $z \neq z_0$ and use the familiar toy contour illustrated in Figure 4.

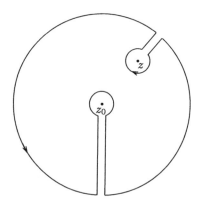

Figure 4. The multiple keyhole contour in the proof of Riemann's theorem

The multiple keyhole avoids the two points z and z_0. Letting the sides of the corridors get closer to each other, and finally overlap, in the limit we get a cancellation:

$$\int_C \frac{f(\zeta)}{\zeta - z}\, d\zeta + \int_{\gamma_\epsilon} \frac{f(\zeta)}{\zeta - z}\, d\zeta + \int_{\gamma_\epsilon'} \frac{f(\zeta)}{\zeta - z}\, d\zeta = 0\,,$$

where γ_ϵ and γ_ϵ' are small circles of radius ϵ with negative orientation and centered at z and z_0 respectively. Copying the argument used in the proof of the Cauchy integral formula in Section 4 of Chapter 2, we find that

$$\int_{\gamma_\epsilon} \frac{f(\zeta)}{\zeta - z}\, d\zeta = -2\pi i f(z).$$

For the second integral, we use the assumption that f is bounded and that since ϵ is small, ζ stays away from z, and therefore

$$\left| \int_{\gamma_\epsilon'} \frac{f(\zeta)}{\zeta - z}\, d\zeta \right| \leq C\epsilon.$$

Letting ϵ tend to 0 proves our contention and concludes the proof of the extension formula (4).

Surprisingly, we may deduce from Riemann's theorem a characterization of poles in terms of the behavior of the function in a neighborhood of a singularity.

Corollary 3.2 *Suppose that f has an isolated singularity at the point z_0. Then z_0 is a pole of f if and only if $|f(z)| \to \infty$ as $z \to z_0$.*

Proof. If z_0 is a pole, then we know that $1/f$ has a zero at z_0, and therefore $|f(z)| \to \infty$ as $z \to z_0$. Conversely, suppose that this condition holds. Then $1/f$ is bounded near z_0, and in fact $1/|f(z)| \to 0$ as $z \to z_0$. Therefore, $1/f$ has a removable singularity at z_0 and must vanish there. This proves the converse, namely that z_0 is a pole.

Isolated singularities belong to one of three categories:

- Removable singularities (f bounded near z_0)

- Pole singularities ($|f(z)| \to \infty$ as $z \to z_0$)

- Essential singularities.

By default, any singularity that is not removable or a pole is defined to be an **essential singularity**. For example, the function $e^{1/z}$ discussed at the very beginning of Section 1 has an essential singularity at

0. We already observed the wild behavior of this function near the origin. Contrary to the controlled behavior of a holomorphic function near a removable singularity or a pole, it is typical for a holomorphic function to behave erratically near an essential singularity. The next theorem clarifies this.

Theorem 3.3 (Casorati-Weierstrass) *Suppose f is holomorphic in the punctured disc $D_r(z_0) - \{z_0\}$ and has an essential singularity at z_0. Then, the image of $D_r(z_0) - \{z_0\}$ under f is dense in the complex plane.*

Proof. We argue by contradiction. Assume that the range of f is not dense, so that there exists $w \in \mathbb{C}$ and $\delta > 0$ such that

$$|f(z) - w| > \delta \quad \text{for all } z \in D_r(z_0) - \{z_0\}.$$

We may therefore define a new function on $D_r(z_0) - \{z_0\}$ by

$$g(z) = \frac{1}{f(z) - w},$$

which is holomorphic on the punctured disc and bounded by $1/\delta$. Hence g has a removable singularity at z_0 by Theorem 3.1. If $g(z_0) \neq 0$, then $f(z) - w$ is holomorphic at z_0, which contradicts the assumption that z_0 is an essential singularity. In the case that $g(z_0) = 0$, then $f(z) - w$ has a pole at z_0 also contradicting the nature of the singularity at z_0. The proof is complete.

In fact, Picard proved a much stronger result. He showed that under the hypothesis of the above theorem, the function f takes on every complex value infinitely many times with at most one exception. Although we shall not give a proof of this remarkable result, a simpler version of it will follow from our study of entire functions in a later chapter. See Exercise 11 in Chapter 5.

We now turn to functions with only isolated singularities that are poles. A function f on an open set Ω is **meromorphic** if there exists a sequence of points $\{z_0, z_1, z_2, \ldots\}$ that has no limit points in Ω, and such that

(i) the function f is holomorphic in $\Omega - \{z_0, z_1, z_2, \ldots\}$, and

(ii) f has poles at the points $\{z_0, z_1, z_2, \ldots\}$.

It is also useful to discuss functions that are meromorphic in the extended complex plane. If a function is holomorphic for all large values of

z, we can describe its behavior at infinity using the tripartite distinction we have used to classify singularities at finite values of z. Thus, if f is holomorphic for all large values of z, we consider $F(z) = f(1/z)$, which is now holomorphic in a deleted neighborhood of the origin. We say that f has a **pole at infinity** if F has a pole at the origin. Similarly, we can speak of f having an **essential singularity at infinity**, or a **removable singularity** (hence holomorphic) **at infinity** in terms of the corresponding behavior of F at 0. A meromorphic function in the complex plane that is either holomorphic at infinity or has a pole at infinity is said to be **meromorphic in the extended complex plane**.

At this stage we return to the principle mentioned at the beginning of the chapter. Here we can see it in its simplest form.

Theorem 3.4 *The meromorphic functions in the extended complex plane are the rational functions.*

Proof. Suppose that f is meromorphic in the extended plane. Then $f(1/z)$ has either a pole or a removable singularity at 0, and in either case it must be holomorphic in a deleted neighborhood of the origin, so that the function f can have only finitely many poles in the plane, say at $z_1, \ldots, z_n$. The idea is to subtract from f its principal parts at all its poles including the one at infinity. Near each pole $z_k \in \mathbb{C}$ we can write

$$f(z) = f_k(z) + g_k(z),$$

where $f_k(z)$ is the principal part of f at z_k and g_k is holomorphic in a (full) neighborhood of z_k. In particular, f_k is a polynomial in $1/(z - z_k)$. Similarly, we can write

$$f(1/z) = \tilde{f}_\infty(z) + \tilde{g}_\infty(z),$$

where $\tilde{g}_\infty$ is holomorphic in a neighborhood of the origin and $\tilde{f}_\infty$ is the principal part of $f(1/z)$ at 0, that is, a polynomial in $1/z$. Finally, let $f_\infty(z) = \tilde{f}_\infty(1/z)$.

We contend that the function $H = f - f_\infty - \sum_{k=1}^n f_k$ is entire and bounded. Indeed, near the pole z_k we subtracted the principal part of f so that the function H has a removable singularity there. Also, $H(1/z)$ is bounded for z near 0 since we subtracted the principal part of the pole at ∞. This proves our contention, and by Liouville's theorem we conclude that H is constant. From the definition of H, we find that f is a rational function, as was to be shown.

Note that as a consequence, a rational function is determined up to a multiplicative constant by prescribing the locations and multiplicities of its zeros and poles.

The Riemann sphere

The extended complex plane, which consists of $\mathbb{C}$ and the point at infinity, has a convenient geometric interpretation, which we briefly discuss here.

Consider the Euclidean space $\mathbb{R}^3$ with coordinates (X, Y, Z) where the XY-plane is identified with $\mathbb{C}$. We denote by $\mathbb{S}$ the sphere centered at $(0, 0, 1/2)$ and of radius $1/2$; this sphere is of unit diameter and lies on top of the origin of the complex plane as pictured in Figure 5. Also, we let $\mathcal{N} = (0, 0, 1)$ denote the north pole of the sphere.

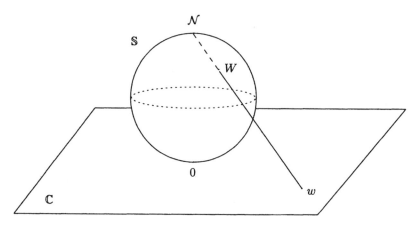

Figure 5. The Riemann sphere $\mathbb{S}$ and stereographic projection

Given any point $W = (X, Y, Z)$ on $\mathbb{S}$ different from the north pole, the line joining $\mathcal{N}$ and W intersects the XY-plane in a single point which we denote by $w = x + iy$; w is called the **stereographic projection** of W (see Figure 5). Conversely, given any point w in $\mathbb{C}$, the line joining $\mathcal{N}$ and $w = (x, y, 0)$ intersects the sphere at $\mathcal{N}$ and another point, which we call W. This geometric construction gives a bijective correspondence between points on the punctured sphere $\mathbb{S} - \{\mathcal{N}\}$ and the complex plane; it is described analytically by the formulas

$$x = \frac{X}{1 - Z} \quad \text{and} \quad y = \frac{Y}{1 - Z},$$

giving w in terms of W, and

$$X = \frac{x}{x^2 + y^2 + 1}, \quad Y = \frac{y}{x^2 + y^2 + 1}, \quad \text{and} \quad Z = \frac{x^2 + y^2}{x^2 + y^2 + 1}$$

giving W in terms of w. Intuitively, we have wrapped the complex plane onto the punctured sphere $\mathbb{S} - \{\mathcal{N}\}$.

As the point w goes to infinity in $\mathbb{C}$ (in the sense that $|w| \to \infty$) the corresponding point W on $\mathbb{S}$ comes arbitrarily close to $\mathcal{N}$. This simple observation makes $\mathcal{N}$ a natural candidate for the so-called "point at infinity." Identifying infinity with the point $\mathcal{N}$ on $\mathbb{S}$, we see that the extended complex plane can be visualized as the full two-dimensional sphere $\mathbb{S}$; this is the **Riemann sphere**. Since this construction takes the unbounded set $\mathbb{C}$ into the compact set $\mathbb{S}$ by adding one point, the Riemann sphere is sometimes called the **one-point compactification** of $\mathbb{C}$.

An important consequence of this interpretation is the following: although the point at infinity required special attention when considered separately from $\mathbb{C}$, it now finds itself on equal footing with all other points on $\mathbb{S}$. In particular, a meromorphic function on the extended complex plane can be thought of as a map from $\mathbb{S}$ to itself, where the image of a pole is now a tractable point on $\mathbb{S}$, namely $\mathcal{N}$. For these reasons (and others) the Riemann sphere provides good geometrical insight into the structure of $\mathbb{C}$ as well as the theory of meromorphic functions.

4 The argument principle and applications

We anticipate our discussion of the logarithm (in Section 6) with a few comments. In general, the function $\log f(z)$ is "multiple-valued" because it cannot be defined unambiguously on the set where $f(z) \neq 0$. However it is to be defined, it must equal $\log|f(z)| + i \arg f(z)$, where $\log|f(z)|$ is the usual real-variable logarithm of the positive quantity $|f(z)|$ (and hence is defined unambiguously), while $\arg f(z)$ is some determination of the argument (up to an additive integral multiple of 2π). Note that in any case, the derivative of $\log f(z)$ is $f'(z)/f(z)$ which is single-valued, and the integral

$$\int_\gamma \frac{f'(z)}{f(z)}\, dz$$

can be interpreted as the change in the argument of f as z traverses the curve γ. Moreover, assuming the curve is closed, this change of argument is determined entirely by the zeros and poles of f inside γ. We now formulate this fact as a precise theorem.

We begin with the observation that while the additivity formula

$$\log(f_1 f_2) = \log f_1 + \log f_2$$

fails in general (as we shall see below), the additivity can be restored to the corresponding derivatives. This is confirmed by the following observation:

$$\frac{(f_1 f_2)'}{f_1 f_2} = \frac{f_1' f_2 + f_1 f_2'}{f_1 f_2} = \frac{f_1'}{f_1} + \frac{f_2'}{f_2},$$

which generalizes to

$$\frac{\left(\prod_{k=1}^{N} f_k\right)'}{\prod_{k=1}^{N} f_k} = \sum_{k=1}^{N} \frac{f_k'}{f_k}.$$

We apply this formula as follows. If f is holomorphic and has a zero of order n at z_0, we can write

$$f(z) = (z - z_0)^n g(z),$$

where g is holomorphic and nowhere vanishing in a neighborhood of z_0, and therefore

$$\frac{f'(z)}{f(z)} = \frac{n}{z - z_0} + G(z)$$

where $G(z) = g'(z)/g(z)$. The conclusion is that if f has a zero of order n at z_0, then f'/f has a simple pole with residue n at z_0. Observe that a similar fact also holds if f has a pole of order n at z_0, that is, if $f(z) = (z - z_0)^{-n} h(z)$. Then

$$\frac{f'(z)}{f(z)} = \frac{-n}{z - z_0} + H(z).$$

Therefore, if f is meromorphic, the function f'/f will have simple poles at the zeros and poles of f, and the residue is simply the order of the zero of f or the negative of the order of the pole of f. As a result, an application of the residue formula gives the following theorem.

Theorem 4.1 (Argument principle) *Suppose f is meromorphic in an open set containing a circle C and its interior. If f has no poles and never vanishes on C, then*

$$\frac{1}{2\pi i} \int_C \frac{f'(z)}{f(z)} \, dz = (number\ of\ zeros\ of\ f\ inside\ C)\ minus$$

$$(number\ of\ poles\ of\ f\ inside\ C),$$

where the zeros and poles are counted with their multiplicities.

Corollary 4.2 *The above theorem holds for toy contours.*

As an application of the argument principle, we shall prove three theorems of interest in the general theory. The first, Rouché's theorem, is in some sense a continuity statement. It says that a holomorphic function can be perturbed slightly without changing the number of its zeros. Then, we prove the open mapping theorem, which states that holomorphic functions map open sets to open sets, an important property that again shows the special nature of holomorphic functions. Finally, the maximum modulus principle is reminiscent of (and in fact implies) the same property for harmonic functions: a non-constant holomorphic function on an open set Ω cannot attain its maximum in the interior of Ω.

Theorem 4.3 (Rouché's theorem) *Suppose that f and g are holomorphic in an open set containing a circle C and its interior. If*

$$|f(z)| > |g(z)| \quad \text{for all } z \in C,$$

then f and $f + g$ have the same number of zeros inside the circle C.

Proof. For $t \in [0, 1]$ define

$$f_t(z) = f(z) + tg(z)$$

so that $f_0 = f$ and $f_1 = f + g$. Let n_t denote the number of zeros of f_t inside the circle counted with multiplicities, so that in particular, n_t is an integer. The condition $|f(z)| > |g(z)|$ for $z \in C$ clearly implies that f_t has no zeros on the circle, and the argument principle implies

$$n_t = \frac{1}{2\pi i} \int_C \frac{f_t'(z)}{f_t(z)} \, dz.$$

To prove that n_t is constant, it suffices to show that it is a continuous function of t. Then we could argue that if n_t were not constant, the intermediate value theorem would guarantee the existence of some $t_0 \in [0, 1]$ with n_{t_0} not integral, contradicting the fact that $n_t \in \mathbb{Z}$ for all t.

To prove the continuity of n_t, we observe that $f_t'(z)/f_t(z)$ is jointly continuous for $t \in [0, 1]$ and $z \in C$. This joint continuity follows from the fact that it holds for both the numerator and denominator, and our assumptions guarantee that $f_t(z)$ does not vanish on C. Hence n_t is integer-valued and continuous, and it must be constant. We conclude that $n_0 = n_1$, which is Rouché's theorem.

We now come to an important geometric property of holomorphic functions that arises when we consider them as mappings (that is, mapping regions in the complex plane to the complex plane).

A mapping is said to be **open** if it maps open sets to open sets.

Theorem 4.4 (Open mapping theorem) *If f is holomorphic and non-constant in a region Ω, then f is open.*

Proof.   Let w_0 belong to the image of f, say $w_0 = f(z_0)$. We must prove that all points w near w_0 also belong to the image of f.

Define $g(z) = f(z) - w$ and write

$$g(z) = (f(z) - w_0) + (w_0 - w)$$
$$= F(z) + G(z).$$

Now choose $\delta > 0$ such that the disc $|z - z_0| \leq \delta$ is contained in Ω and $f(z) \neq w_0$ on the circle $|z - z_0| = \delta$. We then select $\epsilon > 0$ so that we have $|f(z) - w_0| \geq \epsilon$ on the circle $|z - z_0| = \delta$. Now if $|w - w_0| < \epsilon$ we have $|F(z)| > |G(z)|$ on the circle $|z - z_0| = \delta$, and by Rouché's theorem we conclude that $g = F + G$ has a zero inside the circle since F has one.

The next result pertains to the size of a holomorphic function. We shall refer to the **maximum** of a holomorphic function f in an open set Ω as the maximum of its absolute value $|f|$ in Ω.

Theorem 4.5 (Maximum modulus principle) *If f is a non-constant holomorphic function in a region Ω, then f cannot attain a maximum in Ω.*

Proof.   Suppose that f did attain a maximum at z_0. Since f is holomorphic it is an open mapping, and therefore, if $D \subset \Omega$ is a small disc centered at z_0, its image $f(D)$ is open and contains $f(z_0)$. This proves that there are points in $z \in D$ such that $|f(z)| > |f(z_0)|$, a contradiction.

Corollary 4.6 *Suppose that Ω is a region with compact closure $\overline{\Omega}$. If f is holomorphic on Ω and continuous on $\overline{\Omega}$ then*

$$\sup_{z \in \Omega} |f(z)| \leq \sup_{z \in \overline{\Omega} - \Omega} |f(z)|.$$

In fact, since $f(z)$ is continuous on the compact set $\overline{\Omega}$, then $|f(z)|$ attains its maximum in $\overline{\Omega}$; but this cannot be in Ω if f is non-constant. If f is constant, the conclusion is trivial.

Remark. The hypothesis that $\overline{\Omega}$ is compact (that is, bounded) is essential for the conclusion. We give an example related to considerations that we will take up in Chapter 4. Let Ω be the open first quadrant, bounded by the positive half-line $x \geq 0$ and the corresponding imaginary line $y \geq 0$. Consider $F(z) = e^{-iz^2}$. Then F is entire and clearly continuous on $\overline{\Omega}$. Moreover $|F(z)| = 1$ on the two boundary lines $z = x$ and $z = iy$. However, $F(z)$ is unbounded in Ω, since for example, we have $F(z) = e^{r^2}$ if $z = r\sqrt{i} = re^{i\pi/4}$.

5 Homotopies and simply connected domains

The key to the general form of Cauchy's theorem, as well as the analysis of multiple-valued functions, is to understand in what regions we can define the primitive of a given holomorphic function. Note the relevance to the study of the logarithm, which arises as a primitive of $1/z$. The question is not just a local one, but is also global in nature. Its elucidation requires the notion of homotopy, and the resulting idea of simple-connectivity.

Let γ_0 and γ_1 be two curves in an open set Ω with common end-points. So if $\gamma_0(t)$ and $\gamma_1(t)$ are two parametrizations defined on $[a, b]$, we have

$$\gamma_0(a) = \gamma_1(a) = \alpha \quad \text{and} \quad \gamma_0(b) = \gamma_1(b) = \beta.$$

These two curves are said to be **homotopic in Ω** if for each $0 \leq s \leq 1$ there exists a curve $\gamma_s \subset \Omega$, parametrized by $\gamma_s(t)$ defined on $[a, b]$, such that for every s

$$\gamma_s(a) = \alpha \quad \text{and} \quad \gamma_s(b) = \beta,$$

and for all $t \in [a, b]$

$$\gamma_s(t)|_{s=0} = \gamma_0(t) \quad \text{and} \quad \gamma_s(t)|_{s=1} = \gamma_1(t).$$

Moreover, $\gamma_s(t)$ should be jointly continuous in $s \in [0, 1]$ and $t \in [a, b]$.

Loosely speaking, two curves are homotopic if one curve can be deformed into the other by a continuous transformation without ever leaving Ω (Figure 6).

Theorem 5.1 *If f is holomorphic in Ω, then*

$$\int_{\gamma_0} f(z)\, dz = \int_{\gamma_1} f(z)\, dz$$

whenever the two curves γ_0 and γ_1 are homotopic in Ω.

Proof. The key to the proof lies in showing that if two curves are close to each other and have the same end-points, then the integrals over these curves are equal. Recall that by definition, the function $F(s, t) = \gamma_s(t)$ is continuous on $[0, 1] \times [a, b]$. In particular, since the image of F, which we

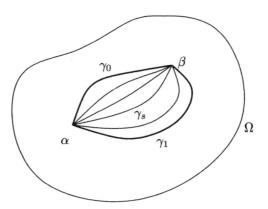

Figure 6. Homotopy of curves

denote by K, is compact, there exists $\epsilon > 0$ such that every disc of radius 3ϵ centered at a point in the image of F is completely contained in Ω. If not, for every $\ell \geq 0$, there exist points $z_\ell \in K$ and w_ℓ in the complement of Ω such that $|z_\ell - w_\ell| < 1/\ell$. By compactness of K, there exists a subsequence of $\{z_\ell\}$, say $\{z_{\ell_k}\}$, that converges to a point $z \in K \subset \Omega$. By construction, we must have $w_{\ell_k} \to z$ as well, and since $\{w_\ell\}$ lies in the complement of Ω which is closed, we must have $z \in \Omega^c$ as well. This is a contradiction.

Having found an ϵ with the desired property, we may, by the uniform continuity of F, select δ so that

$$\sup_{t\in[a,b]} |\gamma_{s_1}(t) - \gamma_{s_2}(t)| < \epsilon \quad \text{whenever } |s_1 - s_2| < \delta.$$

Fix s_1 and s_2 with $|s_1 - s_2| < \delta$. We then choose discs $\{D_0, \ldots, D_n\}$ of radius 2ϵ, and consecutive points $\{z_0, \ldots, z_{n+1}\}$ on γ_{s_1} and $\{w_0, \ldots, w_{n+1}\}$ on γ_{s_2} such that the union of these discs covers both curves, and

$$z_i, z_{i+1}, w_i, w_{i+1} \in D_i.$$

The situation is illustrated in Figure 7.

Also, we choose $z_0 = w_0$ as the beginning end-point of the curves and $z_{n+1} = w_{n+1}$ as the common end-point. On each disc D_i, let F_i denote a primitive of f (Theorem 2.1, Chapter 2). On the intersection of D_i and D_{i+1}, F_i and F_{i+1} are two primitives of the same function, so they must differ by a constant, say c_i. Therefore

$$F_{i+1}(z_{i+1}) - F_i(z_{i+1}) = F_{i+1}(w_{i+1}) - F_i(w_{i+1}),$$

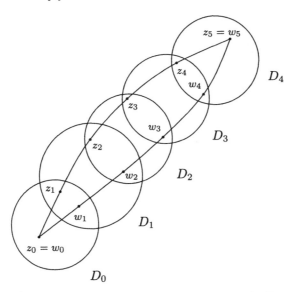

Figure 7. Covering two nearby curves with discs

hence

(5) $$F_{i+1}(z_{i+1}) - F_{i+1}(w_{i+1}) = F_i(z_{i+1}) - F_i(w_{i+1}).$$

This implies

$$\int_{\gamma_{s_1}} f - \int_{\gamma_{s_2}} f = \sum_{i=0}^{n}[F_i(z_{i+1}) - F_i(z_i)] - \sum_{i=0}^{n}[F_i(w_{i+1}) - F_i(w_i)]$$

$$= \sum_{i=0}^{n} F_i(z_{i+1}) - F_i(w_{i+1}) - (F_i(z_i) - F_i(w_i))$$

$$= F_n(z_{n+1}) - F_n(w_{n+1}) - (F_0(z_0) - F_0(w_0)),$$

because of the cancellations due to (5). Since γ_{s_1} and γ_{s_2} have the same beginning and end point, we have proved that

$$\int_{\gamma_{s_1}} f = \int_{\gamma_{s_2}} f.$$

By subdividing $[0, 1]$ into subintervals $[s_i, s_{i+1}]$ of length less than δ, we may go from γ_0 to γ_1 by finitely many applications of the above argument, and the theorem is proved.

A region Ω in the complex plane is **simply connected** if any two pair of curves in Ω with the same end-points are homotopic.

EXAMPLE 1. A disc D is simply connected. In fact, if $\gamma_0(t)$ and $\gamma_1(t)$ are two curves lying in D, we can define $\gamma_s(t)$ by

$$\gamma_s(t) = (1 - s)\gamma_0(t) + s\gamma_1(t).$$

Note that if $0 \leq s \leq 1$, then for each t, the point $\gamma_s(t)$ is on the segment joining $\gamma_0(t)$ and $\gamma_1(t)$, and so is in D. The same argument works if D is replaced by a rectangle, or more generally by any open convex set. (See Exercise 21.)

EXAMPLE 2. The slit plane $\Omega = \mathbb{C} - \{(-\infty, 0]\}$ is simply connected. For a pair of curves γ_0 and γ_1 in Ω, we write $\gamma_j(t) = r_j(t)e^{i\theta_j(t)}$ $(j = 0, 1)$ with $r_j(t)$ continuous and strictly positive, and $\theta_j(t)$ continuous with $|\theta_j(t)| < \pi$. Then, we can define $\gamma_s(t)$ as $r_s(t)e^{i\theta_s(t)}$ where

$$r_s(t) = (1 - s)r_0(t) + sr_1(t) \quad \text{and} \quad \theta_s(t) = (1 - s)\theta_0(t) + s\theta_1(t).$$

We then have $\gamma_s(t) \in \Omega$ whenever $0 \leq s \leq 1$.

EXAMPLE 3. With some effort one can show that the interior of a toy contour is simply connected. This requires that we divide the interior into several subregions. A general form of the argument is given in Problem 4.

EXAMPLE 4. In contrast with the previous examples, the punctured plane $\mathbb{C} - \{0\}$ is not simply connected. Intuitively, consider two curves with the origin between them. It is impossible to continuously pass from one curve to the other without going over 0. A rigorous proof of this fact requires one further result, and will be given shortly.

Theorem 5.2 *Any holomorphic function in a simply connected domain has a primitive.*

Proof. Fix a point z_0 in Ω and define

$$F(z) = \int_{\gamma} f(w)\, dw$$

where the integral is taken over any curve in Ω joining z_0 to z. This definition is independent of the curve chosen, since Ω is simply connected, and if $\tilde{\gamma}$ is another curve in Ω joining z_0 and z, we would have

$$\int_\gamma f(w)\, dw = \int_{\tilde{\gamma}} f(w)\, dw$$

by Theorem 5.1. Now we can write

$$F(z+h) - F(z) = \int_\eta f(w)\, dw$$

where η is the line segment joining z and $z + h$. Arguing as in the proof of Theorem 2.1 in Chapter 2, we find that

$$\lim_{h \to 0} \frac{F(z+h) - F(z)}{h} = f(z).$$

As a result, we obtain the following version of Cauchy's theorem.

Corollary 5.3 *If f is holomorphic in the simply connected region Ω, then*

$$\int_\gamma f(z)\, dz = 0$$

for any closed curve γ in Ω.

This is immediate from the existence of a primitive.

The fact that the punctured plane is not simply connected now follows rigorously from the observation that the integral of $1/z$ over the unit circle is $2\pi i$, and not 0.

6 The complex logarithm

Suppose we wish to define the logarithm of a non-zero complex number. If $z = re^{i\theta}$, and we want the logarithm to be the inverse to the exponential, then it is natural to set

$$\log z = \log r + i\theta.$$

Here and below, we use the convention that $\log r$ denotes the *standard*[1] logarithm of the positive number r. The trouble with the above definition is that θ is unique only up to an integer multiple of 2π. However,

---

[1] By the standard logarithm, we mean the natural logarithm of positive numbers that appears in elementary calculus.

for given z we can fix a choice of θ, and if z varies only a little, this determines the corresponding choice of θ uniquely (assuming we require that θ varies continuously with z). Thus "locally" we can give an unambiguous definition of the logarithm, but this will not work "globally." For example, if z starts at 1, and then winds around the origin and returns to 1, the logarithm does not return to its original value, but rather differs by an integer multiple of $2\pi i$, and therefore is not "single-valued." To make sense of the logarithm as a single-valued function, we must restrict the set on which we define it. This is the so-called choice of a **branch** or **sheet** of the logarithm.

Our discussion of simply connected domains given above leads to a natural global definition of a branch of the logarithm function.

Theorem 6.1 *Suppose that Ω is simply connected with $1 \in \Omega$, and $0 \notin \Omega$. Then in Ω there is a branch of the logarithm $F(z) = \log_\Omega(z)$ so that*

(i) *F is holomorphic in Ω,*

(ii) *$e^{F(z)} = z$ for all $z \in \Omega$,*

(iii) *$F(r) = \log r$ whenever r is a real number and near 1.*

In other words, each branch $\log_\Omega(z)$ is an extension of the standard logarithm defined for positive numbers.

Proof. We shall construct F as a primitive of the function $1/z$. Since $0 \notin \Omega$, the function $f(z) = 1/z$ is holomorphic in Ω. We define

$$\log_\Omega(z) = F(z) = \int_\gamma f(w)\, dw,$$

where γ is any curve in Ω connecting 1 to z. Since Ω is simply connected, this definition does not depend on the path chosen. Arguing as in the proof of Theorem 5.2, we find that F is holomorphic and $F'(z) = 1/z$ for all $z \in \Omega$. This proves (i). To prove (ii), it suffices to show that $ze^{-F(z)} = 1$. For that, we differentiate the left-hand side, obtaining

$$\frac{d}{dz}\left(ze^{-F(z)}\right) = e^{-F(z)} - zF'(z)e^{-F(z)} = (1 - zF'(z))e^{-F(z)} = 0.$$

Since Ω is connected we conclude, by Corollary 3.4 in Chapter 1, that $ze^{-F(z)}$ is constant. Evaluating this expression at $z = 1$, and noting that $F(1) = 0$, we find that this constant must be 1.

Finally, if r is real and close to 1 we can choose as a path from 1 to r a line segment on the real axis so that

$$F(r) = \int_1^r \frac{dx}{x} = \log r,$$

by the usual formula for the standard logarithm. This completes the proof of the theorem.

For example, in the slit plane $\Omega = \mathbb{C} - \{(-\infty, 0]\}$ we have the **principal branch** of the logarithm

$$\log z = \log r + i\theta$$

where $z = re^{i\theta}$ with $|\theta| < \pi$. (Here we drop the subscript Ω, and write simply $\log z$.) To prove this, we use the path of integration γ shown in Figure 8.

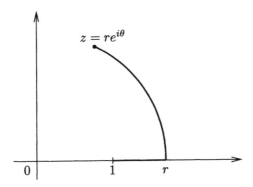

Figure 8. Path of integration for the principal branch of the logarithm

If $z = re^{i\theta}$ with $|\theta| < \pi$, the path consists of the line segment from 1 to r and the arc η from r to z. Then

$$\log z = \int_1^r \frac{dx}{x} + \int_\eta \frac{dw}{w}$$

$$= \log r + \int_0^\theta \frac{ire^{it}}{re^{it}}\, dt$$

$$= \log r + i\theta.$$

An important observation is that in general

$$\log(z_1 z_2) \neq \log z_1 + \log z_2.$$

For example, if $z_1 = e^{2\pi i/3} = z_2$, then for the principal branch of the logarithm, we have

$$\log z_1 = \log z_2 = \frac{2\pi i}{3},$$

and since $z_1 z_2 = e^{-2\pi i/3}$ we have

$$-\frac{2\pi i}{3} = \log(z_1 z_2) \neq \log z_1 + \log z_2.$$

Finally, for the principal branch of the logarithm the following Taylor expansion holds:

(6) $\log(1+z) = z - \dfrac{z^2}{2} + \dfrac{z^3}{3} - \cdots = -\displaystyle\sum_{n=1}^{\infty}(-1)^n \dfrac{z^n}{n}$ for $|z| < 1.$

Indeed, the derivative of both sides equals $1/(1+z)$, so that they differ by a constant. Since both sides are equal to 0 at $z = 0$ this constant must be 0, and we have proved the desired Taylor expansion.

Having defined a logarithm on a simply connected domain, we can now define the powers z^α for any $\alpha \in \mathbb{C}$. If Ω is simply connected with $1 \in \Omega$ and $0 \notin \Omega$, we choose the branch of the logarithm with $\log 1 = 0$ as above, and define

$$z^\alpha = e^{\alpha \log z}.$$

Note that $1^\alpha = 1$, and that if $\alpha = 1/n$, then

$$(z^{1/n})^n = \prod_{k=1}^{n} e^{\frac{1}{n} \log z} = e^{\sum_{k=1}^{n} \frac{1}{n} \log z} = e^{\frac{n}{n} \log z} = e^{\log z} = z.$$

We know that every non-zero complex number w can be written as $w = e^z$. A generalization of this fact is given in the next theorem, which discusses the existence of $\log f(z)$ whenever f does not vanish.

Theorem 6.2 *If f is a nowhere vanishing holomorphic function in a simply connected region Ω, then there exists a holomorphic function g on Ω such that*

$$f(z) = e^{g(z)}.$$

The function $g(z)$ in the theorem can be denoted by $\log f(z)$, and determines a "branch" of that logarithm.

Proof. Fix a point z_0 in Ω, and define a function

$$g(z) = \int_\gamma \frac{f'(w)}{f(w)} \, dw + c_0,$$

where γ is any path in Ω connecting z_0 to z, and c_0 is a complex number so that $e^{c_0} = f(z_0)$. This definition is independent of the path γ since Ω is simply connected. Arguing as in the proof of Theorem 2.1, Chapter 2, we find that g is holomorphic with

$$g'(z) = \frac{f'(z)}{f(z)},$$

and a simple calculation gives

$$\frac{d}{dz}\left(f(z)e^{-g(z)}\right) = 0,$$

so that $f(z)e^{-g(z)}$ is constant. Evaluating this expression at z_0 we find $f(z_0)e^{-c_0} = 1$, so that $f(z) = e^{g(z)}$ for all $z \in \Omega$, and the proof is complete.

7 Fourier series and harmonic functions

In Chapter 4 we shall describe some interesting connections between complex function theory and Fourier analysis on the real line. The motivation for this study comes in part from the simple and direct relation between Fourier series on the circle and power series expansions of holomorphic functions in the disc, which we now investigate.

Suppose that f is holomorphic in a disc $D_R(z_0)$, so that f has a power series expansion

$$f(z) = \sum_{n=0}^{\infty} a_n(z - z_0)^n$$

that converges in that disc.

Theorem 7.1 *The coefficients of the power series expansion of f are given by*

$$a_n = \frac{1}{2\pi r^n} \int_0^{2\pi} f(z_0 + re^{i\theta})e^{-in\theta}\,d\theta$$

for all $n \geq 0$ and $0 < r < R$. Moreover,

$$0 = \frac{1}{2\pi r^n} \int_0^{2\pi} f(z_0 + re^{i\theta})e^{-in\theta}\,d\theta$$

whenever $n < 0$.

Proof. Since $f^{(n)}(z_0) = a_n n!$, the Cauchy integral formula gives

$$a_n = \frac{1}{2\pi i} \int_\gamma \frac{f(\zeta)}{(\zeta - z_0)^{n+1}} \, d\zeta,$$

where γ is a circle of radius $0 < r < R$ centered at z_0 and with the positive orientation. Choosing $\zeta = z_0 + re^{i\theta}$ for the parametrization of this circle, we find that for $n \geq 0$

$$\begin{aligned}
a_n &= \frac{1}{2\pi i} \int_0^{2\pi} \frac{f(z_0 + re^{i\theta})}{(z_0 + re^{i\theta} - z_0)^{n+1}} rie^{i\theta} \, d\theta \\
&= \frac{1}{2\pi r^n} \int_0^{2\pi} f(z_0 + re^{i\theta}) e^{-i(n+1)\theta} e^{i\theta} \, d\theta \\
&= \frac{1}{2\pi r^n} \int_0^{2\pi} f(z_0 + re^{i\theta}) e^{-in\theta} \, d\theta.
\end{aligned}$$

Finally, even when $n < 0$, our calculation shows that we still have the identity

$$\frac{1}{2\pi r^n} \int_0^{2\pi} f(z_0 + re^{i\theta}) e^{-in\theta} \, d\theta = \frac{1}{2\pi i} \int_\gamma \frac{f(\zeta)}{(\zeta - z_0)^{n+1}} \, d\zeta.$$

Since $-n > 0$, the function $f(\zeta)(\zeta - z_0)^{-n-1}$ is holomorphic in the disc, and by Cauchy's theorem the last integral vanishes.

The interpretation of this theorem is as follows. Consider $f(z_0 + re^{i\theta})$ as the restriction to the circle of a holomorphic function on the closure of a disc centered at z_0 with radius r. Then its Fourier coefficients vanish if $n < 0$, while those for $n \geq 0$ are equal (up to a factor of r^n) to coefficients of the power series of the holomorphic function f. The property of the vanishing of the Fourier coefficients for $n < 0$ reveals another special characteristic of holomorphic functions (and in particular their restrictions to any circle).

Next, since $a_0 = f(z_0)$, we obtain the following corollary.

Corollary 7.2 (Mean-value property) *If f is holomorphic in a disc $D_R(z_0)$, then*

$$f(z_0) = \frac{1}{2\pi} \int_0^{2\pi} f(z_0 + re^{i\theta}) \, d\theta, \quad \text{for any } 0 < r < R.$$

Taking the real parts of both sides, we obtain the following consequence.

Corollary 7.3 *If f is holomorphic in a disc $D_R(z_0)$, and $u = \mathrm{Re}(f)$, then*

$$u(z_0) = \frac{1}{2\pi} \int_0^{2\pi} u(z_0 + re^{i\theta})\, d\theta, \quad \text{for any } 0 < r < R.$$

Recall that u is harmonic whenever f is holomorphic, and in fact, the above corollary is a property enjoyed by every harmonic function in the disc $D_R(z_0)$. This follows from Exercise 12 in Chapter 2, which shows that every harmonic function in a disc is the real part of a holomorphic function in that disc.

8 Exercises

1. Using Euler's formula

$$\sin \pi z = \frac{e^{i\pi z} - e^{-i\pi z}}{2i},$$

show that the complex zeros of $\sin \pi z$ are exactly at the integers, and that they are each of order 1.

Calculate the residue of $1/\sin \pi z$ at $z = n \in \mathbb{Z}$.

2. Evaluate the integral

$$\int_{-\infty}^{\infty} \frac{dx}{1 + x^4}.$$

Where are the poles of $1/(1 + z^4)$?

3. Show that

$$\int_{-\infty}^{\infty} \frac{\cos x}{x^2 + a^2}\, dx = \pi \frac{e^{-a}}{a}, \quad \text{for } a > 0.$$

4. Show that

$$\int_{-\infty}^{\infty} \frac{x \sin x}{x^2 + a^2}\, dx = \pi e^{-a}, \quad \text{for all } a > 0.$$

5. Use contour integration to show that

$$\int_{-\infty}^{\infty} \frac{e^{-2\pi i x \xi}}{(1 + x^2)^2}\, dx = \frac{\pi}{2}(1 + 2\pi|\xi|)e^{-2\pi|\xi|}$$

for all ξ real.

6. Show that for $n \geq 1$

$$\int_{-\infty}^{\infty} \frac{dx}{(1+x^2)^{n+1}} = \frac{1 \cdot 3 \cdot 5 \cdots (2n-1)}{2 \cdot 4 \cdot 6 \cdots (2n)} \cdot \pi.$$

7. Prove that

$$\int_{0}^{2\pi} \frac{d\theta}{(a + \cos\theta)^2} = \frac{2\pi a}{(a^2 - 1)^{3/2}}, \qquad \text{whenever } a > 1.$$

8. Prove that

$$\int_{0}^{2\pi} \frac{d\theta}{a + b\cos\theta} = \frac{2\pi}{\sqrt{a^2 - b^2}}$$

if $a > |b|$ and $a, b \in \mathbb{R}$.

9. Show that

$$\int_{0}^{1} \log(\sin \pi x)\, dx = -\log 2.$$

[Hint: Use the contour shown in Figure 9.]

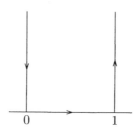

Figure 9. Contour in Exercise 9

10. Show that if $a > 0$, then

$$\int_{0}^{\infty} \frac{\log x}{x^2 + a^2}\, dx = \frac{\pi}{2a} \log a.$$

[Hint: Use the contour in Figure 10.]

11. Show that if $|a| < 1$, then

$$\int_{0}^{2\pi} \log |1 - ae^{i\theta}|\, d\theta = 0.$$

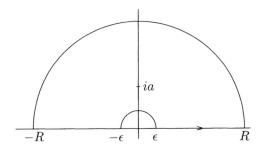

Figure 10. Contour in Exercise 10

Then, prove that the above result remains true if we assume only that $|a| \leq 1$.

12. Suppose u is not an integer. Prove that

$$\sum_{n=-\infty}^{\infty} \frac{1}{(u+n)^2} = \frac{\pi^2}{(\sin \pi u)^2}$$

by integrating

$$f(z) = \frac{\pi \cot \pi z}{(u+z)^2}$$

over the circle $|z| = R_N = N + 1/2$ (N integral, $N \geq |u|$), adding the residues of f inside the circle, and letting N tend to infinity.

Note. Two other derivations of this identity, using Fourier series, were given in Book I.

13. Suppose $f(z)$ is holomorphic in a punctured disc $D_r(z_0) - \{z_0\}$. Suppose also that

$$|f(z)| \leq A|z - z_0|^{-1+\epsilon}$$

for some $\epsilon > 0$, and all z near z_0. Show that the singularity of f at z_0 is removable.

14. Prove that all entire functions that are also injective take the form $f(z) = az + b$ with $a, b \in \mathbb{C}$, and $a \neq 0$.

[Hint: Apply the Casorati-Weierstrass theorem to $f(1/z)$, and use Theorem 4.4.]

15. Use the Cauchy inequalities or the maximum modulus principle to solve the following problems:

(a) Prove that if f is an entire function that satisfies

$$\sup_{|z|=R} |f(z)| \leq AR^k + B$$

for all $R > 0$, and for some integer $k \geq 0$ and some constants $A, B > 0$, then f is a polynomial of degree $\leq k$.

(b) Show that if f is holomorphic in the unit disc, is bounded, and converges uniformly to zero in the sector $\theta < \arg z < \varphi$ as $|z| \to 1$, then $f = 0$.

(c) Let $w_1, \ldots, w_n$ be points on the unit circle in the complex plane. Prove that there exists a point z on the unit circle such that the product of the distances from z to the points w_j, $1 \leq j \leq n$, is at least 1. Conclude that there exists a point w on the unit circle such that the product of the distances from w to the points w_j, $1 \leq j \leq n$, is exactly equal to 1.

(d) Show that if the real part of an entire function f is bounded, then f is constant.

16. Suppose f and g are holomorphic in a region containing the disc $|z| \leq 1$. Suppose that f has a simple zero at $z = 0$ and vanishes nowhere else in $|z| \leq 1$. Let

$$f_\epsilon(z) = f(z) + \epsilon g(z).$$

Show that if ϵ is sufficiently small, then

(a) $f_\epsilon(z)$ has a unique zero in $|z| \leq 1$, and

(b) if z_ϵ is this zero, the mapping $\epsilon \mapsto z_\epsilon$ is continuous.

17. Let f be non-constant and holomorphic in an open set containing the closed unit disc.

(a) Show that if $|f(z)| = 1$ whenever $|z| = 1$, then the image of f contains the unit disc. [Hint: One must show that $f(z) = w_0$ has a root for every $w_0 \in \mathbb{D}$. To do this, it suffices to show that $f(z) = 0$ has a root (why?). Use the maximum modulus principle to conclude.]

(b) If $|f(z)| \geq 1$ whenever $|z| = 1$ and there exists a point $z_0 \in \mathbb{D}$ such that $|f(z_0)| < 1$, then the image of f contains the unit disc.

18. Give another proof of the Cauchy integral formula

$$f(z) = \frac{1}{2\pi i} \int_C \frac{f(\zeta)}{\zeta - z} \, d\zeta$$

using homotopy of curves.

[Hint: Deform the circle C to a small circle centered at z, and note that the quotient $(f(\zeta) - f(z))/(\zeta - z)$ is bounded.]

19. Prove the maximum principle for harmonic functions, that is:

(a) If u is a non-constant real-valued harmonic function in a region Ω, then u cannot attain a maximum (or a minimum) in Ω.

(b) Suppose that Ω is a region with compact closure $\overline{\Omega}$. If u is harmonic in Ω and continuous in $\overline{\Omega}$, then

$$\sup_{z \in \Omega} |u(z)| \leq \sup_{z \in \overline{\Omega} - \Omega} |u(z)|.$$

[Hint: To prove the first part, assume that u attains a local maximum at z_0. Let f be holomorphic near z_0 with $u = \text{Re}(f)$, and show that f is not open. The second part follows directly from the first.]

20. This exercise shows how the mean square convergence dominates the uniform convergence of analytic functions. If U is an open subset of $\mathbb{C}$ we use the notation

$$\|f\|_{L^2(U)} = \left(\int_U |f(z)|^2 \, dx dy \right)^{1/2}$$

for the mean square norm, and

$$\|f\|_{L^\infty(U)} = \sup_{z \in U} |f(z)|$$

for the sup norm.

(a) If f is holomorphic in a neighborhood of the disc $D_r(z_0)$, show that for any $0 < s < r$ there exists a constant $C > 0$ (which depends on s and r) such that

$$\|f\|_{L^\infty(D_s(z_0))} \leq C \|f\|_{L^2(D_r(z_0))}.$$

(b) Prove that if $\{f_n\}$ is a Cauchy sequence of holomorphic functions in the mean square norm $\| \cdot \|_{L^2(U)}$, then the sequence $\{f_n\}$ converges uniformly on every compact subset of U to a holomorphic function.

[Hint: Use the mean-value property.]

21. Certain sets have geometric properties that guarantee they are simply connected.

(a) An open set $\Omega \subset \mathbb{C}$ is **convex** if for any two points in Ω, the straight line segment between them is contained in Ω. Prove that a convex open set is simply connected.

(b) More generally, an open set $\Omega \subset \mathbb{C}$ is **star-shaped** if there exists a point $z_0 \in \Omega$ such that for any $z \in \Omega$, the straight line segment between z and z_0 is contained in Ω. Prove that a star-shaped open set is simply connected. Conclude that the slit plane $\mathbb{C} - \{(-\infty, 0]\}$ (and more generally any sector, convex or not) is simply connected.

(c) What are other examples of open sets that are simply connected?

22. Show that there is no holomorphic function f in the unit disc $\mathbb{D}$ that extends continuously to $\partial\mathbb{D}$ such that $f(z) = 1/z$ for $z \in \partial\mathbb{D}$.

9 Problems

1.* Consider a holomorphic map on the unit disc $f : \mathbb{D} \to \mathbb{C}$ which satisfies $f(0) = 0$. By the open mapping theorem, the image $f(\mathbb{D})$ contains a small disc centered at the origin. We then ask: does there exist $r > 0$ such that for *all* $f : \mathbb{D} \to \mathbb{C}$ with $f(0) = 0$, we have $D_r(0) \subset f(\mathbb{D})$?

(a) Show that with no further restrictions on f, no such r exists. It suffices to find a sequence of functions $\{f_n\}$ holomorphic in $\mathbb{D}$ such that $1/n \notin f(\mathbb{D})$. Compute $f_n'(0)$, and discuss.

(b) Assume in addition that f also satisfies $f'(0) = 1$. Show that despite this new assumption, there exists no $r > 0$ satisfying the desired condition.

[Hint: Try $f_\epsilon(z) = \epsilon(e^{z/\epsilon} - 1)$.]

The Koebe-Bieberbach theorem states that if in addition to $f(0) = 0$ and $f'(0) = 1$ we also assume that f is injective, then such an r exists and the best possible value is $r = 1/4$.

(c) As a first step, show that if $h(z) = \frac{1}{z} + c_0 + c_1 z + c_2 z^2 + \cdots$ is analytic and injective for $0 < |z| < 1$, then $\sum_{n=1}^{\infty} n|c_n|^2 \leq 1$.

[Hint: Calculate the area of the complement of $h(D_\rho(0) - \{0\})$ where $0 < \rho < 1$, and let $\rho \to 1$.]

(d) If $f(z) = z + a_2 z^2 + \cdots$ satisfies the hypotheses of the theorem, show that there exists another function g satisfying the hypotheses of the theorem such that $g^2(z) = f(z^2)$.

[Hint: $f(z)/z$ is nowhere vanishing so there exists ψ such that $\psi^2(z) = f(z)/z$ and $\psi(0) = 1$. Check that $g(z) = z\psi(z^2)$ is injective.]

(e) With the notation of the previous part, show that $|a_2| \leq 2$, and that equality holds if and only if

$$f(z) = \frac{z}{(1 - e^{i\theta}z)^2} \qquad \text{for some } \theta \in \mathbb{R}.$$

[Hint: What is the power series expansion of $1/g(z)$? Use part (c).]

(f) If $h(z) = \frac{1}{z} + c_0 + c_1 z + c_2 z^2 + \cdots$ is injective on $\mathbb{D}$ and avoids the values z_1 and z_2, show that $|z_1 - z_2| \leq 4$.

[Hint: Look at the second coefficient in the power series expansion of $1/(h(z) - z_j)$.]

(g) Complete the proof of the theorem. [Hint: If f avoids w, then $1/f$ avoids 0 and $1/w$.]

2. Let u be a harmonic function in the unit disc that is continuous on its closure. Deduce Poisson's integral formula

$$u(z_0) = \frac{1}{2\pi} \int_0^{2\pi} \frac{1 - |z_0|^2}{|e^{i\theta} - z_0|^2} u(e^{i\theta}) \, d\theta \qquad \text{for } |z_0| < 1$$

from the special case $z_0 = 0$ (the mean value theorem). Show that if $z_0 = re^{i\varphi}$, then

$$\frac{1 - |z_0|^2}{|e^{i\theta} - z_0|^2} = \frac{1 - r^2}{1 - 2r\cos(\theta - \varphi) + r^2} = P_r(\theta - \varphi),$$

and we recover the expression for the Poisson kernel derived in the exercises of the previous chapter.

[Hint: Set $u_0(z) = u(T(z))$ where

$$T(z) = \frac{z_0 - z}{1 - \overline{z_0} z}.$$

Prove that u_0 is harmonic. Then apply the mean value theorem to u_0, and make a change of variables in the integral.]

3. If $f(z)$ is holomorphic in the deleted neighborhood $\{0 < |z - z_0| < r\}$ and has a pole of order k at z_0, then we can write

$$f(z) = \frac{a_{-k}}{(z - z_0)^k} + \cdots + \frac{a_{-1}}{(z - z_0)} + g(z)$$

where g is holomorphic in the disc $\{|z - z_0| < r\}$. There is a generalization of this expansion that holds even if z_0 is an ~~essential singularity~~. ~~This is a special case of~~ the **Laurent series expansion**, which is valid in an even more general setting.

Let f be holomorphic in a region containing the annulus $\{z : r_1 \leq |z - z_0| \leq r_2\}$ where $0 < r_1 < r_2$. Then,

$$f(z) = \sum_{n=-\infty}^{\infty} a_n (z - z_0)^n$$

where the series converges absolutely in the interior of the annulus. To prove this, it suffices to write

$$f(z) = \frac{1}{2\pi i} \int_{C_{r_2}} \frac{f(\zeta)}{\zeta - z} \, d\zeta - \frac{1}{2\pi i} \int_{C_{r_1}} \frac{f(\zeta)}{\zeta - z} \, d\zeta$$

when $r_1 < |z - z_0| < r_2$, and argue as in the proof of Theorem 4.4, Chapter 2. Here C_{r_1} and C_{r_2} are the circles bounding the annulus.

4. * Suppose Ω is a bounded region. Let L be a (two-way infinite) line that intersects Ω. Assume that $\Omega \cap L$ is an interval I. Choosing an orientation for L, we can define Ω_l and Ω_r to be the subregions of Ω lying strictly to the left or right of L, with $\Omega = \Omega_l \cup I \cup \Omega_r$ a disjoint union. If Ω_l and Ω_r are simply connected, then Ω is simply connected.

5. * Let

$$g(z) = \frac{1}{2\pi i} \int_{-M}^{M} \frac{h(x)}{x - z} \, dx$$

where h is continuous and supported in $[-M, M]$.

(a) Prove that the function g is holomorphic in $\mathbb{C} - [-M, M]$, and vanishes at infinity, that is, $\lim_{|z| \to \infty} |g(z)| = 0$. Moreover, the "jump" of g across $[-M, M]$ is h, that is,

$$h(x) = \lim_{\epsilon \to 0, \epsilon > 0} g(x + i\epsilon) - g(x - i\epsilon).$$

[Hint: Express the difference $g(x + i\epsilon) - g(x - i\epsilon)$ in terms of a convolution of h with the Poisson kernel.]

(b) If h satisfies a mild smoothness condition, for instance a Hölder condition with exponent α, that is, $|h(x) - h(y)| \leq C|x - y|^\alpha$ for some $C > 0$ and all $x, y \in [-M, M]$, then $g(x + i\epsilon)$ and $g(x - i\epsilon)$ converge uniformly to functions $g_+(x)$ and $g_-(x)$ as $\epsilon \to 0$. Then, g can be characterized as the unique holomorphic function that satisfies:

(i) g is holomorphic outside $[-M, M]$,

(ii) g vanishes at infinity,

(iii) $g(x + i\epsilon)$ and $g(x - i\epsilon)$ converge uniformly as $\epsilon \to 0$ to functions $g_+(x)$ and $g_-(x)$ with

$$g_+(x) - g_-(x) = h(x).$$

[Hint: If G is another function satisfying these conditions, $g - G$ is entire.]

4 The Fourier Transform

Raymond Edward Alan Christopher Paley, Fellow of
Trinity College, Cambridge, and International Research
Fellow at the Massachusetts Institute of Technology
and at Harvard University, was killed by an avalanche
on April 7, 1933, while skiing in the vicinity of Banff,
Alberta. Although only twenty-six years of age, he
was already recognized as the ablest of the group of
young English mathematicians who have been inspired
by the genius of G. H. Hardy and J. E. Littlewood. In
a group notable for its brilliant technique, no one had
developed this technique to a higher degree than Paley. Nevertheless he should not be thought of as primarily as a technician, for with this ability he combined
creative power of the first order. As he himself was
wont to say, technique without "rugger tactics" will
not get one far, and these rugger tactics he practiced
to a degree that was characteristic of his forthright
and vigorous nature.

Possessed of an extraordinary capacity for making friends and for scientific collaboration, Paley be-
lieved that the inspiration of continual interchange of
ideas stimulates each collaborator to accomplish more
than he would alone. Only the exceptional man works
well with a partner, but Paley had collaborated successfully with many, including Littlewood, Pólya, Zygmund, and Wiener.

N. Wiener, 1933

If f is a function on $\mathbb{R}$ that satisfies appropriate regularity and decay
conditions, then its Fourier transform is defined by

$$\hat{f}(\xi) = \int_{-\infty}^{\infty} f(x)e^{-2\pi i x\xi}\,dx, \quad \xi \in \mathbb{R}$$

and its counterpart, the Fourier inversion formula, holds

$$f(x) = \int_{-\infty}^{\infty} \hat{f}(\xi)e^{2\pi i x\xi}\,d\xi, \quad x \in \mathbb{R}.$$

The Fourier transform (including its d-dimensional variants), plays a basic role in analysis, as the reader of Book I is aware. Here we want to illustrate the intimate and fruitful connection between the one-dimensional theory of the Fourier transform and complex analysis. The main theme (stated somewhat imprecisely) is as follows: for a function f initially defined on the real line, the possibility of extending it to a holomorphic function is closely related to the very rapid (for example, exponential) decay at infinity of its Fourier transform $\hat{f}$. We elaborate on this theme in two stages.

First, we assume that f can be analytically continued in a horizontal strip containing the real axis, and has "moderate decrease" at infinity,[1] so that the integral defining the Fourier transform $\hat{f}$ converges. As a result, we conclude that $\hat{f}$ decreases exponentially at infinity; it also follows directly that the Fourier inversion formula holds. Moreover one can easily obtain from these considerations the Poisson summation formula $\sum_{n\in\mathbb{Z}} f(n) = \sum_{n\in\mathbb{Z}} \hat{f}(n)$. Incidentally, all these theorems are elegant consequences of contour integration.

At a second stage, we take as our starting point the validity of the Fourier inversion formula, which holds if we assume that both f and $\hat{f}$ are of moderate decrease, without making any assumptions on the analyticity of f. We then ask a simple but natural question: What are the conditions on f so that its Fourier transform is supported in a bounded interval, say $[-M, M]$? This is a basic problem that, as one notices, can be stated without any reference to notions of complex analysis. However, it can be resolved only in terms of the holomorphic properties of the function f. The condition, given by the Paley-Wiener theorem, is that there be a holomorphic extension of f to $\mathbb{C}$ that satisfies the growth condition

$$|f(z)| \leq Ae^{2\pi M|z|} \quad \text{for some constant } A > 0.$$

Functions satisfying this condition are said to be of **exponential type**.

Observe that the condition that $\hat{f}$ vanishes outside a compact set can be viewed as an extreme version of a decay property at infinity, and so the above theorem clearly falls within the context of the theme indicated above.

In all these matters a decisive technique will consist in shifting the contour of integration, that is the real line, within the boundaries of a horizontal strip. This will take advantage of the special behavior of $e^{-2\pi i z\xi}$ when z has a non-zero imaginary part. Indeed, when z is real this exponential remains bounded and oscillates, while if $\text{Im}(z) \neq 0$, it will

---

[1] We say that a function f is of **moderate decrease** if f is continuous and there exists $A > 0$ so that $|f(x)| \leq A/(1 + x^2)$ for all $x \in \mathbb{R}$. A more restrictive condition is that $f \in \mathcal{S}$, the Schwartz space of testing functions, which also implies that $\hat{f}$ belongs to $\mathcal{S}$. See Book I for more details.

have exponential decay or exponential increase, depending on whether the product $\xi \text{Im}(z)$ is negative or positive.

1 The class $\mathfrak{F}$

The weakest decay condition imposed on functions in our study of the Fourier transform in Book I was that of moderate decrease. There, we proved the Fourier inversion and Poisson summation formulas under the hypothesis that f and $\hat{f}$ satisfy

$$|f(x)| \leq \frac{A}{1+x^2} \quad \text{and} \quad |\hat{f}(\xi)| \leq \frac{A'}{1+\xi^2}$$

for some positive constants A, A' and all $x, \xi \in \mathbb{R}$. We were led to consider this class of functions because of various examples such as the Poisson kernel

$$P_y(x) = \frac{1}{\pi} \frac{y}{y^2 + x^2}$$

for $y > 0$, which played a fundamental role in the solution of the Dirichlet problem for the steady-state heat equation in the upper half-plane. There we had $\widehat{P_y}(\xi) = e^{-2\pi y |\xi|}$.

In the present context, we introduce a class of functions particularly suited to the goal we have set: proving theorems about the Fourier transform using complex analysis. Moreover, this class will be large enough to contain many of the important applications we have in mind.

For each $a > 0$ we denote by $\mathfrak{F}_a$ the class of all functions f that satisfy the following two conditions:

(i) The function f is holomorphic in the horizontal strip

$$S_a = \{z \in \mathbb{C} : |\text{Im}(z)| < a\}.$$

(ii) There exists a constant $A > 0$ such that

$$|f(x+iy)| \leq \frac{A}{1+x^2} \quad \text{for all } x \in \mathbb{R} \text{ and } |y| < a.$$

In other words, $\mathfrak{F}_a$ consists of those holomorphic functions on S_a that are of moderate decay on each horizontal line $\text{Im}(z) = y$, uniformly in $-a < y < a$. For example, $f(z) = e^{-\pi z^2}$ belongs to $\mathfrak{F}_a$ for all a. Also, the function

$$f(z) = \frac{1}{\pi} \frac{c}{c^2 + z^2},$$

which has simple poles at $z = \pm ci$, belongs to $\mathfrak{F}_a$ for all $0 < a < c$.

Another example is provided by $f(z) = 1/\cosh \pi z$, which belongs to $\mathfrak{F}_a$ whenever $|a| < 1/2$. This function, as well as one of its fundamental properties, was already discussed in Example 3, Section 2.1 of Chapter 3.

Note also that a simple application of the Cauchy integral formulas shows that if $f \in \mathfrak{F}_a$, then for every n, the nth derivative of f belongs to $\mathfrak{F}_b$ for all b with $0 < b < a$ (Exercise 2).

Finally, we denote by $\mathfrak{F}$ the class of all functions that belong to $\mathfrak{F}_a$ for some a.

Remark. The condition of moderate decrease can be weakened somewhat by replacing the order of decrease of $A/(1 + x^2)$ by $A/(1 + |x|^{1+\epsilon})$ for any $\epsilon > 0$. As the reader will observe, many of the results below remain unchanged with this less restrictive condition.

2 Action of the Fourier transform on $\mathfrak{F}$

Here we prove three theorems, including the Fourier inversion and Poisson summation formulas, for functions in $\mathfrak{F}$. The idea behind all three proofs is the same: contour integration. Thus the approach used will be different from that of the corresponding results in Book I.

Theorem 2.1 *If f belongs to the class $\mathfrak{F}_a$ for some $a > 0$, then $|\hat{f}(\xi)| \le Be^{-2\pi b|\xi|}$ for any $0 \le b < a$.*

Proof. Recall that $\hat{f}(\xi) = \int_{-\infty}^{\infty} f(x)e^{-2\pi i x\xi}\, dx$. The case $b = 0$ simply says that $\hat{f}$ is bounded, which follows at once from the integral defining $\hat{f}$, the assumption that f is of moderate decrease, and the fact that the exponential is bounded by 1.

Now suppose $0 < b < a$ and assume first that $\xi > 0$. The main step consists of shifting the contour of integration, that is the real line, down by b. More precisely, consider the contour in Figure 1 as well as the function $g(z) = f(z)e^{-2\pi i z\xi}$.

We claim that as R tends to infinity, the integrals of g over the two vertical sides converge to zero. For example, the integral over the vertical segment on the left can be estimated by

$$\left| \int_{-R-ib}^{-R} g(z)\, dz \right| \le \int_0^b |f(-R-it)e^{-2\pi i(-R-it)\xi}|\, dt$$

$$\le \int_0^b \frac{A}{R^2} e^{-2\pi t\xi}\, dt$$

$$= O(1/R^2).$$

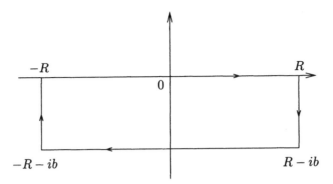

Figure 1. The contour in the proof of Theorem 2.1 when $\xi > 0$

A similar estimate for the other side proves our claim. Therefore, by Cauchy's theorem applied to the large rectangle, we find in the limit as R tends to infinity that

$$(1) \qquad \hat{f}(\xi) = \int_{-\infty}^{\infty} f(x - ib)e^{-2\pi i(x-ib)\xi}\, dx,$$

which leads to the estimate

$$|\hat{f}(\xi)| \leq \int_{-\infty}^{\infty} \frac{A}{1 + x^2}e^{-2\pi b\xi}\, dx \leq Be^{-2\pi b\xi},$$

where B is a suitable constant. A similar argument for $\xi < 0$, but this time shifting the real line up by b, allows us to finish the proof of the theorem.

This result says that whenever $f \in \mathfrak{F}$, then $\hat{f}$ has rapid decay at infinity. We remark that the further we can extend f (that is, the larger a), then the larger we can choose b, hence the better the decay. We shall return to this circle of ideas in Section 3, where we describe those f for which $\hat{f}$ has the ultimate decay condition: compact support.

Since $\hat{f}$ decreases rapidly on $\mathbb{R}$, the integral in the Fourier inversion formula makes sense, and we now turn to the complex analytic proof of this identity.

Theorem 2.2 *If $f \in \mathfrak{F}$, then the Fourier inversion holds, namely*

$$f(x) = \int_{-\infty}^{\infty} \hat{f}(\xi)e^{2\pi i x\xi}\, d\xi \qquad \text{for all } x \in \mathbb{R}.$$

Besides contour integration, the proof of the theorem requires a simple identity, which we isolate.

Lemma 2.3 *If A is positive and B is real, then $\int_0^\infty e^{-(A+iB)\xi} \, d\xi = \frac{1}{A+iB}$.*

Proof. Since $A > 0$ and $B \in \mathbb{R}$, we have $|e^{-(A+iB)\xi}| = e^{-A\xi}$, and the integral converges. By definition

$$\int_0^\infty e^{-(A+iB)\xi} \, d\xi = \lim_{R \to \infty} \int_0^R e^{-(A+iB)\xi} \, d\xi.$$

However,

$$\int_0^R e^{-(A+iB)\xi} \, d\xi = \left[-\frac{e^{-(A+iB)\xi}}{A+iB} \right]_0^R,$$

which tends to $1/(A+iB)$ as R tends to infinity.

We can now prove the inversion theorem. Once again, the sign of ξ matters, so we begin by writing

$$\int_{-\infty}^\infty \hat{f}(\xi)e^{2\pi i x \xi} \, d\xi = \int_{-\infty}^0 \hat{f}(\xi)e^{2\pi i x \xi} \, d\xi + \int_0^\infty \hat{f}(\xi)e^{2\pi i x \xi} \, d\xi.$$

For the second integral we argue as follows. Say $f \in \mathfrak{F}_a$ and choose $0 < b < a$. Arguing as the proof of Theorem 2.1, or simply using equation (1), we get

$$\hat{f}(\xi) = \int_{-\infty}^\infty f(u - ib)e^{-2\pi i (u-ib)\xi} \, du,$$

so that with an application of the lemma and the convergence of the integration in ξ, we find

$$
\begin{aligned}
\int_0^\infty \hat{f}(\xi)e^{2\pi i x \xi} \, d\xi &= \int_0^\infty \int_{-\infty}^\infty f(u - ib)e^{-2\pi i (u-ib)\xi}e^{2\pi i x \xi} \, du \, d\xi \\
&= \int_{-\infty}^\infty f(u - ib) \int_0^\infty e^{-2\pi i (u-ib-x)\xi} \, d\xi \, du \\
&= \int_{-\infty}^\infty f(u - ib) \frac{1}{2\pi b + 2\pi i (u - x)} \, du \\
&= \frac{1}{2\pi i} \int_{-\infty}^\infty \frac{f(u - ib)}{u - ib - x} \, du \\
&= \frac{1}{2\pi i} \int_{L_1} \frac{f(\zeta)}{\zeta - x} \, d\zeta,
\end{aligned}
$$

where L_1 denotes the line $\{u - ib : u \in \mathbb{R}\}$ traversed from left to right. (In other words, L_1 is the real line shifted down by b.) For the integral when $\xi < 0$, a similar calculation gives

$$\int_{-\infty}^{0} \hat{f}(\xi)e^{2\pi ix\xi}\,d\xi = -\frac{1}{2\pi i}\int_{L_2}\frac{f(\zeta)}{\zeta - x}\,d\zeta,$$

where L_2 is the real line shifted up by b, with orientation from left to right. Now given $x \in \mathbb{R}$, consider the contour γ_R in Figure 2.

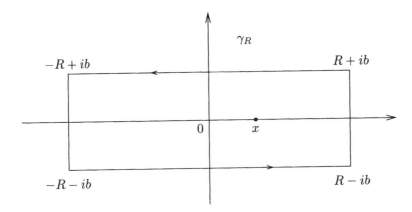

Figure 2. The contour γ_R in the proof of Theorem 2.2

The function $f(\zeta)/(\zeta - x)$ has a simple pole at x with residue $f(x)$, so the residue formula gives

$$f(x) = \frac{1}{2\pi i}\int_{\gamma_R}\frac{f(\zeta)}{\zeta - x}\,d\zeta.$$

Letting R tend to infinity, one checks easily that the integral over the vertical sides goes to 0 and therefore, combining with the previous results, we get

$$\begin{aligned}
f(x) &= \frac{1}{2\pi i}\int_{L_1}\frac{f(\zeta)}{\zeta - x}\,d\zeta - \frac{1}{2\pi i}\int_{L_2}\frac{f(\zeta)}{\zeta - x}\,d\zeta \\
&= \int_{0}^{\infty}\hat{f}(\xi)e^{2\pi ix\xi}\,d\xi + \int_{-\infty}^{0}\hat{f}(\xi)e^{2\pi ix\xi}\,d\xi \\
&= \int_{-\infty}^{\infty}\hat{f}(\xi)e^{2\pi ix\xi}\,d\xi,
\end{aligned}$$

and the theorem is proved.

The last of our three theorems is the Poisson summation formula.

Theorem 2.4 *If $f \in \mathfrak{F}$, then*

$$\sum_{n \in \mathbb{Z}} f(n) = \sum_{n \in \mathbb{Z}} \hat{f}(n).$$

Proof. Say $f \in \mathfrak{F}_a$ and choose some b satisfying $0 < b < a$. The function $1/(e^{2\pi i z} - 1)$ has simple poles with residue $1/(2\pi i)$ at the integers. Thus $f(z)/(e^{2\pi i z} - 1)$ has simple poles at the integers n, with residues $f(n)/2\pi i$. We may therefore apply the residue formula to the contour γ_N in Figure 3 where N is an integer.

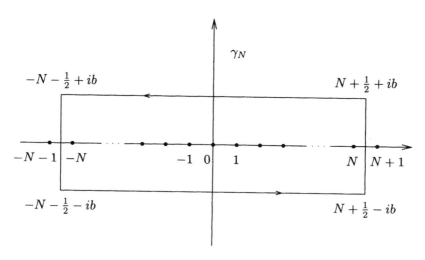

Figure 3. The contour γ_N in the proof of Theorem 2.4

This yields

$$\sum_{|n| \leq N} f(n) = \int_{\gamma_N} \frac{f(z)}{e^{2\pi i z} - 1} \, dz.$$

Letting N tend to infinity, and recalling that f has moderate decrease, we see that the sum converges to $\sum_{n \in \mathbb{Z}} f(n)$, and also that the integral over the vertical segments goes to 0. Therefore, in the limit we get

$$(2) \qquad \sum_{n \in \mathbb{Z}} f(n) = \int_{L_1} \frac{f(z)}{e^{2\pi i z} - 1} \, dz - \int_{L_2} \frac{f(z)}{e^{2\pi i z} - 1} \, dz$$

where L_1 and L_2 are the real line shifted down and up by b, respectively.
Now we use the fact that if $|w| > 1$, then

$$\frac{1}{w-1} = w^{-1} \sum_{n=0}^{\infty} w^{-n}$$

to see that on L_1 (where $|e^{2\pi i z}| > 1$) we have

$$\frac{1}{e^{2\pi i z} - 1} = e^{-2\pi i z} \sum_{n=0}^{\infty} e^{-2\pi i n z}.$$

Also if $|w| < 1$, then

$$\frac{1}{w-1} = -\sum_{n=0}^{\infty} w^{n}$$

so that on L_2

$$\frac{1}{e^{2\pi i z} - 1} = -\sum_{n=0}^{\infty} e^{2\pi i n z}.$$

Substituting these observations in (2) we find that

$$\sum_{n \in \mathbb{Z}} f(n) = \int_{L_1} f(z) \left(e^{-2\pi i z} \sum_{n=0}^{\infty} e^{-2\pi i n z} \right) dz + \int_{L_2} f(z) \left(\sum_{n=0}^{\infty} e^{2\pi i n z} \right) dz$$

$$= \sum_{n=0}^{\infty} \int_{L_1} f(z) e^{-2\pi i (n+1) z} \, dz + \sum_{n=0}^{\infty} \int_{L_2} f(z) e^{2\pi i n z} \, dz$$

$$= \sum_{n=0}^{\infty} \int_{-\infty}^{\infty} f(x) e^{-2\pi i (n+1) x} \, dx + \sum_{n=0}^{\infty} \int_{-\infty}^{\infty} f(x) e^{2\pi i n x} \, dz$$

$$= \sum_{n=0}^{\infty} \hat{f}(n+1) + \sum_{n=0}^{\infty} \hat{f}(-n)$$

$$= \sum_{n \in \mathbb{Z}} \hat{f}(n),$$

where we have shifted L_1 and L_2 back to the real line according to
equation (1) and its analogue for the shift down.

The Poisson summation formula has many far-reaching consequences,
and we close this section by deriving several interesting identities that
are of importance for later applications.

First, we recall the calculation in Example 1, Chapter 2, which showed that the function $e^{-\pi x^2}$ was its own Fourier transform:

$$\int_{-\infty}^{\infty} e^{-\pi x^2} e^{-2\pi i x \xi}\, dx = e^{-\pi \xi^2}.$$

For fixed values of $t > 0$ and $a \in \mathbb{R}$, the change of variables $x \mapsto t^{1/2}(x + a)$ in the above integral shows that the Fourier transform of the function $f(x) = e^{-\pi t(x+a)^2}$ is $\hat{f}(\xi) = t^{-1/2} e^{-\pi \xi^2 / t} e^{2\pi i a \xi}$. Applying the Poisson summation formula to the pair f and $\hat{f}$ (which belong to $\mathfrak{F}$) provides the following relation:

$$(3) \qquad \sum_{n=-\infty}^{\infty} e^{-\pi t(n+a)^2} = \sum_{n=-\infty}^{\infty} t^{-1/2} e^{-\pi n^2 / t} e^{2\pi i n a}.$$

This identity has noteworthy consequences. For instance, the special case $a = 0$ is the transformation law for a version of the "theta function": if we define ϑ for $t > 0$ by the series $\vartheta(t) = \sum_{n=-\infty}^{\infty} e^{-\pi n^2 t}$, then the relation (3) says precisely that

$$(4) \qquad\qquad \vartheta(t) = t^{-1/2} \vartheta(1/t) \quad \text{for } t > 0.$$

This equation will be used in Chapter 6 to derive the key functional equation of the Riemann zeta function, and this leads to its analytic continuation. Also, the general case $a \in \mathbb{R}$ will be used in Chapter 10 to establish a corresponding law for the more general Jacobi theta function Θ.

For another application of the Poisson summation formula we recall that we proved in Example 3, Chapter 3, that the function $1/\cosh \pi x$ was also its own Fourier transform:

$$\int_{-\infty}^{\infty} \frac{e^{-2\pi i x \xi}}{\cosh \pi x}\, dx = \frac{1}{\cosh \pi \xi}.$$

This implies that if $t > 0$ and $a \in \mathbb{R}$, then the Fourier transform of the function $f(x) = e^{-2\pi i a x} / \cosh(\pi x / t)$ is $\hat{f}(\xi) = t / \cosh(\pi(\xi + a)t)$, and the Poisson summation formula yields

$$(5) \qquad \sum_{n=-\infty}^{\infty} \frac{e^{-2\pi i a n}}{\cosh(\pi n / t)} = \sum_{n=-\infty}^{\infty} \frac{t}{\cosh(\pi(n + a)t)}.$$

This formula will be used in Chapter 10 in the context of the two-squares theorem.

3 Paley-Wiener theorem

In this section we change our point of view somewhat: we do not suppose any analyticity of f, but we do assume the validity of the Fourier inversion formula

$$f(x) = \int_{-\infty}^{\infty} \hat{f}(\xi) e^{2\pi i x \xi} \, d\xi \quad \text{if} \quad \hat{f}(\xi) = \int_{-\infty}^{\infty} f(x) e^{-2\pi i x \xi} \, dx,$$

under the conditions $|f(x)| \leq A/(1+x^2)$ and $|\hat{f}(\xi)| \leq A'/(1+\xi^2)$. For a proof of the inversion formula under these conditions, we refer the reader to Chapter 5 in Book I.

We start by pointing out a partial converse to Theorem 2.1.

Theorem 3.1 *Suppose $\hat{f}$ satisfies the decay condition $|\hat{f}(\xi)| \leq Ae^{-2\pi a|\xi|}$ for some constants $a, A > 0$. Then $f(x)$ is the restriction to $\mathbb{R}$ of a function $f(z)$ holomorphic in the strip $S_b = \{z \in \mathbb{C}: |\mathrm{Im}(z)| < b\}$, for any $0 < b < a$.*

Proof. Define

$$f_n(z) = \int_{-n}^{n} \hat{f}(\xi) e^{2\pi i \xi z} \, d\xi,$$

and note that f_n is entire by Theorem 5.4 in Chapter 2. Observe also that $f(z)$ may be defined for all z in the strip S_b by

$$f(z) = \int_{-\infty}^{\infty} \hat{f}(\xi) e^{2\pi i \xi z} \, d\xi,$$

because the integral converges absolutely by our assumption on $\hat{f}$: it is majorized by

$$A \int_{-\infty}^{\infty} e^{-2\pi a|\xi|} e^{2\pi b|\xi|} \, d\xi,$$

which is finite if $b < a$. Moreover, for $z \in S_b$

$$|f(z) - f_n(z)| \leq A \int_{|\xi| \geq n} e^{-2\pi a|\xi|} e^{2\pi b|\xi|} \, d\xi$$

$$\to 0 \quad \text{as } n \to \infty,$$

and thus the sequence $\{f_n\}$ converges to f uniformly in S_b, which, by Theorem 5.2 in Chapter 2, proves the theorem.

We digress briefly to make the following observation.

Corollary 3.2 If $\hat{f}(\xi) = O(e^{-2\pi a|\xi|})$ for some $a > 0$, and f vanishes in a non-empty open interval, then $f = 0$.

Since by the theorem f is analytic in a region containing the real line, the corollary is a consequence of Theorem 4.8 in Chapter 2. In particular, we recover the fact proved in Exercise 21, Chapter 5 in Book I, namely that f and $\hat{f}$ cannot both have compact support unless $f = 0$.

The Paley-Wiener theorem goes a step further than the previous theorem, and describes the nature of those functions whose Fourier transforms are supported in a given interval $[-M, M]$.

Theorem 3.3 Suppose f is continuous and of moderate decrease on $\mathbb{R}$. Then, f has an extension to the complex plane that is entire with $|f(z)| \leq Ae^{2\pi M|z|}$ for some $A > 0$, if and only if $\hat{f}$ is supported in the interval $[-M, M]$.

One direction is simple. Suppose $\hat{f}$ is supported in $[-M, M]$. Then both f and $\hat{f}$ have moderate decrease, and the Fourier inversion formula applies

$$f(x) = \int_{-M}^{M} \hat{f}(\xi)e^{2\pi i \xi x} \, d\xi.$$

Since the range of integration is finite, we may replace x by the complex variable z in the integral, thereby defining a complex-valued function on $\mathbb{C}$ by

$$g(z) = \int_{-M}^{M} \hat{f}(\xi)e^{2\pi i \xi z} \, d\xi.$$

By construction $g(z) = f(z)$ if z is real, and g is holomorphic by Theorem 5.4 in Chapter 2. Finally, if $z = x + iy$, we have

$$|g(z)| \leq \int_{-M}^{M} |\hat{f}(\xi)|e^{-2\pi \xi y} \, d\xi$$

$$\leq Ae^{2\pi M|z|}.$$

The converse result requires a little more work. It starts with the observation that if $\hat{f}$ were supported in $[-M, M]$, then the argument above would give the stronger bound $|f(z)| \leq Ae^{2\pi|y|}$ instead of what we assume, that is $|f(z)| \leq Ae^{2\pi|z|}$. The idea is then to try to reduce to the better situation, where this stronger bound holds. However, this is not quite enough because we need in addition a (moderate) decay as $x \to \infty$

(when $y \neq 0$) to deal with the convergence of certain integrals at infinity. Thus we begin by also assuming this further property of f, and then we remove the additional assumptions, one step at a time.

Step 1. We first assume that f is holomorphic in the complex plane, and satisfies the following condition regarding decay in x and growth in y:

$$(6) \qquad\qquad |f(x+iy)| \leq A' \frac{e^{2\pi M|y|}}{1+x^2}.$$

We then prove under this stronger assumption that $\hat{f}(\xi) = 0$ if $|\xi| > M$. To see this, we first suppose that $\xi > M$ and write

$$\hat{f}(\xi) = \int_{-\infty}^{\infty} f(x)e^{-2\pi i\xi x} \, dx$$
$$= \int_{-\infty}^{\infty} f(x-iy)e^{-2\pi i\xi(x-iy)} \, dx.$$

Here we have shifted the real line down by an amount $y > 0$ using the standard argument (equation (1)). Putting absolute values gives the bound

$$|\hat{f}(\xi)| \leq A' \int_{-\infty}^{\infty} \frac{e^{2\pi My - 2\pi\xi y}}{1+x^2} \, dx$$
$$\leq Ce^{-2\pi y(\xi - M)}.$$

Letting y tend to infinity, and recalling that $\xi - M > 0$, proves that $\hat{f}(\xi) = 0$. A similar argument, shifting the contour up by $y > 0$, proves that $\hat{f}(\xi) = 0$ whenever $\xi < -M$.

Step 2. We relax condition (6) by assuming only that f satisfies

$$(7) \qquad\qquad |f(x+iy)| \leq Ae^{2\pi M|y|}.$$

This is still a stronger condition than in the theorem, but it is weaker than (6). Suppose first that $\xi > M$, and for $\epsilon > 0$ consider the following auxiliary function

$$f_\epsilon(z) = \frac{f(z)}{(1-i\epsilon z)^2}.$$

We observe that the quantity $1/(1-i\epsilon z)^2$ has absolute value less than or equal to 1 in the closed lower half-plane (including the real line) and converges to 1 as ϵ tends to 0. In particular, this shows that $\widehat{f_\epsilon}(\xi) \to \hat{f}(\xi)$ as $\epsilon \to 0$ since we may write

$$|\widehat{f_\epsilon}(\xi) - \hat{f}(\xi)| \leq \int_{-\infty}^{\infty} |f(x)| \left[\frac{1}{(1 - i\epsilon x)^2} - 1 \right] dx,$$

and recall that f has moderate decrease on $\mathbb{R}$.

But for each fixed ϵ, we have

$$|f_\epsilon(x + iy)| \leq A'' \frac{e^{2\pi M|y|}}{1 + x^2},$$

so by Step 1 we must have $\widehat{f_\epsilon}(\xi) = 0$, and hence $\hat{f}(\xi) = 0$ after passing to the limit as $\epsilon \to 0$. A similar argument applies if $\xi < -M$, although we must now argue in the upper half-plane, and use the factor $1/(1 - i\epsilon z)^2$ instead.

Step 3. To conclude the proof, it suffices to show that the conditions in the theorem imply condition (7) in Step 2. In fact, after dividing by an appropriate constant, it suffices to show that if $|f(x)| \leq 1$ for all real x, and $|f(z)| \leq e^{2\pi M|z|}$ for all complex z, then

$$|f(x + iy)| \leq e^{2\pi M|y|}.$$

This will follow from an ingenious and very useful idea of Phragmén and Lindelöf that allows one to adapt the maximum modulus principle to various unbounded regions. The particular result we need is as follows.

Theorem 3.4 *Suppose F is a holomorphic function in the sector*

$$S = \{z : -\pi/4 < \arg z < \pi/4\}$$

that is continuous on the closure of S. Assume $|F(z)| \leq 1$ on the boundary of the sector, and that there are constants $C, c > 0$ such that $|F(z)| \leq C e^{c|z|}$ for all z in the sector. Then

$$|F(z)| \leq 1 \quad \textit{for all } z \in S.$$

In other words, if F is bounded by 1 on the boundary of S and has no more than a reasonable amount of growth, then F is actually bounded everywhere by 1. That some restriction on the growth of F is necessary follows from a simple observation. Consider the function $F(z) = e^{z^2}$. Then F is bounded by 1 on the boundary of S, but if x is real, $F(x)$ is unbounded as $x \to \infty$. We now give the proof of Theorem 3.4.

Proof. The idea is to subdue the "enemy" function e^{z^2} and turn it to our advantage: in brief, one modifies e^{z^2} by replacing it by e^{z^α} with $\alpha < 2$. For simplicity we take the case $\alpha = 3/2$.

If $\epsilon > 0$, let

$$F_\epsilon(z) = F(z)e^{-\epsilon z^{3/2}}.$$

Here we have chosen the principal branch of the logarithm to define $z^{3/2}$ so that if $z = re^{i\theta}$ (with $-\pi < \theta < \pi$), then $z^{3/2} = r^{3/2}e^{3i\theta/2}$. Hence F_ϵ is holomorphic in S and continuous up to its boundary. Also

$$\left|e^{-\epsilon z^{3/2}}\right| = e^{-\epsilon r^{3/2}\cos(3\theta/2)};$$

and since $-\pi/4 < \theta < \pi/4$ in the sector, we get the inequalities

$$-\frac{\pi}{2} < -\frac{3\pi}{8} < \frac{3\theta}{2} < \frac{3\pi}{8} < \frac{\pi}{2},$$

and therefore $\cos(3\theta/2)$ is strictly positive in the sector. This, together with the fact that $|F(z)| \le Ce^{c|z|}$, shows that $F_\epsilon(z)$ decreases rapidly in the closed sector as $|z| \to \infty$, and in particular F_ϵ is bounded. We claim that in fact $|F_\epsilon(z)| \le 1$ for all $z \in \bar{S}$, where $\bar{S}$ denotes the closure of S. To prove this, we define

$$M = \sup_{z \in \bar{S}} |F_\epsilon(z)|.$$

Assuming F is not identically zero, let $\{w_j\}$ be a sequence of points such that $|F_\epsilon(w_j)| \to M$. Since $M \ne 0$ and F_ϵ decays to 0 as $|z|$ becomes large in the sector, w_j cannot escape to infinity, and we conclude that this sequence accumulates to a point $w \in \bar{S}$. By the maximum principle, w cannot be an interior point of S, so w lies on its boundary. But on the boundary, we have first $|F(z)| \le 1$ by assumption, and also $|e^{-\epsilon z^{3/2}}| \le 1$, so that $M \le 1$, and the claim is proved.

Finally, we may let ϵ tend to 0 to conclude the proof of the theorem.

Further generalizations of the Phragmén-Lindelöf theorem are included in Exercise 9 and Problem 3.

We must now use this result to conclude the proof of the Paley-Wiener theorem, that is, show that if $|f(x)| \le 1$ and $|f(z)| \le e^{2\pi M|z|}$, then $|f(z)| \le e^{2\pi M|y|}$. First, note that the sector in the Phragmén-Lindelöf theorem can be rotated, say to the first quadrant $Q = \{z = x + iy : x > 0, y > 0\}$, and the result remains the same. Then, we consider

$$F(z) = f(z)e^{2\pi iMz},$$

and note that F is bounded by 1 on the positive real and positive imaginary axes. Since we also have $|F(z)| \leq Ce^{c|z|}$ in the quadrant, we conclude by the Phragmén-Lindelöf theorem that $|F(z)| \leq 1$ for all z in Q, which implies $|f(z)| \leq e^{2\pi My}$. A similar argument for the other quadrants concludes Step 3 as well as the proof of the Paley-Wiener theorem.

We conclude with another version of the idea behind the Paley-Wiener theorem, this time characterizing the functions whose Fourier transform vanishes for all negative ξ.

Theorem 3.5 *Suppose f and $\hat{f}$ have moderate decrease. Then $\hat{f}(\xi) = 0$ for all $\xi < 0$ if and only if f can be extended to a continuous and bounded function in the closed upper half-plane $\{z = x + iy : y \geq 0\}$ with f holomorphic in the interior.*

Proof. First assume $\hat{f}(\xi) = 0$ for $\xi < 0$. By the Fourier inversion formula

$$f(x) = \int_0^\infty \hat{f}(\xi)e^{2\pi i x\xi}\, d\xi,$$

and we can extend f when $z = x + iy$ with $y \geq 0$ by

$$f(z) = \int_0^\infty \hat{f}(\xi)e^{2\pi i z\xi}\, d\xi.$$

Notice that the integral converges and that

$$|f(z)| \leq A \int_0^\infty \frac{d\xi}{1+\xi^2} < \infty,$$

which proves the boundedness of f. The uniform convergence of

$$f_n(z) = \int_0^n \hat{f}(\xi)e^{2\pi i x\xi}\, d\xi$$

to $f(z)$ in the closed half-plane establishes the continuity of f there, and its holomorphicity in the interior.

For the converse, we argue in the spirit of the proof of Theorem 3.3. For ϵ and δ positive, we set

$$f_{\epsilon,\delta}(z) = \frac{f(z + i\delta)}{(1 - i\epsilon z)^2}.$$

Then $f_{\epsilon,\delta}$ is holomorphic in a region containing the closed upper half-plane. One also shows as before, using Cauchy's theorem, that $\widehat{f_{\epsilon,\delta}}(\xi) = 0$

for all $\xi < 0$. Then, passing to the limit successively, one has $\widehat{f_{\epsilon,0}}(\xi) = 0$ for $\xi < 0$, and finally $\hat{f}(\xi) = \widehat{f_{0,0}}(\xi) = 0$ for all $\xi < 0$.

Remark. The reader should note a certain analogy between the above theorem and Theorem 7.1 in Chapter 3. Here we are dealing with a function holomorphic in the upper half-plane, and there with a function holomorphic in a disc. In the present case the Fourier transform vanishes when $\xi < 0$, and in the earlier case, the Fourier coefficients vanish when $n < 0$.

4 Exercises

1. Suppose f is continuous and of moderate decrease, and $\hat{f}(\xi) = 0$ for all $\xi \in \mathbb{R}$. Show that $f = 0$ by completing the following outline:

(a) For each fixed real number t consider the two functions

$$A(z) = \int_{-\infty}^{t} f(x)e^{-2\pi i z(x-t)}\,dx \quad \text{and} \quad B(z) = -\int_{t}^{\infty} f(x)e^{-2\pi i z(x-t)}\,dx.$$

Show that $A(\xi) = B(\xi)$ for all $\xi \in \mathbb{R}$.

(b) Prove that the function F equal to A in the closed upper half-plane, and B in the lower half-plane, is entire and bounded, thus constant. In fact, show that $F = 0$.

(c) Deduce that

$$\int_{-\infty}^{t} f(x)\,dx = 0,$$

for all t, and conclude that $f = 0$.

2. If $f \in \mathfrak{F}_a$ with $a > 0$, then for any positive integer n one has $f^{(n)} \in \mathfrak{F}_b$ whenever $0 \le b < a$.

[Hint: Modify the solution to Exercise 8 in Chapter 2.]

3. Show, by contour integration, that if $a > 0$ and $\xi \in \mathbb{R}$ then

$$\frac{1}{\pi} \int_{-\infty}^{\infty} \frac{a}{a^2 + x^2} e^{-2\pi i x \xi}\,dx = e^{-2\pi a|\xi|},$$

and check that

$$\int_{-\infty}^{\infty} e^{-2\pi a|\xi|} e^{2\pi i \xi x}\,d\xi = \frac{1}{\pi}\frac{a}{a^2 + x^2}.$$

4. Suppose Q is a polynomial of degree ≥ 2 with distinct roots, none lying on the real axis. Calculate

$$\int_{-\infty}^{\infty} \frac{e^{-2\pi i x \xi}}{Q(x)} \, dx, \qquad \xi \in \mathbb{R}$$

in terms of the roots of Q. What happens when several roots coincide?

[Hint: Consider separately the cases $\xi < 0$, $\xi = 0$, and $\xi > 0$. Use residues.]

5. More generally, let $R(x) = P(x)/Q(x)$ be a rational function with (degree Q) $\geq$ (degree P)$+2$ and $Q(x) \neq 0$ on the real axis.

(a) Prove that if $\alpha_1, \ldots, \alpha_k$ are the roots of Q in the upper half-plane, then there exists polynomials $P_j(\xi)$ of degree less than the multiplicity of α_j so that

$$\int_{-\infty}^{\infty} R(x) e^{-2\pi i x \xi} \, dx = \sum_{j=1}^{k} P_j(\xi) e^{-2\pi i \alpha_j \xi}, \qquad \text{when } \xi < 0.$$

(b) In particular, if $Q(z)$ has no zeros in the upper half-plane, then $\int_{-\infty}^{\infty} R(x) e^{-2\pi i x \xi} \, dx = 0$ for $\xi < 0$.

(c) Show that similar results hold in the case $\xi > 0$.

(d) Show that

$$\int_{-\infty}^{\infty} R(x) e^{-2\pi i x \xi} \, dx = O(e^{-a|\xi|}), \qquad \xi \in \mathbb{R}$$

as $|\xi| \to \infty$ for some $a > 0$. Determine the best possible a's in terms of the roots of Q.

[Hint: For part (a), use residues. The powers of ξ appear when one differentiates the function $f(z) = R(z) e^{-2\pi i z \xi}$ (as in the formula of Theorem 1.4 in the previous chapter). For part (c) argue in the lower half-plane.]

6. Prove that

$$\frac{1}{\pi} \sum_{n=-\infty}^{\infty} \frac{a}{a^2 + n^2} = \sum_{n=-\infty}^{\infty} e^{-2\pi a |n|}$$

whenever $a > 0$. Hence show that the sum equals $\coth \pi a$.

7. The Poisson summation formula applied to specific examples often provides interesting identities.

(a) Let τ be fixed with $\mathrm{Im}(\tau) > 0$. Apply the Poisson summation formula to

$$f(z) = (\tau + z)^{-k},$$

where k is an integer ≥ 2, to obtain

$$\sum_{n=-\infty}^{\infty} \frac{1}{(\tau + n)^k} = \frac{(-2\pi i)^k}{(k-1)!} \sum_{m=1}^{\infty} m^{k-1} e^{2\pi i m \tau}.$$

(b) Set $k = 2$ in the above formula to show that if $\text{Im}(\tau) > 0$, then

$$\sum_{n=-\infty}^{\infty} \frac{1}{(\tau + n)^2} = \frac{\pi^2}{\sin^2(\pi\tau)}.$$

(c) Can one conclude that the above formula holds true whenever τ is any complex number that is not an integer?

[Hint: For (a), use residues to prove that $\hat{f}(\xi) = 0$, if $\xi < 0$, and

$$\hat{f}(\xi) = \frac{(-2\pi i)^k}{(k-1)!} \xi^{k-1} e^{2\pi i \xi \tau}, \qquad \text{when } \xi > 0.]$$

8. Suppose $\hat{f}$ has compact support contained in $[-M, M]$ and let $f(z) = \sum_{n=0}^{\infty} a_n z^n$. Show that

$$a_n = \frac{(2\pi i)^n}{n!} \int_{-M}^{M} \hat{f}(\xi) \xi^n \, d\xi,$$

and as a result

$$\limsup_{n \to \infty} (n! |a_n|)^{1/n} \leq 2\pi M.$$

In the converse direction, let f be *any* power series $f(z) = \sum_{n=0}^{\infty} a_n z^n$ with $\limsup_{n \to \infty} (n! |a_n|)^{1/n} \leq 2\pi M$. Then, f is holomorphic in the complex plane, and for every $\epsilon > 0$ there exists $A_\epsilon > 0$ such that

$$|f(z)| \leq A_\epsilon e^{2\pi(M+\epsilon)|z|}.$$

9. Here are further results similar to the Phragmén-Lindelöf theorem.

(a) Let F be a holomorphic function in the right half-plane that extends continuously to the boundary, that is, the imaginary axis. Suppose that $|F(iy)| \leq 1$ for all $y \in \mathbb{R}$, and

$$|F(z)| \leq Ce^{c|z|^\gamma}$$

for some $c, C > 0$ and $\gamma < 1$. Prove that $|F(z)| \leq 1$ for all z in the right half-plane.

(b) More generally, let S be a sector whose vertex is the origin, and forming an angle of π/β. Let F be a holomorphic function in S that is continuous on the closure of S, so that $|F(z)| \leq 1$ on the boundary of S and

$$|F(z)| \leq Ce^{c|z|^\alpha} \text{ for all } z \in S$$

for some $c, C > 0$ and $0 < \alpha < \beta$. Prove that $|F(z)| \leq 1$ for all $z \in S$.

10. This exercise generalizes some of the properties of $e^{-\pi x^2}$ related to the fact that it is its own Fourier transform.

Suppose $f(z)$ is an entire function that satisfies

$$|f(x + iy)| \leq ce^{-ax^2 + by^2}$$

for some $a, b, c > 0$. Let

$$\hat{f}(\zeta) = \int_{-\infty}^{\infty} f(x)e^{-2\pi i x \zeta}\, dx.$$

Then, $\hat{f}$ is an entire function of ζ that satisfies

$$|\hat{f}(\xi + i\eta)| \leq c'e^{-a'\xi^2 + b'\eta^2}$$

for some $a', b', c' > 0$.

[Hint: To prove $\hat{f}(\xi) = O(e^{-a'\xi^2})$, assume $\xi > 0$ and change the contour of integration to $x - iy$ for some $y > 0$ fixed, and $-\infty < x < \infty$. Then

$$\hat{f}(\xi) = O(e^{-2\pi y\xi}e^{by^2}).$$

Finally, choose $y = d\xi$ where d is a small constant.]

11. One can give a neater formulation of the result in Exercise 10 by proving the following fact.

Suppose $f(z)$ is an entire function of strict order 2, that is,

$$f(z) = O(e^{c_1|z|^2})$$

for some $c_1 > 0$. Suppose also that for x real,

$$f(x) = O(e^{-c_2|x|^2})$$

for some $c_2 > 0$. Then

$$|f(x + iy)| = O(e^{-ax^2 + by^2})$$

for some $a, b > 0$. The converse is obviously true.

12. The principle that a function and its Fourier transform cannot both be too small at infinity is illustrated by the following theorem of Hardy.

If f is a function on $\mathbb{R}$ that satisfies

$$f(x) = O(e^{-\pi x^2}) \quad \text{and} \quad \hat{f}(\xi) = O(e^{-\pi \xi^2}),$$

then f is a constant multiple of $e^{-\pi x^2}$. As a result, if $f(x) = O(e^{-\pi A x^2})$, and $\hat{f}(\xi) = O(e^{-\pi B \xi^2})$, with $AB > 1$ and $A, B > 0$, then f is identically zero.

(a) If f is even, show that $\hat{f}$ extends to an even entire function. Moreover, if $g(z) = \hat{f}(z^{1/2})$, then g satisfies

$$|g(x)| \le ce^{-\pi x} \quad \text{and} \quad |g(z)| \le ce^{\pi R \sin^2(\theta/2)} \le ce^{\pi |z|}$$

when $x \in \mathbb{R}$ and $z = Re^{i\theta}$ with $R \ge 0$ and $\theta \in \mathbb{R}$.

(b) Apply the Phragmén-Lindelöf principle to the function

$$F(z) = g(z)e^{\gamma z} \quad \text{where } \gamma = i\pi \frac{e^{-i\pi/(2\beta)}}{\sin \pi/(2\beta)}$$

and the sector $0 \le \theta \le \pi/\beta < \pi$, and let $\beta \to \pi$ to deduce that $e^{\pi z}g(z)$ is bounded in the closed upper half-plane. The same result holds in the lower half-plane, so by Liouville's theorem $e^{\pi z}g(z)$ is constant, as desired.

(c) If f is odd, then $\hat{f}(0) = 0$, and apply the above argument to $\hat{f}(z)/z$ to deduce that $f = \hat{f} = 0$. Finally, write an arbitrary f as an appropriate sum of an even function and an odd function.

5 Problems

1. Suppose $\hat{f}(\xi) = O(e^{-a|\xi|^p})$ as $|\xi| \to \infty$, for some $p > 1$. Then f is holomorphic for all z and satisfies the growth condition

$$|f(z)| \le Ae^{a|z|^q}$$

where $1/p + 1/q = 1$.

Note that on the one hand, when $p \to \infty$ then $q \to 1$, and this limiting case can be interpreted as part of Theorem 3.3. On the other hand, when $p \to 1$ then $q \to \infty$, and this limiting case in a sense brings us back to Theorem 2.1.

[Hint: To prove the result, use the inequality $-\xi^p + \xi u \le u^q$, which is valid when ξ and u are non-negative. To establish this inequality, examine separately the cases $\xi^p \ge \xi u$ and $\xi^p < \xi u$; note also that the functions $\xi = u^{q-1}$ and $u = \xi^{p-1}$ are inverses of each other because $(p-1)(q-1) = 1$.]

2. The problem is to solve the differential equation

$$a_n \frac{d^n}{dt^n} u(t) + a_{n-1} \frac{d^{n-1}}{dt^{n-1}} u(t) + \cdots + a_0 u(t) = f(t),$$

where $a_0, a_1, \ldots, a_n$ are complex constants, and f is a given function. Here we suppose that f has bounded support and is smooth (say of class C^2).

(a) Let

$$\hat{f}(z) = \int_{-\infty}^{\infty} f(t) e^{-2\pi i z t} \, dt.$$

Observe that $\hat{f}$ is an entire function, and using integration by parts show that

$$|\hat{f}(x + iy)| \leq \frac{A}{1 + x^2}$$

if $|y| \leq a$ for any fixed $a \geq 0$.

(b) Write

$$P(z) = a_n (2\pi i z)^n + a_{n-1} (2\pi i z)^{n-1} + \cdots + a_0.$$

Find a real number c so that $P(z)$ does not vanish on the line

$$L = \{z : z = x + ic, \ x \in \mathbb{R}\}.$$

(c) Set

$$u(t) = \int_L \frac{e^{2\pi i z t}}{P(z)} \hat{f}(z) \, dz.$$

Check that

$$\sum_{j=0}^{n} a_j \left(\frac{d}{dt}\right)^j u(t) = \int_L e^{2\pi i z t} \hat{f}(z) \, dz$$

and

$$\int_L e^{2\pi i z t} \hat{f}(z) \, dz = \int_{-\infty}^{\infty} e^{2\pi i x t} \hat{f}(x) \, dx.$$

Conclude by the Fourier inversion theorem that

$$\sum_{j=0}^{n} a_j \left(\frac{d}{dt}\right)^j u(t) = f(t).$$

Note that the solution u depends on the choice c.

3.[*] In this problem, we investigate the behavior of certain bounded holomorphic functions in an infinite strip. The particular result described here is sometimes called the three-lines lemma.

(a) Suppose $F(z)$ is holomorphic and bounded in the strip $0 < \text{Im}(z) < 1$ and continuous on its closure. If $|F(z)| \leq 1$ on the boundary lines, then $|F(z)| \leq 1$ throughout the strip.

(b) For the more general F, let $\sup_{x \in \mathbb{R}} |F(x)| = M_0$ and $\sup_{x \in \mathbb{R}} |F(x+i)| = M_1$. Then,

$$\sup_{x \in \mathbb{R}} |F(x+iy)| \leq M_0^{1-y} M_1^y, \qquad \text{if } 0 \leq y \leq 1.$$

(c) As a consequence, prove that $\log \sup_{x \in \mathbb{R}} |F(x+iy)|$ is a convex function of y when $0 \leq y \leq 1$.

[Hint: For part (a), apply the maximum modulus principle to $F_\epsilon(z) = F(z)e^{-\epsilon z^2}$. For part (b), consider $M_0^{z-1} M_1^{-z} F(z)$.]

4.[*] There is a relation between the Paley-Wiener theorem and an earlier representation due to E. Borel.

(a) A function $f(z)$, holomorphic for all z, satisfies $|f(z)| \leq A_\epsilon e^{2\pi(M+\epsilon)|z|}$ for all ϵ if and only if it is representable in the form

$$f(z) = \int_C e^{2\pi i z w} g(w) \, dw$$

where g is holomorphic outside the circle of radius M centered at the origin, and g vanishes at infinity. Here C is any circle centered at the origin of radius larger than M. In fact, if $f(z) = \sum a_n z^n$, then $g(w) = \sum_{n=0}^{\infty} A_n w^{-n-1}$ with $a_n = A_n (2\pi i)^{n+1}/n!$.

(b) The connection with Theorem 3.3 is as follows. For these functions f (for which in addition f and $\hat{f}$ are of moderate decrease on the real axis), one can assert that the g above is holomorphic in the larger region, which consists of the slit plane $\mathbb{C} - [-M, M]$. Moreover, the relation between g and the Fourier transform $\hat{f}$ is

$$g(z) = \frac{1}{2\pi i} \int_{-M}^{M} \frac{\hat{f}(\xi)}{\xi - z} \, d\xi$$

so that $\hat{f}$ represents the jump of g across the segment $[-M, M]$; that is,

$$\hat{f}(x) = \lim_{\epsilon \to 0, \epsilon > 0} g(x+i\epsilon) - g(x-i\epsilon).$$

See Problem 5 in Chapter 3.

5 Entire Functions

...but after the 15th of October I felt myself a free man, with such longing for mathematical work, that the last two months flew by quickly, and that only today I found the letter of the 19th of October that I had not answered. The result of my work, with which I am not entirely satisfied, I want to share with you.

Firstly, in looking back at my lectures, a gap in function theory needed to be filled. As you know, up to now the following question had been unresolved. Given an arbitrary sequence of complex numbers, $a_1, a_2, \ldots$, can one construct an entire (transcendental) function that vanishes at these values, with prescribed multiplicities, and nowhere else?...

K. Weierstrass, 1874

In this chapter, we will study functions that are holomorphic in the whole complex plane; these are called **entire** functions. Our presentation will be organized around the following three questions:

1. Where can such functions vanish? We shall see that the obvious necessary condition is also sufficient: if $\{z_n\}$ is any sequence of complex numbers having no limit point in $\mathbb{C}$, then there exists an entire function vanishing exactly at the points of this sequence. The construction of the desired function is inspired by Euler's product formula for $\sin \pi z$ (the prototypical case when $\{z_n\}$ is $\mathbb{Z}$), but requires an additional refinement: the Weierstrass canonical factors.

2. How do these functions grow at infinity? Here, matters are controlled by an important principle: the larger a function is, the more zeros it can have. This principle already manifests itself in the simple case of polynomials. By the fundamental theorem of algebra, the number of zeros of a polynomial P of degree d is precisely d, which is also the exponent in the order of (polynomial) growth of P, namely

$$\sup_{|z|=R} |P(z)| \approx R^d \quad \text{as } R \to \infty.$$

A precise version of this general principle is contained in Jensen's formula, which we prove in the first section. This formula, central to much of the theory developed in this chapter, exhibits a deep connection between the number of zeros of a function in a disc and the (logarithmic) average of the function over the circle. In fact, Jensen's formula not only constitutes a natural starting point for us, but also leads to the fruitful theory of value distributions, also called Nevanlinna theory (which, however, we do not take up here).

3. To what extent are these functions determined by their zeros? It turns out that if an entire function has a finite (exponential) order of growth, then it can be specified by its zeros up to multiplication by a simple factor. The precise version of this assertion is the Hadamard factorization theorem. It may be viewed as another instance of the general rule that was formulated in Chapter 3, that is, that under appropriate conditions, a holomorphic function is essentially determined by its zeros.

1 Jensen's formula

In this section, we denote by D_R and C_R the open disc and circle of radius R centered at the origin. We shall also, in the rest of this chapter, exclude the trivial case of the function that vanishes identically.

Theorem 1.1 *Let Ω be an open set that contains the closure of a disc D_R and suppose that f is holomorphic in Ω, $f(0) \neq 0$, and f vanishes nowhere on the circle C_R. If $z_1, \ldots, z_N$ denote the zeros of f inside the disc (counted with multiplicities),[1] then*

$$(1) \qquad \log|f(0)| = \sum_{k=1}^{N} \log\left(\frac{|z_k|}{R}\right) + \frac{1}{2\pi} \int_0^{2\pi} \log|f(Re^{i\theta})| \, d\theta.$$

The proof of the theorem consists of several steps.

Step 1. First, we observe that if f_1 and f_2 are two functions satisfying the hypotheses and the conclusion of the theorem, then the product $f_1 f_2$ also satisfies the hypothesis of the theorem and formula (1). This observation is a simple consequence of the fact that $\log xy = \log x + \log y$ whenever x and y are positive numbers, and that the set of zeros of $f_1 f_2$ is the union of the sets of zeros of f_1 and f_2.

---

[1]That is, each zero appears in the sequence as many times as its order.

Step 2. The function

$$g(z) = \frac{f(z)}{(z - z_1) \cdots (z - z_N)}$$

initially defined on $\Omega - \{z_1, \ldots, z_N\}$, is bounded near each z_j. Therefore each z_j is a removable singularity, and hence we can write

$$f(z) = (z - z_1) \cdots (z - z_N) g(z)$$

where g is holomorphic in Ω and nowhere vanishing in the closure of D_R. By Step 1, it suffices to prove Jensen's formula for functions like g that vanish nowhere, and for functions of the form $z - z_j$.

Step 3. We first prove (1) for a function g that vanishes nowhere in the closure of D_R. More precisely, we must establish the following identity:

$$\log |g(0)| = \frac{1}{2\pi} \int_0^{2\pi} \log |g(Re^{i\theta})| \, d\theta.$$

In a slightly larger disc, we can write $g(z) = e^{h(z)}$ where h is holomorphic in that disc. This is possible since discs are simply connected, and we can define $h = \log g$ (see Theorem 6.2 in Chapter 3). Now we observe that

$$|g(z)| = |e^{h(z)}| = |e^{\mathrm{Re}(h(z)) + i \, \mathrm{Im}(h(z))}| = e^{\mathrm{Re}(h(z))},$$

so that $\log |g(z)| = \mathrm{Re}(h(z))$. The mean value property (Corollary 7.3 in Chapter 3) then immediately implies the desired formula for g.

Step 4. The last step is to prove the formula for functions of the form $f(z) = z - w$, where $w \in D_R$. That is, we must show that

$$\log |w| = \log \left(\frac{|w|}{R} \right) + \frac{1}{2\pi} \int_0^{2\pi} \log |Re^{i\theta} - w| \, d\theta.$$

Since $\log(|w|/R) = \log |w| - \log R$ and $\log |Re^{i\theta} - w| = \log R + \log |e^{i\theta} - w/R|$, it suffices to prove that

$$\int_0^{2\pi} \log |e^{i\theta} - a| \, d\theta = 0, \quad \text{whenever } |a| < 1.$$

This in turn is equivalent (after the change of variables $\theta \mapsto -\theta$) to

$$\int_0^{2\pi} \log |1 - ae^{i\theta}| \, d\theta = 0, \quad \text{whenever } |a| < 1.$$

To prove this, we use the function $F(z) = 1 - az$, which vanishes nowhere in the closure of the unit disc. As a consequence, there exists a holomorphic function G in a disc of radius greater than 1 such that $F(z) = e^{G(z)}$. Then $|F| = e^{\mathrm{Re}(G)}$, and therefore $\log|F| = \mathrm{Re}(G)$. Since $F(0) = 1$ we have $\log|F(0)| = 0$, and an application of the mean value property (Corollary 7.3 in Chapter 3) to the harmonic function $\log|F(z)|$ concludes the proof of the theorem.

From Jensen's formula we can derive an identity linking the growth of a holomorphic function with its number of zeros inside a disc. If f is a holomorphic function on the closure of a disc D_R, we denote by $\mathfrak{n}(r)$ (or $\mathfrak{n}_f(r)$ when it is necessary to keep track of the function in question) the number of zeros of f (counted with their multiplicities) inside the disc D_r, with $0 < r < R$. A simple but useful observation is that $\mathfrak{n}(r)$ is a non-decreasing function of r.

We claim that if $f(0) \neq 0$, and f does not vanish on the circle C_R, then

$$(2) \qquad \int_0^R \mathfrak{n}(r) \, \frac{dr}{r} = \frac{1}{2\pi} \int_0^{2\pi} \log|f(Re^{i\theta})| \, d\theta - \log|f(0)|.$$

This formula is immediate from Jensen's equality and the next lemma.

Lemma 1.2 *If $z_1, \ldots, z_N$ are the zeros of f inside the disc D_R, then*

$$\int_0^R \mathfrak{n}(r) \, \frac{dr}{r} = \sum_{k=1}^N \log\left|\frac{R}{z_k}\right|.$$

Proof. First we have

$$\sum_{k=1}^N \log\left|\frac{R}{z_k}\right| = \sum_{k=1}^N \int_{|z_k|}^R \frac{dr}{r}.$$

If we define the characteristic function

$$\eta_k(r) = \begin{cases} 1 & \text{if } r > |z_k|, \\ 0 & \text{if } r \leq |z_k|, \end{cases}$$

then $\sum_{k=1}^N \eta_k(r) = \mathfrak{n}(r)$, and the lemma is proved using

$$\sum_{k=1}^N \int_{|z_k|}^R \frac{dr}{r} = \sum_{k=1}^N \int_0^R \eta_k(r) \, \frac{dr}{r} = \int_0^R \left(\sum_{k=1}^N \eta_k(r)\right) \frac{dr}{r} = \int_0^R \mathfrak{n}(r) \, \frac{dr}{r}.$$

2 Functions of finite order

Let f be an entire function. If there exist a positive number ρ and constants $A, B > 0$ such that

$$|f(z)| \le Ae^{B|z|^\rho} \qquad \text{for all } z \in \mathbb{C},$$

then we say that f has **an order of growth** $\le \rho$. We define *the* **order of growth of** f as

$$\rho_f = \inf \rho \,,$$

where the infimum is over all $\rho > 0$ such that f has an order of growth $\le \rho$.

For example, the order of growth of the function e^{z^2} is 2.

Theorem 2.1 *If f is an entire function that has an order of growth $\le \rho$, then:*

(i) $n(r) \le Cr^\rho$ *for some $C > 0$ and all sufficiently large r.*

(ii) *If $z_1, z_2, \ldots$ denote the zeros of f, with $z_k \ne 0$, then for all $s > \rho$ we have*

$$\sum_{k=1}^{\infty} \frac{1}{|z_k|^s} < \infty.$$

Proof. It suffices to prove the estimate for $n(r)$ when $f(0) \ne 0$. Indeed, consider the function $F(z) = f(z)/z^\ell$ where ℓ is the order of the zero of f at the origin. Then $n_f(r)$ and $n_F(r)$ differ only by a constant, and F also has an of order of growth $\le \rho$.

If $f(0) \ne 0$ we may use formula (2), namely

$$\int_0^R n(x) \frac{dx}{x} = \frac{1}{2\pi} \int_0^{2\pi} \log|f(Re^{i\theta})| \, d\theta - \log|f(0)|.$$

Choosing $R = 2r$, this formula implies

$$\int_r^{2r} n(x) \frac{dx}{x} \le \frac{1}{2\pi} \int_0^{2\pi} \log|f(Re^{i\theta})| \, d\theta - \log|f(0)|.$$

On the one hand, since $n(r)$ is increasing, we have

$$\int_r^{2r} n(x) \frac{dx}{x} \ge n(r) \int_r^{2r} \frac{dx}{x} = n(r)[\log 2r - \log r] = n(r) \log 2,$$

and on the other hand, the growth condition on f gives

$$\int_0^{2\pi} \log|f(Re^{i\theta})|\,d\theta \leq \int_0^{2\pi} \log|Ae^{BR^\rho}|\,d\theta \leq C'r^\rho$$

for all large r. Consequently, $\mathfrak{n}(r) \leq Cr^\rho$ for an appropriate $C > 0$ and all sufficiently large r.

The following estimates prove the second part of the theorem:

$$\sum_{|z_k|\geq 1}|z_k|^{-s} = \sum_{j=0}^{\infty}\left(\sum_{2^j \leq |z_k| < 2^{j+1}}|z_k|^{-s}\right)$$

$$\leq \sum_{j=0}^{\infty}2^{-js}\mathfrak{n}(2^{j+1})$$

$$\leq c\sum_{j=0}^{\infty}2^{-js}2^{(j+1)\rho}$$

$$\leq c'\sum_{j=0}^{\infty}(2^{\rho-s})^j$$

$$< \infty.$$

The last series converges because $s > \rho$.

Part (ii) of the theorem is a noteworthy fact, which we shall use in a later part of this chapter.

We give two simple examples of the theorem; each of these shows that the condition $s > \rho$ cannot be improved.

EXAMPLE 1. Consider $f(z) = \sin\pi z$. Recall Euler's identity, namely

$$f(z) = \frac{e^{i\pi z} - e^{-i\pi z}}{2i},$$

which implies that $|f(z)| \leq e^{\pi|z|}$, and f has an order of growth ≤ 1. By taking $z = ix$, where $x \in \mathbb{R}$, it is clear that the order of growth of f is actually equal to 1. However, f vanishes to order 1 at $z = n$ for each $n \in \mathbb{Z}$, and $\sum_{n\neq 0}1/|n|^s < \infty$ precisely when $s > 1$.

EXAMPLE 2. Consider $f(z) = \cos z^{1/2}$, which we *define* by

$$\cos z^{1/2} = \sum_{n=0}^{\infty}(-1)^n\frac{z^n}{(2n)!}.$$

Then f is entire, and it is easy to see that

$$|f(z)| \leq e^{|z|^{1/2}},$$

and the order of growth of f is $1/2$. Moreover, $f(z)$ vanishes when $z_n = ((n+1/2)\pi)^2$, while $\sum_n 1/|z_n|^s < \infty$ exactly when $s > 1/2$.

A natural question is whether or not, given any sequence of complex numbers $z_1, z_2, \ldots$, there exists an entire function f with zeros precisely at the points of this sequence. A necessary condition is that $z_1, z_2, \ldots$ do not accumulate, in other words we must have

$$\lim_{k \to \infty} |z_k| = \infty,$$

otherwise f would vanish identically by Theorem 4.8 in Chapter 2. Weierstrass proved that this condition is also sufficient by explicitly constructing a function with these prescribed zeros. A first guess is of course the product

$$(z - z_1)(z - z_2) \cdots,$$

which provides a solution in the special case when the sequence of zeros is finite. In general, Weierstrass showed how to insert factors in this product so that the convergence is guaranteed, yet no new zeros are introduced.

Before coming to the general construction, we review infinite products and study a basic example.

3 Infinite products

3.1 Generalities

Given a sequence $\{a_n\}_{n=1}^{\infty}$ of complex numbers, we say that the product

$$\prod_{n=1}^{\infty} (1 + a_n)$$

converges if the limit

$$\lim_{N \to \infty} \prod_{n=1}^{N} (1 + a_n)$$

of the partial products exists.

A useful necessary condition that guarantees the existence of a product is contained in the following proposition.

Proposition 3.1 *If $\sum |a_n| < \infty$, then the product $\prod_{n=1}^{\infty}(1 + a_n)$ converges. Moreover, the product converges to 0 if and only if one of its factors is 0.*

This is simply Proposition 1.9 of Chapter 8 in Book I. We repeat the proof here.

Proof. If $\sum |a_n|$ converges, then for all large n we must have $|a_n| < 1/2$. Disregarding if necessary finitely many terms, we may assume that this inequality holds for all n. In particular, we can define $\log(1 + a_n)$ by the usual power series (see (6) in Chapter 3), and this logarithm satisfies the property that $1 + z = e^{\log(1+z)}$ whenever $|z| < 1$. Hence we may write the partial products as follows:

$$\prod_{n=1}^{N}(1 + a_n) = \prod_{n=1}^{N} e^{\log(1+a_n)} = e^{B_N},$$

where $B_N = \sum_{n=1}^{N} b_n$ with $b_n = \log(1 + a_n)$. By the power series expansion we see that $|\log(1 + z)| \leq 2|z|$, if $|z| < 1/2$. Hence $|b_n| \leq 2|a_n|$, so B_N converges as $N \to \infty$ to a complex number, say B. Since the exponential function is continuous, we conclude that e^{B_N} converges to e^B as $N \to \infty$, proving the first assertion of the proposition. Observe also that if $1 + a_n \neq 0$ for all n, then the product converges to a non-zero limit since it is expressed as e^B.

More generally, we can consider products of holomorphic functions.

Proposition 3.2 *Suppose $\{F_n\}$ is a sequence of holomorphic functions on the open set Ω. If there exist constants $c_n > 0$ such that*

$$\sum c_n < \infty \quad and \quad |F_n(z) - 1| \leq c_n \quad for \ all \ z \in \Omega,$$

then:

(i) *The product $\prod_{n=1}^{\infty} F_n(z)$ converges uniformly in Ω to a holomorphic function $F(z)$.*

(ii) *If $F_n(z)$ does not vanish for any n, then*

$$\frac{F'(z)}{F(z)} = \sum_{n=1}^{\infty} \frac{F_n'(z)}{F_n(z)}.$$

Proof. To prove the first statement, note that for each z we may argue as in the previous proposition if we write $F_n(z) = 1 + a_n(z)$, with

$|a_n(z)| \leq c_n$. Then, we observe that the estimates are actually uniform in z because the c_n's are constants. It follows that the product converges uniformly to a holomorphic function, which we denote by $F(z)$.

To establish the second part of the theorem, suppose that K is a compact subset of Ω, and let

$$G_N(z) = \prod_{n=1}^{N} F_n(z).$$

We have just proved that $G_N \to F$ uniformly in Ω, so by Theorem 5.3 in Chapter 2, the sequence $\{G'_N\}$ converges uniformly to F' in K. Since G_N is uniformly bounded from below on K, we conclude that $G'_N/G_N \to F'/F$ uniformly on K, and because K is an arbitrary compact subset of Ω, the limit holds for every point of Ω. Moreover, as we saw in Section 4 of Chapter 3

$$\frac{G'_N}{G_N} = \sum_{n=1}^{N} \frac{F'_n}{F_n},$$

so part (ii) of the proposition is also proved.

3.2 Example: the product formula for the sine function

Before proceeding with the general theory of Weierstrass products, we consider the key example of the product formula for the sine function:

$$(3) \qquad \frac{\sin \pi z}{\pi} = z \prod_{n=1}^{\infty} \left(1 - \frac{z^2}{n^2}\right).$$

This identity will in turn be derived from the sum formula for the cotangent function ($\cot \pi z = \cos \pi z / \sin \pi z$):

$$(4) \quad \pi \cot \pi z = \sum_{n=-\infty}^{\infty} \frac{1}{z+n} = \lim_{N \to \infty} \sum_{|n| \leq N} \frac{1}{z+n} = \frac{1}{z} + \sum_{n=1}^{\infty} \frac{2z}{z^2 - n^2}.$$

The first formula holds for all complex numbers z, and the second whenever z is not an integer. The sum $\sum_{n=-\infty}^{\infty} 1/(z+n)$ needs to be properly understood, because the separate halves corresponding to positive and negative values of n do not converge. Only when interpreted symmetrically, as $\lim_{N \to \infty} \sum_{|n| \leq N} 1/(z+n)$, does the cancellation of terms lead to a convergent series as in (4) above.

We prove (4) by showing that both $\pi \cot \pi z$ and the series have the same structural properties. In fact, observe that if $F(z) = \pi \cot \pi z$, then F has the following three properties:

(i) $F(z + 1) = F(z)$ whenever z is not an integer.

(ii) $F(z) = \dfrac{1}{z} + F_0(z)$, where F_0 is analytic near 0.

(iii) $F(z)$ has simple poles at the integers, and no other singularities.

Then, we note that the function

$$\sum_{n=-\infty}^{\infty} \frac{1}{z+n} = \lim_{N \to \infty} \sum_{|n| \leq N} \frac{1}{z+n}$$

also satisfies these same three properties. In fact, property (i) is simply the observation that the passage from z to $z + 1$ merely shifts the terms in the infinite sum. To be precise,

$$\sum_{|n| \leq N} \frac{1}{z+1+n} = \frac{1}{z+1+N} - \frac{1}{z-N} + \sum_{|n| \leq N} \frac{1}{z+n}.$$

Letting N tend to infinity proves the assertion. Properties (ii) and (iii) are evident from the representation $\frac{1}{z} + \sum_{n=1}^{\infty} \frac{2z}{z^2 - n^2}$ of the sum.

Therefore, the function defined by

$$\Delta(z) = F(z) - \sum_{n=-\infty}^{\infty} \frac{1}{z+n}$$

is periodic in the sense that $\Delta(z + 1) = \Delta(z)$, and by (ii) the singularity of Δ at the origin is removable, and hence by periodicity the singularities at all the integers are also removable; this implies that Δ is entire. To prove our formula, it will suffice to show that the function Δ is bounded in the complex plane. By the periodicity above, it is enough to do so in the strip $|\text{Re}(z)| \leq 1/2$. This is because every $z' \in \mathbb{C}$ is of the form $z' = z + k$, where z is in the strip and k is an integer. Since Δ is holomorphic, it is bounded in the rectangle $|\text{Im}(z)| \leq 1$, and we need only control the behavior of that function for $|\text{Im}(z)| > 1$. If $\text{Im}(z) > 1$ and $z = x + iy$, then

$$\cot \pi z = i \frac{e^{i\pi z} + e^{-i\pi z}}{e^{i\pi z} - e^{-i\pi z}} = i \frac{e^{-2\pi y} + e^{-2\pi i x}}{e^{-2\pi y} - e^{-2\pi i x}},$$

and in absolute value this quantity is bounded. Also

$$\frac{1}{z} + \sum_{n=1}^{\infty} \frac{2z}{z^2 - n^2} = \frac{1}{x + iy} + \sum_{n=1}^{\infty} \frac{2(x + iy)}{x^2 - y^2 - n^2 + 2ixy} ;$$

therefore if $y > 1$, we have

$$\left| \frac{1}{z} + \sum_{n=1}^{\infty} \frac{2z}{z^2 - n^2} \right| \leq C + C \sum_{n=1}^{\infty} \frac{y}{y^2 + n^2}.$$

Now the sum on the right-hand side is majorized by

$$\int_0^{\infty} \frac{y}{y^2 + x^2} \, dx,$$

because the function $y/(y^2 + x^2)$ is decreasing in x; moreover, as the change of variables $x \mapsto yx$ shows, the integral is independent of y, and hence bounded. By a similar argument Δ is bounded in the strip where $\operatorname{Im}(z) < -1$, hence is bounded throughout the whole strip $|\operatorname{Re}(z)| \leq 1/2$. Therefore Δ is bounded in $\mathbb{C}$, and by Liouville's theorem, $\Delta(z)$ is constant. The observation that Δ is odd shows that this constant must be 0, and concludes the proof of formula (4).

To prove (3), we now let

$$G(z) = \frac{\sin \pi z}{\pi} \quad \text{and} \quad P(z) = z \prod_{n=1}^{\infty} \left(1 - \frac{z^2}{n^2} \right).$$

Proposition 3.2 and the fact that $\sum 1/n^2 < \infty$ guarantee that the product $P(z)$ converges, and that away from the integers we have

$$\frac{P'(z)}{P(z)} = \frac{1}{z} + \sum_{n=1}^{\infty} \frac{2z}{z^2 - n^2}.$$

Since $G'(z)/G(z) = \pi \cot \pi z$, the cotangent formula (4) gives

$$\left(\frac{P(z)}{G(z)} \right)' = \frac{P(z)}{G(z)} \left[\frac{P'(z)}{P(z)} - \frac{G'(z)}{G(z)} \right] = 0,$$

and so $P(z) = cG(z)$ for some constant c. Dividing this identity by z, and taking the limit as $z \to 0$, we find $c = 1$.

Remark. Other proofs of (4) and (3) can be given by integrating analogous identities for $\pi^2/(\sin \pi z)^2$ derived in Exercise 12, Chapter 3, and Exercise 7, Chapter 4. Still other proofs using Fourier series can be found in the exercises of Chapters 3 and 5 of Book I.

4 Weierstrass infinite products

We now turn to Weierstrass's construction of an entire function with prescribed zeros.

Theorem 4.1 *Given any sequence $\{a_n\}$ of complex numbers with $|a_n| \to \infty$ as $n \to \infty$, there exists an entire function f that vanishes at all $z = a_n$ and nowhere else. Any other such entire function is of the form $f(z)e^{g(z)}$, where g is entire.*

Recall that if a holomorphic function f vanishes at $z = a$, then the multiplicity of the zero a is the integer m so that

$$f(z) = (z - a)^m g(z),$$

where g is holomorphic and nowhere vanishing in a neighborhood of a. Alternatively, m is the first non-zero power of $z - a$ in the power series expansion of f at a. Since, as before, we allow for repetitions in the sequence $\{a_n\}$, the theorem actually guarantees the existence of entire functions with prescribed zeros and with desired multiplicities.

To begin the proof, note first that if f_1 and f_2 are two entire functions that vanish at all $z = a_n$ and nowhere else, then f_1/f_2 has removable singularities at all the points a_n. Hence f_1/f_2 is entire and vanishes nowhere, so that there exists an entire function g with $f_1(z)/f_2(z) = e^{g(z)}$, as we showed in Section 6 of Chapter 3. Therefore $f_1(z) = f_2(z)e^{g(z)}$ and the last statement of the theorem is verified.

Hence we are left with the task of constructing a function that vanishes at all the points of the sequence $\{a_n\}$ and nowhere else. A naive guess, suggested by the product formula for $\sin \pi z$, is the product $\prod_n (1 - z/a_n)$. The problem is that this product converges only for suitable sequences $\{a_n\}$, so we correct this by inserting exponential factors. These factors will make the product converge without adding new zeros.

For each integer $k \geq 0$ we define **canonical factors** by

$$E_0(z) = 1 - z \quad \text{and} \quad E_k(z) = (1 - z)e^{z + z^2/2 + \cdots + z^k/k}, \quad \text{for } k \geq 1.$$

The integer k is called the **degree** of the canonical factor.

Lemma 4.2 *If $|z| \leq 1/2$, then $|1 - E_k(z)| \leq c|z|^{k+1}$ for some $c > 0$.*

Proof. If $|z| \leq 1/2$, then with the logarithm defined in terms of the power series, we have $1 - z = e^{\log(1-z)}$, and therefore

$$E_k(z) = e^{\log(1-z) + z + z^2/2 + \cdots + z^k/k} = e^w,$$

where $w = -\sum_{n=k+1}^{\infty} z^n/n$. Observe that since $|z| \leq 1/2$ we have

$$|w| \leq |z|^{k+1} \sum_{n=k+1}^{\infty} |z|^{n-k-1}/n \leq |z|^{k+1} \sum_{j=0}^{\infty} 2^{-j} \leq 2|z|^{k+1}.$$

In particular, we have $|w| \leq 1$ and this implies that

$$|1 - E_k(z)| = |1 - e^w| \leq c'|w| \leq c|z|^{k+1}.$$

Remark. An important technical point is that the constant c in the statement of the lemma can be chosen to be independent of k. In fact, an examination of the proof shows that we may take $c' = e$ and then $c = 2e$.

Suppose that we are given a zero of order m at the origin, and that $a_1, a_2 \ldots$ are all non-zero. Then we define the Weierstrass product by

$$f(z) = z^m \prod_{n=1}^{\infty} E_n(z/a_n).$$

We claim that this function has the required properties; that is, f is entire with a zero of order m at the origin, zeros at each point of the sequence $\{a_n\}$, and f vanishes nowhere else.

Fix $R > 0$, and suppose that z belongs to the disc $|z| < R$. We shall prove that f has all the desired properties in this disc, and since R is arbitrary, this will prove the theorem.

We can consider two types of factors in the formula defining f, with the choice depending on whether $|a_n| \leq 2R$ or $|a_n| > 2R$. There are only finitely many terms of the first kind (since $|a_n| \to \infty$), and we see that the finite product vanishes at all $z = a_n$ with $|a_n| < R$. If $|a_n| \geq 2R$, we have $|z/a_n| \leq 1/2$, hence the previous lemma implies

$$|1 - E_n(z/a_n)| \leq c \left|\frac{z}{a_n}\right|^{n+1} \leq \frac{c}{2^{n+1}}.$$

Note that by the above remark, c does not depend on n. Therefore, the product

$$\prod_{|a_n| \geq 2R} E_n(z/a_n)$$

defines a holomorphic function when $|z| < R$, and does not vanish in that disc by the propositions in Section 3. This shows that the function

f has the desired properties, and the proof of Weierstrass's theorem is complete.

5 Hadamard's factorization theorem

The theorem of this section combines the results relating the growth of a function to the number of zeros it possesses, and the above product theorem. Weierstrass's theorem states that a function that vanishes at the points $a_1, a_2, \ldots$ takes the form

$$e^{g(z)} z^m \prod_{n=1}^{\infty} E_n(z/a_n).$$

Hadamard refined this result by showing that in the case of functions of finite order, the degree of the canonical factors can be taken to be constant, and g is then a polynomial.

Recall that an entire function has an order of growth $\leq \rho$ if

$$|f(z)| \leq A e^{B|z|^\rho},$$

and that *the* order of growth ρ_0 of f is the infimum of all such ρ's.

A basic result we proved earlier was that if f has order of growth $\leq \rho$, then

$$\mathfrak{n}(r) \leq C r^\rho, \quad \text{for all large } r,$$

and if $a_1, a_2, \ldots$ are the non-zero zeros of f, and $s > \rho$, then

$$\sum |a_n|^{-s} < \infty.$$

Theorem 5.1 *Suppose f is entire and has growth order ρ_0. Let k be the integer so that $k \leq \rho_0 < k + 1$. If $a_1, a_2, \ldots$ denote the (non-zero) zeros of f, then*

$$f(z) = e^{P(z)} z^m \prod_{n=1}^{\infty} E_k(z/a_n),$$

where P is a polynomial of degree $\leq k$, and m is the order of the zero of f at $z = 0$.

Main lemmas

Here we gather a few lemmas needed in the proof of Hadamard's theorem.

Lemma 5.2 *The canonical products satisfy*

$$|E_k(z)| \geq e^{-c|z|^{k+1}} \qquad \text{if } |z| \leq 1/2$$

and

$$|E_k(z)| \geq |1 - z| \, e^{-c'|z|^k} \qquad \text{if } |z| \geq 1/2.$$

Proof. If $|z| \leq 1/2$ we can use the power series to define the logarithm of $1 - z$, so that

$$E_k(z) = e^{\log(1-z) + \sum_{n=1}^{k} z^n/n} = e^{-\sum_{n=k+1}^{\infty} z^n/n} = e^w.$$

Since $|e^w| \geq e^{-|w|}$ and $|w| \leq c|z|^{k+1}$, the first part of the lemma follows. For the second part, simply observe that if $|z| \geq 1/2$, then

$$|E_k(z)| = |1 - z| |e^{z + z^2/2 + \cdots + z^k/k}|,$$

and that there exists $c' > 0$ such that

$$|e^{z + z^2/2 + \cdots + z^k/k}| \geq e^{-|z + z^2/2 + \cdots + z^k/k|} \geq e^{-c'|z|^k}.$$

The inequality in the lemma when $|z| \geq 1/2$ then follows from these observations.

The key to the proof of Hadamard's theorem consists of finding a lower bound for the product of the canonical factors when z stays away from the zeros $\{a_n\}$. Therefore, we shall first estimate the product from below, in the complement of small discs centered at these points.

Lemma 5.3 *For any s with $\rho_0 < s < k + 1$, we have*

$$\left| \prod_{n=1}^{\infty} E_k(z/a_n) \right| \geq e^{-c|z|^s},$$

except possibly when z belongs to the union of the discs centered at a_n of radius $|a_n|^{-k-1}$, for $n = 1, 2, 3, \dots$.

Proof. The proof this lemma is a little subtle. First, we write

$$\prod_{n=1}^{\infty} E_k(z/a_n) = \prod_{|a_n| \leq 2|z|} E_k(z/a_n) \prod_{|a_n| > 2|z|} E_k(z/a_n).$$

For the second product the estimate asserted above holds with no restriction on z. Indeed, by the previous lemma

$$\left| \prod_{|a_n|>2|z|} E_k(z/a_n) \right| = \prod_{|a_n|>2|z|} |E_k(z/a_n)|$$

$$\geq \prod_{|a_n|>2|z|} e^{-c|z/a_n|^{k+1}}$$

$$\geq e^{-c|z|^{k+1} \sum_{|a_n|>2|z|} |a_n|^{-k-1}}.$$

But $|a_n| > 2|z|$ and $s < k+1$, so we must have

$$|a_n|^{-k-1} = |a_n|^{-s}|a_n|^{s-k-1} \leq C|a_n|^{-s}|z|^{s-k-1}.$$

Therefore, the fact that $\sum |a_n|^{-s}$ converges implies that

$$\left| \prod_{|a_n|>2|z|} E_k(z/a_n) \right| \geq e^{-c|z|^s}$$

for some $c > 0$.

To estimate the first product, we use the second part of Lemma 5.2, and write

$$(5) \qquad \left| \prod_{|a_n|\leq 2|z|} E_k(z/a_n) \right| \geq \prod_{|a_n|\leq 2|z|} \left| 1 - \frac{z}{a_n} \right| \prod_{|a_n|\leq 2|z|} e^{-c'|z/a_n|^k}.$$

We now note that

$$\prod_{|a_n|\leq 2|z|} e^{-c'|z/a_n|^k} = e^{-c'|z|^k \sum_{|a_n|\leq 2|z|} |a_n|^{-k}},$$

and again, we have $|a_n|^{-k} = |a_n|^{-s}|a_n|^{s-k} \leq C|a_n|^{-s}|z|^{s-k}$, thereby proving that

$$\prod_{|a_n|\leq 2|z|} e^{-c'|z/a_n|^k} \geq e^{-c|z|^s}.$$

It is the estimate on the first product on the right-hand side of (5) which requires the restriction on z imposed in the statement of the lemma. Indeed, whenever z does not belong to a disc of radius $|a_n|^{-k-1}$ centered at a_n, we must have $|a_n - z| \geq |a_n|^{-k-1}$. Therefore

$$\prod_{|a_n| \leq 2|z|} \left| 1 - \frac{z}{a_n} \right| = \prod_{|a_n| \leq 2|z|} \left| \frac{a_n - z}{a_n} \right|$$

$$\geq \prod_{|a_n| \leq 2|z|} |a_n|^{-k-1} |a_n|^{-1}$$

$$= \prod_{|a_n| \leq 2|z|} |a_n|^{-k-2}.$$

Finally, the estimate for the first product follows from the fact that

$$(k+2) \sum_{|a_n| \leq 2|z|} \log |a_n| \leq (k+2)\mathbf{n}(2|z|) \log 2|z|$$

$$\leq c|z|^\rho \log 2|z|$$

$$\leq c'|z|^s,$$

if we take $\rho_0 < \rho < s$. The second inequality then follows because $\mathbf{n}(2|z|) \leq c|z|^\rho$ by Theorem 2.1. This completes the proof.

Corollary 5.4 *There exists a sequence of radii, $r_1, r_2, \ldots,$ with $r_m \to \infty$, such that*

$$\left| \prod_{n=1}^{\infty} E_k(z/a_n) \right| \geq e^{-c|z|^s} \qquad \text{for } |z| = r_m.$$

Proof. Since $\sum |a_n|^{-k-1} < \infty$, there exists an integer N so that

$$\sum_{n=N}^{\infty} |a_n|^{-k-1} < 1/2.$$

Therefore, given any two consecutive large integers L and $L+1$, we can find a positive number r with $L \leq r \leq L+1$, such that the circle of radius r centered at the origin does not intersect the forbidden discs of Lemma 5.3. For otherwise, the union of the intervals

$$I_n = \left[|a_n| - \frac{1}{|a_n|^{k+1}}, \; |a_n| + \frac{1}{|a_n|^{k+1}} \right]$$

(which are of length $2|a_n|^{-k-1}$) would cover all the interval $[L, L+1]$. (See Figure 1.) This would imply $2\sum_{n=N}^{\infty} |a_n|^{-k-1} \geq 1$, which is a contradiction. We can then apply the previous lemma with $|z| = r$ to conclude the proof of the corollary.

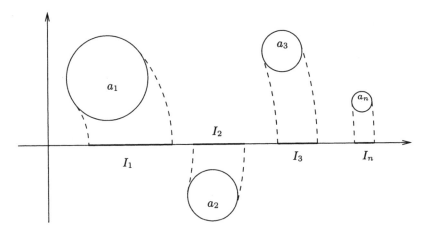

Figure 1. The intervals I_n

Proof of Hadamard's theorem

Let

$$E(z) = z^m \prod_{n=1}^{\infty} E_k(z/a_n).$$

To prove that E is entire, we repeat the argument in the proof of Theorem 4.1; we take into account that by Lemma 4.2

$$|1 - E_k(z/a_n)| \leq c \left| \frac{z}{a_n} \right|^{k+1} \qquad \text{for all large } n,$$

and that the series $\sum |a_n|^{-k-1}$ converges. (Recall $\rho_0 < s < k+1$.) Moreover, E has the zeros of f, therefore f/E is holomorphic and nowhere vanishing. Hence

$$\frac{f(z)}{E(z)} = e^{g(z)}$$

for some entire function g. By the fact that f has growth order ρ_0, and because of the estimate from below for E obtained in Corollary 5.4, we have

$$e^{\text{Re}(g(z))} = \left| \frac{f(z)}{E(z)} \right| \leq c' e^{c|z|^s}$$

for $|z| = r_m$. This proves that

$$\text{Re}(g(z)) \leq C|z|^s, \qquad \text{for } |z| = r_m.$$

The proof of Hadamard's theorem is therefore complete if we can establish the following final lemma.

Lemma 5.5 *Suppose g is entire and $u = \text{Re}(g)$ satisfies*

$$u(z) \leq Cr^s \qquad \text{whenever } |z| = r$$

for a sequence of positive real numbers r that tends to infinity. Then g is a polynomial of degree $\leq s$.

Proof. We can expand g in a power series centered at the origin

$$g(z) = \sum_{n=0}^{\infty} a_n z^n.$$

We have already proved in the last section of Chapter 3 (as a simple application of Cauchy's integral formulas) that

(6) $$\frac{1}{2\pi} \int_0^{2\pi} g(re^{i\theta}) e^{-in\theta} \, d\theta = \begin{cases} a_n r^n & \text{if } n \geq 0 \\ 0 & \text{if } n < 0. \end{cases}$$

By taking complex conjugates we find that

(7) $$\frac{1}{2\pi} \int_0^{2\pi} \overline{g(re^{i\theta})} e^{-in\theta} \, d\theta = 0$$

whenever $n > 0$, and since $2u = g + \bar{g}$ we add equations (6) and (7) to obtain

$$a_n r^n = \frac{1}{\pi} \int_0^{2\pi} u(re^{i\theta}) e^{-in\theta} \, d\theta, \qquad \text{whenever } n > 0.$$

For $n = 0$ we can simply take real parts of both sides of (6) to find that

$$2\text{Re}(a_0) = \frac{1}{\pi} \int_0^{2\pi} u(re^{i\theta}) \, d\theta.$$

Now we recall the simple fact that whenever $n \neq 0$, the integral of $e^{-in\theta}$ over any circle centered at the origin vanishes. Therefore

$$a_n = \frac{1}{\pi r^n} \int_0^{2\pi} [u(re^{i\theta}) - Cr^s]e^{-in\theta}\, d\theta \quad \text{when } n > 0,$$

hence

$$|a_n| \leq \frac{1}{\pi r^n} \int_0^{2\pi} [Cr^s - u(re^{i\theta})]\, d\theta \leq 2Cr^{s-n} - 2\mathrm{Re}(a_0)r^{-n}.$$

Letting r tend to infinity along the sequence given in the hypothesis of the lemma proves that $a_n = 0$ for $n > s$. This completes the proof of the lemma and of Hadamard's theorem.

6 Exercises

1. Give another proof of Jensen's formula in the unit disc using the functions (called Blaschke factors)

$$\psi_\alpha(z) = \frac{\alpha - z}{1 - \bar{\alpha}z}.$$

[Hint: The function $f/(\psi_{z_1} \cdots \psi_{z_N})$ is nowhere vanishing.]

2. Find the order of growth of the following entire functions:

(a) $p(z)$ where p is a polynomial.

(b) e^{bz^n} for $b \neq 0$.

(c) e^{e^z}.

3. Show that if τ is fixed with $\mathrm{Im}(\tau) > 0$, then the Jacobi theta function

$$\Theta(z|\tau) = \sum_{n=-\infty}^{\infty} e^{\pi i n^2 \tau} e^{2\pi i n z}$$

is of order 2 as a function of z. Further properties of Θ will be studied in Chapter 10.

[Hint: $-n^2 t + 2n|z| \leq -n^2 t/2$ when $t > 0$ and $n \geq 4|z|/t$.]

4. Let $t > 0$ be given and fixed, and define $F(z)$ by

$$F(z) = \prod_{n=1}^{\infty} (1 - e^{-2\pi n t} e^{2\pi i z}).$$

Note that the product defines an entire function of z.

(a) Show that $|F(z)| \le Ae^{a|z|^2}$, hence F is of order 2.

(b) F vanishes exactly when $z = -int + m$ for $n \ge 1$ and n, m integers. Thus, if z_n is an enumeration of these zeros we have

$$\sum \frac{1}{|z_n|^2} = \infty \quad \text{but} \quad \sum \frac{1}{|z_n|^{2+\epsilon}} < \infty.$$

[Hint: To prove (a), write $F(z) = F_1(z)F_2(z)$ where

$$F_1(z) = \prod_{n=1}^{N} (1 - e^{-2\pi nt}e^{2\pi iz}) \quad \text{and} \quad F_2(z) = \prod_{n=N+1}^{\infty} (1 - e^{-2\pi nt}e^{2\pi iz}).$$

Choose $N \approx c|z|$ with c appropriately large. Then, since

$$\left(\sum_{N+1}^{\infty} e^{-2\pi nt} \right) e^{2\pi|z|} \le 1,$$

one has $|F_2(z)| \le A$. However,

$$|1 - e^{-2\pi nt}e^{2\pi iz}| \le 1 + e^{2\pi|z|} \le 2e^{2\pi|z|}.$$

Thus $|F_1(z)| \le 2^N e^{2\pi N|z|} \le e^{c'|z|^2}$. Note that a simple variant of the function F arises as a factor in the triple product formula for the Jacobi theta function Θ, taken up in Chapter 10.]

5. Show that if $\alpha > 1$, then

$$F_\alpha(z) = \int_{-\infty}^{\infty} e^{-|t|^\alpha}e^{2\pi izt}\, dt$$

is an entire function of growth order $\alpha/(\alpha - 1)$.
[Hint: Show that

$$-\frac{|t|^\alpha}{2} + 2\pi|z||t| \le c|z|^{\alpha/(\alpha-1)}$$

by considering the two cases $|t|^{\alpha-1} \le A|z|$ and $|t|^{\alpha-1} \ge A|z|$, for an appropriate constant A.]

6. Prove Wallis's product formula

$$\frac{\pi}{2} = \frac{2 \cdot 2}{1 \cdot 3} \cdot \frac{4 \cdot 4}{3 \cdot 5} \cdots \frac{2m \cdot 2m}{(2m - 1) \cdot (2m + 1)} \cdots .$$

[Hint: Use the product formula for $\sin z$ at $z = \pi/2$.]

7. Establish the following properties of infinite products.

(a) Show that if $\sum |a_n|^2$ converges, and $a_n \neq -1$, then the product $\prod(1 + a_n)$ converges to a non-zero limit if and only if $\sum a_n$ converges.

(b) Find an example of a sequence of complex numbers $\{a_n\}$ such that $\sum a_n$ converges but $\prod(1 + a_n)$ diverges.

(c) Also find an example such that $\prod(1 + a_n)$ converges and $\sum a_n$ diverges.

8. Prove that for every z the product below converges, and

$$\cos(z/2) \cos(z/4) \cos(z/8) \cdots = \prod_{k=1}^{\infty} \cos(z/2^k) = \frac{\sin z}{z}.$$

[Hint: Use the fact that $\sin 2z = 2 \sin z \cos z.$]

9. Prove that if $|z| < 1$, then

$$(1+z)(1+z^2)(1+z^4)(1+z^8) \cdots = \prod_{k=0}^{\infty} (1 + z^{2^k}) = \frac{1}{1-z}.$$

10. Find the Hadamard products for:

(a) $e^z - 1$;

(b) $\cos \pi z$.

[Hint: The answers are $e^{z/2} z \prod_{n=1}^{\infty} (1 + z^2/4n^2\pi^2)$ and $\prod_{n=0}^{\infty} (1 - 4z^2/(2n+1)^2)$, respectively.]

11. Show that if f is an entire function of finite order that omits two values, then f is constant. This result remains true for any entire function and is known as Picard's little theorem.

[Hint: If f misses a, then $f(z) - a$ is of the form $e^{p(z)}$ where p is a polynomial.]

12. Suppose f is entire and never vanishes, and that none of the higher derivatives of f ever vanish. Prove that if f is also of finite order, then $f(z) = e^{az+b}$ for some constants a and b.

13. Show that the equation $e^z - z = 0$ has infinitely many solutions in $\mathbb{C}$.

[Hint: Apply Hadamard's theorem.]

14. Deduce from Hadamard's theorem that if F is entire and of growth order ρ that is non-integral, then F has infinitely many zeros.

15. Prove that every meromorphic function in $\mathbb{C}$ is the quotient of two entire functions. Also, if $\{a_n\}$ and $\{b_n\}$ are two disjoint sequences having no finite limit points, then there exists a meromorphic function in the whole complex plane that vanishes exactly at $\{a_n\}$ and has poles exactly at $\{b_n\}$.

16. Suppose that

$$Q_n(z) = \sum_{k=1}^{N_n} c_k^n \, z^k$$

are given polynomials for $n = 1, 2, \ldots$. Suppose also that we are given a sequence of complex numbers $\{a_n\}$ without limit points. Prove that there exists a meromorphic function $f(z)$ whose only poles are at $\{a_n\}$, and so that for each n, the difference

$$f(z) - Q_n\left(\frac{1}{z - a_n}\right)$$

is holomorphic near a_n. In other words, f has a prescribed poles and principal parts at each of these poles. This result is due to Mittag-Leffler.

17. Given two countably infinite sequences of complex numbers $\{a_k\}_{k=0}^{\infty}$ and $\{b_k\}_{k=0}^{\infty}$, with $\lim_{k \to \infty} |a_k| = \infty$, it is always possible to find an entire function F that satisfies $F(a_k) = b_k$ for all k.

(a) Given n distinct complex numbers $a_1, \ldots, a_n$, and another n complex numbers $b_1, \ldots, b_n$, construct a polynomial P of degree $\leq n - 1$ with

$$P(a_i) = b_i \qquad \text{for } i = 1, \ldots, n.$$

(b) Let $\{a_k\}_{k=0}^{\infty}$ be a sequence of distinct complex numbers such that $a_0 = 0$ and $\lim_{k \to \infty} |a_k| = \infty$, and let $E(z)$ denote a Weierstrass product associated with $\{a_k\}$. Given complex numbers $\{b_k\}_{k=0}^{\infty}$, show that there exist integers $m_k \geq 1$ such that the series

$$F(z) = \frac{b_0}{E'(z)} \frac{E(z)}{z} + \sum_{k=1}^{\infty} \frac{b_k}{E'(a_k)} \frac{E(z)}{z - a_k} \left(\frac{z}{a_k}\right)^{m_k}$$

defines an entire function that satisfies

$$F(a_k) = b_k \qquad \text{for all } k \geq 0.$$

This is known as the Pringsheim interpolation formula.

7 Problems

1. Prove that if f is holomorphic in the unit disc, bounded and not identically zero, and $z_1, z_2, \ldots, z_n, \ldots$ are its zeros ($|z_k| < 1$), then

$$\sum_n (1 - |z_n|) < \infty.$$

[Hint: Use Jensen's formula.]

2.* In this problem, we discuss Blaschke products, which are bounded analogues in the disc of the Weierstrass products for entire functions.

(a) Show that for $0 < |\alpha| < 1$ and $|z| \leq r < 1$ the inequality

$$\left| \frac{\alpha + |\alpha|z}{(1 - \bar{\alpha}z)\alpha} \right| \leq \frac{1+r}{1-r}$$

holds.

(b) Let $\{\alpha_n\}$ be a sequence in the unit disc such that $\alpha_n \neq 0$ for all n and

$$\sum_{n=1}^{\infty} (1 - |\alpha_n|) < \infty.$$

Note that this will be the case if $\{\alpha_n\}$ are the zeros of a bounded holomorphic function on the unit disc (see Problem 1). Show that the product

$$f(z) = \prod_{n=1}^{\infty} \frac{\alpha_n - z}{1 - \bar{\alpha}_n z} \frac{|\alpha_n|}{\alpha_n}$$

converges uniformly for $|z| \leq r < 1$, and defines a holomorphic function on the unit disc having precisely the zeros α_n and no other zeros. Show that $|f(z)| \leq 1$.

3.* Show that $\sum \dfrac{z^n}{(n!)^\alpha}$ is an entire function of order $1/\alpha$.

4.* Let $F(z) = \sum_{n=0}^{\infty} a_n z^n$ be an entire function of finite order. Then the growth order of F is intimately linked with the growth of the coefficients a_n as $n \to \infty$. In fact:

(a) Suppose $|F(z)| \leq Ae^{a|z|^\rho}$. Then

(8) $\limsup\limits_{n\to\infty} |a_n|^{1/n} n^{1/\rho} < \infty.$

(b) Conversely, if (8) holds, then $|F(z)| \leq A_\epsilon e^{a_\epsilon |z|^{\rho+\epsilon}}$, for every $\epsilon > 0$.

[Hint: To prove (a), use Cauchy's inequality

$$|a_n| \leq \frac{A}{r^n} e^{ar^\rho},$$

and the fact that the function $u^{-n}e^{u^\rho}$, $0 < u < \rho$, attains its minimum value $e^{n/\rho}(\rho/n)^{n/\rho}$ at $u = n^{1/\rho}/\rho^{1/\rho}$. Then, choose r in terms of n to achieve this minimum.

To establish (b), note that for $|z| = r$,

$$|F(z)| \leq \sum \frac{c^n r^n}{n^{n/\rho}} \leq \sum \frac{c^n r^n}{(n!)^{1/\rho}}$$

for some constant c, since $n^n \geq n!$. This yields a reduction to Problem 3.]

6 The Gamma and Zeta Functions

It is no exaggeration to say that the gamma and zeta functions are among the most important nonelementary functions in mathematics. The gamma function Γ is ubiquitous in nature. It arises in a host of calculations and is featured in a large number of identities that occur in analysis. Part of the explanation for this probably lies in the basic structural property of the gamma function, which essentially characterizes it: $1/\Gamma(s)$ is the (simplest) entire function[1] which has zeros at exactly $s = 0, -1, -2, \ldots$.

The zeta function ζ (whose study, like that of the gamma function, was initiated by Euler) plays a fundamental role in the analytic theory of numbers. Its intimate connection with prime numbers comes about via the identity for $\zeta(s)$:

$$\prod_{p} \frac{1}{1 - p^{-s}} = \sum_{n=1}^{\infty} \frac{1}{n^s},$$

where the product is over all primes. The behavior of $\zeta(s)$ for real $s > 1$, with s tending to 1, was used by Euler to prove that $\sum_{p} 1/p$ diverges, and a similar reasoning for L-functions is at the starting point of the proof of Dirichlet's theorem on primes in arithmetic progression, as we saw in Book I.

While there is no difficulty in seeing that $\zeta(s)$ is well-defined (and analytic) when $\mathrm{Re}(s) > 1$, it was Riemann who realized that the further study of primes was bound up with the analytic (in fact, meromorphic) continuation of ζ into the rest of the complex plane. Beyond this, we also consider its remarkable functional equation, which reveals a symmetry about the line $\mathrm{Re}(s) = 1/2$, and whose proof is based on a corresponding identity for the theta function. We also make a more detailed study of the growth of $\zeta(s)$ near the line $\mathrm{Re}(s) = 1$, which will be required in the proof of the prime number theorem given in the next chapter.

---

[1] In keeping with the standard notation of the subject, we denote by s (instead of z) the argument of the functions Γ and ζ.

1 The gamma function

For $s > 0$, the **gamma function** is defined by

$$\Gamma(s) = \int_0^\infty e^{-t} t^{s-1}\, dt. \tag{1}$$

The integral converges for each positive s because near $t = 0$ the function t^{s-1} is integrable, and for t large the convergence is guaranteed by the exponential decay of the integrand. These observations allow us to extend the domain of definition of Γ as follows.

Proposition 1.1 *The gamma function extends to an analytic function in the half-plane* $\mathrm{Re}(s) > 0$, *and is still given there by the integral formula* (1).

Proof. It suffices to show that the integral defines a holomorphic function in every strip

$$S_{\delta,M} = \{\delta < \mathrm{Re}(s) < M\},$$

where $0 < \delta < M < \infty$. Note that if σ denotes the real part of s, then $|e^{-t} t^{s-1}| = e^{-t} t^{\sigma-1}$, so that the integral

$$\Gamma(s) = \int_0^\infty e^{-t} t^{s-1}\, dt,$$

which is defined by the limit $\lim_{\epsilon \to 0} \int_\epsilon^{1/\epsilon} e^{-t} t^{s-1}\, dt$, converges for each $s \in S_{\delta,M}$. For $\epsilon > 0$, let

$$F_\epsilon(s) = \int_\epsilon^{1/\epsilon} e^{-t} t^{s-1}\, dt.$$

By Theorem 5.4 in Chapter 2, the function F_ϵ is holomorphic in the strip $S_{\delta,M}$. By Theorem 5.2, also of Chapter 2, it suffices to show that F_ϵ converges uniformly to Γ on the strip $S_{\delta,M}$. To see this, we first observe that

$$|\Gamma(s) - F_\epsilon(s)| \leq \int_0^\epsilon e^{-t} t^{\sigma-1}\, dt + \int_{1/\epsilon}^\infty e^{-t} t^{\sigma-1}\, dt.$$

The first integral converges uniformly to 0, as ϵ tends to 0 since it can be easily estimated by ϵ^δ/δ whenever $0 < \epsilon < 1$. The second integral converges uniformly to 0 as well, since

$$\left| \int_{1/\epsilon}^\infty e^{-t} t^{\sigma-1}\, dt \right| \leq \int_{1/\epsilon}^\infty e^{-t} t^{M-1}\, dt \leq C \int_{1/\epsilon}^\infty e^{-t/2}\, dt \to 0,$$

and the proof is complete.

1.1 Analytic continuation

Despite the fact that the integral defining Γ is not absolutely convergent for other values of s, we can go further and prove that there exists a meromorphic function defined on all of $\mathbb{C}$ that equals Γ in the half-plane $\mathrm{Re}(s) > 0$. In the same sense as in Chapter 2, we say that this function is the analytic continuation[2] of Γ, and we therefore continue to denote it by Γ.

To prove the asserted analytic extension to a meromorphic function, we need a lemma, which incidentally exhibits an important property of Γ.

Lemma 1.2 *If* $\mathrm{Re}(s) > 0$, *then*

$$(2) \qquad\qquad \Gamma(s+1) = s\Gamma(s).$$

As a consequence $\Gamma(n+1) = n!$ *for* $n = 0, 1, 2, \ldots$.

Proof. Integrating by parts in the finite integrals gives

$$\int_{\epsilon}^{1/\epsilon} \frac{d}{dt}(e^{-t}t^s)\, dt = -\int_{\epsilon}^{1/\epsilon} e^{-t}t^s\, dt + s\int_{\epsilon}^{1/\epsilon} e^{-t}t^{s-1}\, dt,$$

and the desired formula (2) follows by letting ϵ tend to 0, and noting that the left-hand side vanishes because $e^{-t}t^s \to 0$ as t tends to 0 or ∞. Now it suffices to check that

$$\Gamma(1) = \int_0^\infty e^{-t}\, dt = \left[-e^{-t}\right]_0^\infty = 1,$$

and to apply (2) successively to find that $\Gamma(n+1) = n!$.

Formula (2) in the lemma is all we need to give a proof of the following theorem.

Theorem 1.3 *The function* $\Gamma(s)$ *initially defined for* $\mathrm{Re}(s) > 0$ *has an analytic continuation to a meromorphic function on* $\mathbb{C}$ *whose only singularities are simple poles at the negative integers* $s = 0, -1, \ldots$. *The residue of* Γ *at* $s = -n$ *is* $(-1)^n/n!$.

---

[2]Uniqueness of the analytic continuation is guaranteed since the complement of the poles of a meromorphic function forms a connected set.

Proof. It suffices to extend Γ to each half-plane $\mathrm{Re}(s) > -m$, where $m \geq 1$ is an integer. For $\mathrm{Re}(s) > -1$, we define

$$F_1(s) = \frac{\Gamma(s+1)}{s}.$$

Since $\Gamma(s+1)$ is holomorphic in $\mathrm{Re}(s) > -1$, we see that F_1 is meromorphic in that half-plane, with the only possible singularity a simple pole at $s = 0$. The fact that $\Gamma(1) = 1$ shows that F_1 does in fact have a simple pole at $s = 0$ with residue 1. Moreover, if $\mathrm{Re}(s) > 0$, then

$$F_1(s) = \frac{\Gamma(s+1)}{s} = \Gamma(s)$$

by the previous lemma. So F_1 extends Γ to a meromorphic function on the half-plane $\mathrm{Re}(s) > -1$. We can now continue in this fashion by defining a meromorphic F_m for $\mathrm{Re}(s) > -m$ that agrees with Γ on $\mathrm{Re}(s) > 0$. For $\mathrm{Re}(s) > -m$, where m is an integer ≥ 1, define

$$F_m(s) = \frac{\Gamma(s+m)}{(s+m-1)(s+m-2)\cdots s}.$$

The function F_m is meromorphic in $\mathrm{Re}(s) > -m$ and has simple poles at $s = 0, -1, -2, \ldots, -m+1$ with residues

$$
\begin{aligned}
\mathrm{res}_{s=-n} F_m(s) &= \frac{\Gamma(-n+m)}{(m-1-n)!(-1)(-2)\cdots(-n)} \\
&= \frac{(m-n-1)!}{(m-1-n)!(-1)(-2)\cdots(-n)} \\
&= \frac{(-1)^n}{n!}.
\end{aligned}
$$

Successive applications of the lemma show that $F_m(s) = \Gamma(s)$ for $\mathrm{Re}(s) > 0$. By uniqueness, this also means that $F_m = F_k$ for $1 \leq k \leq m$ on the domain of definition of F_k. Therefore, we have obtained the desired continuation of Γ.

Remark. We have already proved that $\Gamma(s+1) = s\Gamma(s)$ whenever $\mathrm{Re}(s) > 0$. In fact, by analytic continuation, this formula remains true whenever $s \neq 0, -1, -2, \ldots$, that is, whenever s is not a pole of Γ. This is because both sides of the formula are holomorphic in the complement of the poles of Γ and are equal when $\mathrm{Re}(s) > 0$. Actually, one can go further, and note that if s is a negative integer $s = -n$ with $n \geq 1$, then both sides of the formula are infinite and moreover

$$\mathrm{res}_{s=-n}\Gamma(s+1) = -n\ \mathrm{res}_{s=-n}\Gamma(s).$$

Finally, note that when $s = 0$ we have $\Gamma(1) = \lim_{s \to 0} s\Gamma(s)$.

An alternate proof of Theorem 1.3, which is interesting in its own right and whose ideas recur later, is obtained by splitting the integral for $\Gamma(s)$ defined on $\text{Re}(s) > 0$ as follows:

$$\Gamma(s) = \int_0^1 e^{-t} t^{s-1} \, dt + \int_1^\infty e^{-t} t^{s-1} \, dt.$$

The integral on the far right defines an entire function; also expanding e^{-t} in a power series and integrating term by term gives

$$\int_0^1 e^{-t} t^{s-1} \, dt = \sum_{n=0}^\infty \frac{(-1)^n}{n!(n+s)}.$$

Therefore

$$(3) \qquad \Gamma(s) = \sum_{n=0}^\infty \frac{(-1)^n}{n!(n+s)} + \int_1^\infty e^{-t} t^{s-1} \, dt \qquad \text{for } \text{Re}(s) > 0.$$

Finally, the series defines a meromorphic function on $\mathbb{C}$ with poles at the negative integers and residue $(-1)^n/n!$ at $s = -n$. To prove this, we argue as follows. For a fixed $R > 0$ we may split the sum in two parts

$$\sum_{n=0}^\infty \frac{(-1)^n}{n!(n+s)} = \sum_{n=0}^N \frac{(-1)^n}{n!(n+s)} + \sum_{n=N+1}^\infty \frac{(-1)^n}{n!(n+s)},$$

where N is an integer chosen so that $N > 2R$. The first sum, which is finite, defines a meromorphic function in the disc $|s| < R$ with poles at the desired points and the correct residues. The second sum converges uniformly in that disc, hence defines a holomorphic function there, since $n > N > 2R$ and $|n + s| \geq R$ imply

$$\left| \frac{(-1)^n}{n!(n+s)} \right| \leq \frac{1}{n!R}.$$

Since R was arbitrary, we conclude that the series in (3) has the desired properties.

In particular, the relation (3) now holds on all of $\mathbb{C}$.

1.2 Further properties of Γ

The following identity reveals the symmetry of Γ about the line $\text{Re}(s) = 1/2$.

Theorem 1.4 *For all $s \in \mathbb{C}$,*

$$(4) \qquad\qquad \Gamma(s)\Gamma(1-s) = \frac{\pi}{\sin \pi s}.$$

Observe that $\Gamma(1-s)$ has simple poles at the positive integers $s = 1, 2, 3, \ldots$, so that $\Gamma(s)\Gamma(1-s)$ is a meromorphic function on $\mathbb{C}$ with simple poles at *all* the integers, a property also shared by $\pi/\sin \pi s$.

To prove the identity, it suffices to do so for $0 < s < 1$ since it then holds on all of $\mathbb{C}$ by analytic continuation.

Lemma 1.5 *For $0 < a < 1$,* $\displaystyle \int_0^\infty \frac{v^{a-1}}{1+v}\, dv = \frac{\pi}{\sin \pi a}.$

Proof. We observe first that

$$\int_0^\infty \frac{v^{a-1}}{1+v}\, dv = \int_{-\infty}^\infty \frac{e^{ax}}{1+e^x}\, dx \,,$$

which follows by making the change of variables $v = e^x$. However, using contour integration, we saw in Example 2 of Section 2.1 in Chapter 3, that the second integral equals $\pi/\sin \pi a$, as desired.

To establish the theorem, we first note that for $0 < s < 1$ we may write

$$\Gamma(1-s) = \int_0^\infty e^{-u} u^{-s}\, du = t \int_0^\infty e^{-vt}(vt)^{-s}\, dv\,,$$

where for $t > 0$ we made the change of variables $vt = u$. This trick then gives

$$\begin{aligned}
\Gamma(1-s)\Gamma(s) &= \int_0^\infty e^{-t} t^{s-1} \Gamma(1-s)\, dt \\
&= \int_0^\infty e^{-t} t^{s-1} \left(t \int_0^\infty e^{-vt}(vt)^{-s}\, dv \right)\, dt \\
&= \int_0^\infty \int_0^\infty e^{-t[1+v]} v^{-s}\, dv\, dt \\
&= \int_0^\infty \frac{v^{-s}}{1+v}\, dv \\
&= \frac{\pi}{\sin \pi(1-s)} \\
&= \frac{\pi}{\sin \pi s},
\end{aligned}$$

and the theorem is proved.

In particular, by putting $s = 1/2$, and noting that $\Gamma(s) > 0$ whenever $s > 0$, we find that

$$\Gamma(1/2) = \sqrt{\pi}.$$

We continue our study of the gamma function by considering its reciprocal, which turns out to be an entire function with remarkably simple properties.

Theorem 1.6 *The function Γ has the following properties:*

(i) $1/\Gamma(s)$ *is an entire function of s with simple zeros at $s=0,-1,-2,\ldots$ and it vanishes nowhere else.*

(ii) $1/\Gamma(s)$ *has growth*

$$\left| \frac{1}{\Gamma(s)} \right| \leq c_1 e^{c_2 |s| \log |s|}.$$

Therefore, $1/\Gamma$ is of order 1 in the sense that for every $\epsilon > 0$, there exists a bound $c(\epsilon)$ so that

$$\left| \frac{1}{\Gamma(s)} \right| \leq c(\epsilon) e^{c_2 |s|^{1+\epsilon}}.$$

Proof. By the theorem we may write

$$(5) \qquad \frac{1}{\Gamma(s)} = \Gamma(1-s) \frac{\sin \pi s}{\pi},$$

so the simple poles of $\Gamma(1-s)$, which are at $s = 1, 2, 3, \ldots$ are cancelled by the simple zeros of $\sin \pi s$, and therefore $1/\Gamma$ is entire with simple zeros at $s = 0, -1, -2, -3, \ldots$.

To prove the estimate, we begin by showing that

$$\int_1^\infty e^{-t} t^\sigma \, dt \leq e^{(\sigma+1) \log(\sigma+1)}$$

whenever $\sigma = \mathrm{Re}(s)$ is positive. Choose n so that $\sigma \leq n \leq \sigma + 1$. Then

$$\int_1^\infty e^{-t} t^\sigma \, dt \leq \int_0^\infty e^{-t} t^n \, dt$$

$$= n!$$

$$\leq n^n$$

$$= e^{n \log n}$$

$$\leq e^{(\sigma+1) \log(\sigma+1)}.$$

Since the relation (3) holds on all of $\mathbb{C}$, we see from (5) that

$$\frac{1}{\Gamma(s)} = \left(\sum_{n=0}^{\infty} \frac{(-1)^n}{n!(n+1-s)}\right) \frac{\sin \pi s}{\pi} + \left(\int_1^{\infty} e^{-t} t^{-s}\, dt\right) \frac{\sin \pi s}{\pi}.$$

However, from our previous observation,

$$\left|\int_1^{\infty} e^{-t} t^{-s}\, dt\right| \leq \int_1^{\infty} e^{-t} t^{|\sigma|}\, dt \leq e^{(|\sigma|+1)\log(|\sigma|+1)},$$

and because $|\sin \pi s| \leq e^{\pi|s|}$ (by Euler's formula for the sine function) we find that the second term in the formula for $1/\Gamma(s)$ is dominated by $ce^{(|s|+1)\log(|s|+1)}e^{\pi|s|}$, which is itself majorized by $c_1 e^{c_2|s|\log|s|}$. Next, we consider the term

$$\sum_{n=0}^{\infty} \frac{(-1)^n}{n!(n+1-s)} \frac{\sin \pi s}{\pi}.$$

There are two cases: $|\text{Im}(s)| > 1$ and $|\text{Im}(s)| \leq 1$. In the first case, this expression is dominated in absolute value by $ce^{\pi|s|}$. If $|\text{Im}(s)| \leq 1$, we choose k to be the integer so that $k - 1/2 \leq \text{Re}(s) < k + 1/2$. Then if $k \geq 1$,

$$\sum_{n=0}^{\infty} \frac{(-1)^n}{n!(n+1-s)} \frac{\sin \pi s}{\pi} = (-1)^{k-1} \frac{\sin \pi s}{(k-1)!(k-s)\pi} +$$

$$+ \sum_{n \neq k-1} (-1)^n \frac{\sin \pi s}{n!(n+1-s)\pi}.$$

Both terms on the right are bounded; the first because $\sin \pi s$ vanishes at $s = k$, and the second because the sum is majorized by $c \sum 1/n!$.

   When $k \leq 0$, then $\text{Re}(s) < 1/2$ by our supposition, and $\sum_{n=0}^{\infty} \frac{(-1)^n}{n!(n+1-s)}$ is again bounded by $c \sum 1/n!$. This concludes the proof of the theorem.

   The fact that $1/\Gamma$ satisfies the type of growth conditions discussed in Chapter 5 leads naturally to the product formula for the function $1/\Gamma$, which we treat next.

Theorem 1.7 *For all $s \in \mathbb{C}$,*

$$\frac{1}{\Gamma(s)} = e^{\gamma s} s \prod_{n=1}^{\infty} \left(1 + \frac{s}{n}\right) e^{-s/n}.$$

The real number γ, which is known as **Euler's constant**, is defined by

$$\gamma = \lim_{N \to \infty} \sum_{n=1}^{N} \frac{1}{n} - \log N.$$

The existence of the limit was already proved in Proposition 3.10, Chapter 8 of Book I, but we shall repeat the argument here for completeness. Observe that

$$\sum_{n=1}^{N} \frac{1}{n} - \log N = \sum_{n=1}^{N} \frac{1}{n} - \int_{1}^{N} \frac{1}{x} \, dx = \sum_{n=1}^{N-1} \int_{n}^{n+1} \left[\frac{1}{n} - \frac{1}{x} \right] dx + \frac{1}{N},$$

and by the mean value theorem applied to $f(x) = 1/x$ we have

$$\left| \frac{1}{n} - \frac{1}{x} \right| \leq \frac{1}{n^2} \qquad \text{for all } n \leq x \leq n+1.$$

Hence

$$\sum_{n=1}^{\infty} \frac{1}{n} - \log N = \sum_{n=1}^{N-1} a_n + \frac{1}{N}$$

where $|a_n| \leq 1/n^2$. Therefore $\sum a_n$ converges, which proves that the limit defining γ exists. We may now proceed with the proof of the factorization of $1/\Gamma$.

Proof. By the Hadamard factorization theorem and the fact that $1/\Gamma$ is entire, of growth order 1, and has simple zeros at $s = 0, -1, -2, \ldots$, we can expand $1/\Gamma$ in a Weierstrass product of the form

$$\frac{1}{\Gamma(s)} = e^{As+B} s \prod_{n=1}^{\infty} \left(1 + \frac{s}{n} \right) e^{-s/n}.$$

Here A and B are two constants that are to be determined. Remembering that $s\Gamma(s) \to 1$ as $s \to 0$, we find that $B = 0$ (or some integer multiple of $2\pi i$, which of course gives the same result). Putting $s = 1$, and using the fact that $\Gamma(1) = 1$ yields

$$
\begin{aligned}
e^{-A} &= \prod_{n=1}^{\infty} \left(1 + \frac{1}{n}\right) e^{-1/n} \\
&= \lim_{N \to \infty} \prod_{n=1}^{N} \left(1 + \frac{1}{n}\right) e^{-1/n} \\
&= \lim_{N \to \infty} e^{\sum_{n=1}^{N}[\log(1+1/n) - 1/n]} \\
&= \lim_{N \to \infty} e^{-\left(\sum_{n=1}^{N} 1/n\right) + \log N + \log(1+1/N)} \\
&= e^{-\gamma}.
\end{aligned}
$$

Therefore $A = \gamma + 2\pi i k$ for some integer k. Since $\Gamma(s)$ is real whenever s is real, we must have $k = 0$, and the argument is complete.

Note that the proof shows that the function $1/\Gamma$ is essentially characterized (up to two normalizing constants) as the entire function that has:

(i) simple zeros at $s = 0, -1, -2, \ldots$ and vanishes nowhere else, and

(ii) order of growth ≤ 1.

Observe that $\sin \pi s$ has a similar characterization (except the zeros are now at *all* the integers). However, while $\sin \pi s$ has a stricter growth estimate of the form $\sin \pi s = O\left(e^{c|s|}\right)$, this estimate (without the logarithm in the exponent) does not hold for $1/\Gamma(s)$ as Exercise 12 demonstrates.

2 The zeta function

The Riemann **zeta function** is initially defined for real $s > 1$ by the convergent series

$$
\zeta(s) = \sum_{n=1}^{\infty} \frac{1}{n^s}.
$$

As in the case of the gamma function, ζ can be continued into the complex plane. There are several proofs of this fact, and we present in the next section the one that relies on the functional equation of ζ.

2.1 Functional equation and analytic continuation

In parallel to the gamma function, we first provide a simple extension of ζ to a half-plane in $\mathbb{C}$.

Proposition 2.1 *The series defining $\zeta(s)$ converges for $\mathrm{Re}(s) > 1$, and the function ζ is holomorphic in this half-plane.*

Proof. If $s = \sigma + it$ where σ and t are real, then

$$|n^{-s}| = |e^{-s \log n}| = e^{-\sigma \log n} = n^{-\sigma}.$$

As a consequence, if $\sigma > 1 + \delta > 1$ the series defining ζ is uniformly bounded by $\sum_{n=1}^{\infty} 1/n^{1+\delta}$, which converges. Therefore, the series $\sum 1/n^s$ converges uniformly on every half-plane $\mathrm{Re}(s) > 1 + \delta > 1$, and therefore defines a holomorphic function in $\mathrm{Re}(s) > 1$.

The analytic continuation of ζ to a meromorphic function in $\mathbb{C}$ is more subtle than in the case of the gamma function. The proof we present here relates ζ to Γ and another important function.

Consider the **theta function**, already introduced in Chapter 4, which is defined for real $t > 0$ by

$$\vartheta(t) = \sum_{n=-\infty}^{\infty} e^{-\pi n^2 t}.$$

An application of the Poisson summation formula (Theorem 2.4 in Chapter 4) gave the functional equation satisfied by ϑ, namely

$$\vartheta(t) = t^{-1/2} \vartheta(1/t).$$

The growth and decay of ϑ we shall need are

$$\vartheta(t) \leq C t^{-1/2} \quad \text{as } t \to 0,$$

and

$$|\vartheta(t) - 1| \leq C e^{-\pi t} \quad \text{for some } C > 0, \text{ and all } t \geq 1.$$

The inequality for t tending to zero follows from the functional equation, while the behavior as t tends to infinity follows from the fact that

$$\sum_{n \geq 1} e^{-\pi n^2 t} \leq \sum_{n \geq 1} e^{-\pi n t} \leq C e^{-\pi t}$$

for $t \geq 1$.

We are now in a position to prove an important relation among ζ, Γ and ϑ.

Theorem 2.2 *If* $\mathrm{Re}(s) > 1$, *then*

$$\pi^{-s/2}\Gamma(s/2)\zeta(s) = \frac{1}{2}\int_0^\infty u^{(s/2)-1}[\vartheta(u) - 1]\, du.$$

Proof. This and further arguments are based on the observation that

(6) $$\int_0^\infty e^{-\pi n^2 u} u^{(s/2)-1}\, du = \pi^{-s/2}\Gamma(s/2)n^{-s}, \quad \text{if } n \geq 1.$$

Indeed, if we make the change of variables $u = t/\pi n^2$ in the integral, the left-hand side becomes

$$\left(\int_0^\infty e^{-t} t^{(s/2)-1}\, dt\right)(\pi n^2)^{-s/2},$$

which is precisely $\pi^{-s/2}\Gamma(s/2)n^{-s}$. Next, note that

$$\frac{\vartheta(u) - 1}{2} = \sum_{n=1}^\infty e^{-\pi n^2 u}.$$

The estimates for ϑ given before the statement of the theorem justify an interchange of the infinite sum with the integral, and thus

$$\frac{1}{2}\int_0^\infty u^{(s/2)-1}[\vartheta(u) - 1]\, du = \sum_{n=1}^\infty \int_0^\infty u^{(s/2)-1} e^{-\pi n^2 u}\, du$$

$$= \pi^{-s/2}\Gamma(s/2)\sum_{n=1}^\infty n^{-s}$$

$$= \pi^{-s/2}\Gamma(s/2)\zeta(s),$$

as was to be shown.

In view of this, we now consider the modification of the ζ function called the **xi function**, which makes the former appear more symmetric. It is defined for $\mathrm{Re}(s) > 1$ by

(7) $$\xi(s) = \pi^{-s/2}\Gamma(s/2)\zeta(s).$$

Theorem 2.3 *The function* ξ *is holomorphic for* $\mathrm{Re}(s) > 1$ *and has an analytic continuation to all of* $\mathbb{C}$ *as a meromorphic function with simple poles at* $s = 0$ *and* $s = 1$. *Moreover,*

$$\xi(s) = \xi(1 - s) \quad \text{for all } s \in \mathbb{C}.$$

Proof. The idea of the proof is to use the functional equation for ϑ, namely

$$\sum_{n=-\infty}^{\infty} e^{-\pi n^2 u} = u^{-1/2} \sum_{n=-\infty}^{\infty} e^{-\pi n^2/u}, \quad u > 0.$$

We then could multiply both sides by $u^{(s/2)-1}$ and try to integrate in u. Disregarding the terms corresponding to $n = 0$ (which produce infinities in both sums), we would get the desired equality once we invoked formula (6), and the parallel formula obtained by making the change of variables $u \mapsto 1/u$. The actual proof requires a little more work and goes as follows.

Let $\psi(u) = [\vartheta(u) - 1]/2$. The functional equation for the theta function, namely $\vartheta(u) = u^{-1/2}\vartheta(1/u)$, implies

$$\psi(u) = u^{-1/2}\psi(1/u) + \frac{1}{2u^{1/2}} - \frac{1}{2}.$$

Now, by Theorem 2.2 for $\text{Re}(s) > 1$, we have

$$
\begin{aligned}
\pi^{-s/2}\Gamma(s/2)\zeta(s) &= \int_0^{\infty} u^{(s/2)-1}\psi(u)\,du \\
&= \int_0^1 u^{(s/2)-1}\psi(u)\,du + \int_1^{\infty} u^{(s/2)-1}\psi(u)\,du \\
&= \int_0^1 u^{(s/2)-1}\left[u^{-1/2}\psi(1/u) + \frac{1}{2u^{1/2}} - \frac{1}{2}\right]du + \\
&\quad + \int_1^{\infty} u^{(s/2)-1}\psi(u)\,du \\
&= \frac{1}{s-1} - \frac{1}{s} + \int_1^{\infty}\left(u^{(-s/2)-1/2} + u^{(s/2)-1}\right)\psi(u)\,du
\end{aligned}
$$

whenever $\text{Re}(s) > 1$. Therefore

$$\xi(s) = \frac{1}{s-1} - \frac{1}{s} + \int_1^{\infty}\left(u^{(-s/2)-1/2} + u^{(s/2)-1}\right)\psi(u)\,du.$$

Since the function ψ has exponential decay at infinity, the integral above defines an entire function, and we conclude that ξ has an analytic continuation to all of $\mathbb{C}$ with simple poles at $s = 0$ and $s = 1$. Moreover, it is immediate that the integral remains unchanged if we replace s by $1 - s$, and the same is true for the sum of the two terms $1/(s - 1) - 1/s$. We conclude that $\xi(s) = \xi(1 - s)$ as was to be shown.

From the identity we have proved for ξ we obtain the desired result for the zeta function: its analytic continuation and its functional equation.

Theorem 2.4 *The zeta function has a meromorphic continuation into the entire complex plane, whose only singularity is a simple pole at $s = 1$.*

Proof.   A look at (7) provides the meromorphic continuation of ζ, namely

$$\zeta(s) = \pi^{s/2} \frac{\xi(s)}{\Gamma(s/2)}.$$

Recall that $1/\Gamma(s/2)$ is entire with simple zeros at $0, -2, -4, \ldots$, so the simple pole of $\xi(s)$ at the origin is cancelled by the corresponding zero of $1/\Gamma(s/2)$. As a consequence, the only singularity of ζ is a simple pole at $s = 1$.

We shall now present a more elementary approach to the analytic continuation of the zeta function, which easily leads to its extension in the half-plane $\text{Re}(s) > 0$. This method will be useful in studying the growth properties of ζ near the line $\text{Re}(s) = 1$ (which will be needed in the next chapter). The idea behind it is to compare the sum $\sum_{n=1}^{\infty} n^{-s}$ with the integral $\int_{1}^{\infty} x^{-s} \, dx$.

Proposition 2.5 *There is a sequence of entire functions $\{\delta_n(s)\}_{n=1}^{\infty}$ that satisfy the estimate $|\delta_n(s)| \leq |s|/n^{\sigma+1}$, where $s = \sigma + it$, and such that*

$$(8) \qquad \sum_{1 \leq n < N} \frac{1}{n^s} - \int_{1}^{N} \frac{dx}{x^s} = \sum_{1 \leq n < N} \delta_n(s),$$

whenever N is an integer > 1.

This proposition has the following consequence.

Corollary 2.6 *For $\text{Re}(s) > 0$ we have*

$$\zeta(s) - \frac{1}{s - 1} = H(s),$$

where $H(s) = \sum_{n=1}^{\infty} \delta_n(s)$ is holomorphic in the half-plane $\text{Re}(s) > 0$.

To prove the proposition we compare $\sum_{1 \leq n < N} n^{-s}$ with $\sum_{1 \leq n < N} \int_n^{n+1} x^{-s}\, dx$, and set

$$(9) \qquad \delta_n(s) = \int_n^{n+1} \left[\frac{1}{n^s} - \frac{1}{x^s} \right] dx.$$

The mean-value theorem applied to $f(x) = x^{-s}$ yields

$$\left| \frac{1}{n^s} - \frac{1}{x^s} \right| \leq \frac{|s|}{n^{\sigma+1}}, \qquad \text{whenever } n \leq x \leq n+1.$$

Therefore $|\delta_n(s)| \leq |s|/n^{\sigma+1}$, and since

$$\int_1^N \frac{dx}{x^s} = \sum_{1 \leq n < N} \int_n^{n+1} \frac{dx}{x^s},$$

the proposition is proved.

Turning to the corollary, we assume first that $\mathrm{Re}(s) > 1$. We let N tend to infinity in formula (8) of the proposition, and observe that by the estimate $|\delta_n(s)| \leq |s|/n^{\sigma+1}$ we have the uniform convergence of the series $\sum \delta_n(s)$ (in any half-plane $\mathrm{Re}(s) \geq \delta$ when $\delta > 0$). Since $\mathrm{Re}(s) > 1$, the series $\sum n^{-s}$ converges to $\zeta(s)$, and this proves the assertion when $\mathrm{Re}(s) > 1$. The uniform convergence also shows that $\sum \delta_n(s)$ is holomorphic when $\mathrm{Re}(s) > 0$, and thus shows that $\zeta(s)$ is extendable to that half-plane, and that the identity continues to hold there.

Remark. The idea described above can be developed step by step to yield the continuation of ζ into the entire complex plane, as shown in Problems 2 and 3. Another argument giving the full analytic continuation of ζ is outlined in Exercises 15 and 16.

As an application of the proposition we can show that the growth of $\zeta(s)$ near the line $\mathrm{Re}(s) = 1$ is "mild." Recall that when $\mathrm{Re}(s) > 1$, we have $|\zeta(s)| \leq \sum_{n=1}^{\infty} n^{-\sigma}$, and so $\zeta(s)$ is bounded in any half-plane $\mathrm{Re}(s) \geq 1 + \delta$, with $\delta > 0$. We shall see that on the line $\mathrm{Re}(s) = 1$, $|\zeta(s)|$ is majorized by $|t|^\epsilon$, for every $\epsilon > 0$, and that the growth near the line is not much worse. The estimates below are not optimal. In fact, they are rather crude but suffice for what is needed later on.

Proposition 2.7 *Suppose $s = \sigma + it$ with $\sigma, t \in \mathbb{R}$. Then for each σ_0, $0 \leq \sigma_0 \leq 1$, and every $\epsilon > 0$, there exists a constant c_ϵ so that*

(i) $|\zeta(s)| \leq c_\epsilon |t|^{1-\sigma_0+\epsilon}$, *if $\sigma_0 \leq \sigma$ and $|t| \geq 1$.*

(ii) $|\zeta'(s)| \leq c_\epsilon |t|^\epsilon$, if $1 \leq \sigma$ and $|t| \geq 1$.

In particular, the proposition implies that $\zeta(1 + it) = O(|t|^\epsilon)$ as $|t|$ tends to infinity,[3] and the same estimate also holds for ζ'. For the proof, we use Corollary 2.6. Recall the estimate $|\delta_n(s)| \leq |s|/n^{\sigma+1}$. We also have the estimate $|\delta_n(s)| \leq 2/n^\sigma$, which follows from the expression for $\delta_n(s)$ given by (9) and the fact that $|n^{-s}| = n^{-\sigma}$ and $|x^{-s}| \leq n^{-\sigma}$ if $x \geq n$. We then combine these two estimates for $|\delta_n(s)|$ via the observation that $A = A^\delta A^{1-\delta}$, to obtain the bound

$$|\delta_n(s)| \leq \left(\frac{|s|}{n^{\sigma_0+1}}\right)^\delta \left(\frac{2}{n^{\sigma_0}}\right)^{1-\delta} \leq \frac{2|s|^\delta}{n^{\sigma_0+\delta}},$$

as long as $\delta \geq 0$. Now choose $\delta = 1 - \sigma_0 + \epsilon$ and apply the identity in Corollary 2.6. Then, with $\sigma = \mathrm{Re}(s) \geq \sigma_0$, we find

$$|\zeta(s)| \leq \left|\frac{1}{s-1}\right| + 2|s|^{1-\sigma_0+\epsilon} \sum_{n=1}^\infty \frac{1}{n^{1+\epsilon}},$$

and conclusion (i) is proved. The second conclusion is actually a consequence of the first by a slight modification of Exercise 8 in Chapter 2. For completeness we sketch the argument. By the Cauchy integral formula,

$$\zeta'(s) = \frac{1}{2\pi r} \int_0^{2\pi} \zeta(s + re^{i\theta})e^{i\theta}\, d\theta,$$

where the integration is taken over a circle of radius r centered at the point s. Now choose $r = \epsilon$ and observe that this circle lies in the half-plane $\mathrm{Re}(s) \geq 1 - \epsilon$, and so (ii) follows as a consequence of (i) on replacing 2ϵ by ϵ.

3 Exercises

1. Prove that

$$\Gamma(s) = \lim_{n\to\infty} \frac{n^s n!}{s(s+1)\cdots(s+n)}$$

whenever $s \neq 0, -1, -2, \ldots$.

[Hint: Use the product formula for $1/\Gamma$, and the definition of the Euler constant γ.]

2. Prove that

$$\prod_{n=1}^\infty \frac{n(n+a+b)}{(n+a)(n+b)} = \frac{\Gamma(a+1)\Gamma(b+1)}{\Gamma(a+b+1)}$$

---

[3]The reader should recall the O notation which was introduced at the end of Chapter 1.

whenever a and b are positive. Using the product formula for $\sin \pi s$, give another proof that $\Gamma(s)\Gamma(1-s) = \pi/\sin \pi s$.

3. Show that Wallis's product formula can be written as

$$\sqrt{\frac{\pi}{2}} = \lim_{n \to \infty} \frac{2^{2n}(n!)^2}{(2n+1)!}(2n+1)^{1/2}.$$

As a result, prove the following identity:

$$\Gamma(s)\Gamma(s+1/2) = \sqrt{\pi}2^{1-2s}\Gamma(2s).$$

4. Prove that if we take

$$f(z) = \frac{1}{(1-z)^\alpha}, \qquad \text{for } |z| < 1$$

(defined in terms of the principal branch of the logarithm), where α is a fixed complex number, then

$$f(z) = \sum_{n=0}^{\infty} a_n(\alpha)z^n$$

with

$$a_n(\alpha) \sim \frac{1}{\Gamma(\alpha)} n^{\alpha-1} \qquad \text{as } n \to \infty.$$

5. Use the fact that $\Gamma(s)\Gamma(1-s) = \pi/\sin \pi s$ to prove that

$$|\Gamma(1/2 + it)| = \sqrt{\frac{2\pi}{e^{\pi t} + e^{-\pi t}}}, \qquad \text{whenever } t \in \mathbb{R}.$$

6. Show that

$$1 + \frac{1}{3} + \frac{1}{5} + \cdots + \frac{1}{2n-1} - \frac{1}{2}\log n \to \frac{\gamma}{2} + \log 2,$$

where γ is Euler's constant.

7. The **Beta function** is defined for $\mathrm{Re}(\alpha) > 0$ and $\mathrm{Re}(\beta) > 0$ by

$$B(\alpha, \beta) = \int_0^1 (1-t)^{\alpha-1}t^{\beta-1}\, dt.$$

(a) Prove that $B(\alpha, \beta) = \dfrac{\Gamma(\alpha)\Gamma(\beta)}{\Gamma(\alpha+\beta)}$.

(b) Show that $B(\alpha, \beta) = \displaystyle\int_0^\infty \dfrac{u^{\alpha-1}}{(1+u)^{\alpha+\beta}}\, du.$

[Hint: For part (a), note that

$$\Gamma(\alpha)\Gamma(\beta) = \int_0^\infty \int_0^\infty t^{\alpha-1} s^{\beta-1} e^{-t-s}\, dt ds,$$

and make the change of variables $s = ur$, $t = u(1 - r)$.]

8. The Bessel functions arise in the study of spherical symmetries and the Fourier transform. See Chapter 6 in Book I. Prove that the following power series identity holds for Bessel functions of real order $\nu > -1/2$:

$$J_\nu(x) = \frac{(x/2)^\nu}{\Gamma(\nu + 1/2)\sqrt{\pi}} \int_{-1}^1 e^{ixt}(1 - t^2)^{\nu-(1/2)}\, dt = \left(\frac{x}{2}\right)^\nu \sum_{m=0}^\infty \frac{(-1)^m \left(\frac{x^2}{4}\right)^m}{m!\,\Gamma(\nu + m + 1)}$$

whenever $x > 0$. In particular, the Bessel function J_ν satisfies the ordinary differential equation

$$\frac{d^2 J_\nu}{dx^2} + \frac{1}{x}\frac{d J_\nu}{dx} + \left(1 - \frac{\nu^2}{x^2}\right) J_\nu = 0.$$

[Hint: Expand the exponential e^{ixt} in a power series, and express the remaining integrals in terms of the gamma function, using Exercise 7.]

9. The hypergeometric series $F(\alpha, \beta, \gamma; z)$ was defined in Exercise 16 of Chapter 1. Show that

$$F(\alpha, \beta, \gamma; z) = \frac{\Gamma(\gamma)}{\Gamma(\beta)\Gamma(\gamma - \beta)} \int_0^1 t^{\beta-1}(1 - t)^{\gamma-\beta-1}(1 - zt)^{-\alpha}\, dt.$$

Here $\alpha > 0, \beta > 0, \gamma > \beta$, and $|z| < 1$.

Show as a result that the hypergeometric function, initially defined by a power series convergent in the unit disc, can be continued analytically to the complex plane slit along the half-line $[1, \infty)$.

Note that

$$\log(1 - z) \quad = \quad -zF(1, 1, 2; z),$$

$$e^z \quad = \quad \lim_{\beta \to \infty} F(1, \beta, 1; z/\beta),$$

$$(1 - z)^{-\alpha} \quad = \quad F(\alpha, 1, 1; z).$$

[Hint: To prove the integral identity, expand $(1 - zt)^{-\alpha}$ as a power series.]

10. An integral of the form

$$F(z) = \int_0^\infty f(t)t^{z-1}\, dt$$

is called a **Mellin transform**, and we shall write $\mathcal{M}(f)(z) = F(z)$. For example, the gamma function is the Mellin transform of the function e^{-t}.

(a) Prove that

$$\mathcal{M}(\cos)(z) = \int_0^\infty \cos(t) t^{z-1} \, dt = \Gamma(z) \cos\left(\pi \frac{z}{2}\right) \qquad \text{for } 0 < \mathrm{Re}(z) < 1,$$

and

$$\mathcal{M}(\sin)(z) = \int_0^\infty \sin(t) t^{z-1} \, dt = \Gamma(z) \sin\left(\pi \frac{z}{2}\right) \qquad \text{for } 0 < \mathrm{Re}(z) < 1.$$

(b) Show that the second of the above identities is valid in the larger strip $-1 < \mathrm{Re}(z) < 1$, and that as a consequence, one has

$$\int_0^\infty \frac{\sin x}{x} \, dx = \frac{\pi}{2} \qquad \text{and} \qquad \int_0^\infty \frac{\sin x}{x^{3/2}} \, dx = \sqrt{2\pi}.$$

This generalizes the calculation in Exercise 2 of Chapter 2.

[Hint: For the first part, consider the integral of the function $f(w) = e^{-w} w^{z-1}$ around the contour illustrated in Figure 1. Use analytic continuation to prove the second part.]

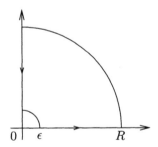

Figure 1. The contour in Exercise 10

11. Let $f(z) = e^{az} e^{-e^z}$ where $a > 0$. Observe that in the strip $\{x + iy : |y| < \pi/2\}$ the function $f(x + iy)$ is exponentially decreasing as $|x|$ tends to infinity. Prove that

$$\hat{f}(\xi) = \Gamma(a - 2\pi i \xi), \qquad \text{for all } \xi \in \mathbb{R}.$$

12. This exercise gives two simple observations about $1/\Gamma$.

(a) Show that $1/|\Gamma(s)|$ is not $O(c^{c|s|})$ for any $c > 0$. [Hint: If $s = -k - 1/2$, where k is a positive integer, then $|1/\Gamma(s)| \geq k!/\pi$.]

(b) Show that there is no entire function $F(s)$ with $F(s) = O(e^{c|s|})$ that has simple zeros at $s = 0, -1, -2, \ldots, -n, \ldots$, and that vanishes nowhere else.

13. Prove that

$$\frac{d^2 \log \Gamma(s)}{ds^2} = \sum_{n=0}^{\infty} \frac{1}{(s+n)^2}$$

whenever s is a positive number. Show that if the left-hand side is interpreted as $(\Gamma'/\Gamma)'$, then the above formula also holds for all complex numbers s with $s \neq 0, -1, -2, \ldots$.

14. This exercise gives an asymptotic formula for $\log n!$. A more refined asymptotic formula for $\Gamma(s)$ as $s \to \infty$ (Stirling's formula) is given in Appendix A.

(a) Show that

$$\frac{d}{dx} \int_x^{x+1} \log \Gamma(t) \, dt = \log x, \qquad \text{for } x > 0,$$

and as a result

$$\int_x^{x+1} \log \Gamma(t) \, dt = x \log x - x + c.$$

(b) Show as a consequence that $\log \Gamma(n) \sim n \log n$ as $n \to \infty$. In fact, prove that $\log \Gamma(n) \sim n \log n + O(n)$ as $n \to \infty$. [Hint: Use the fact that $\Gamma(x)$ is monotonically increasing for all large x.]

15. Prove that for $\mathrm{Re}(s) > 1$,

$$\zeta(s) = \frac{1}{\Gamma(s)} \int_0^\infty \frac{x^{s-1}}{e^x - 1} \, dx.$$

[Hint: Write $1/(e^x - 1) = \sum_{n=1}^{\infty} e^{-nx}$.]

16. Use the previous exercise to give another proof that $\zeta(s)$ is continuable in the complex plane with only singularity a simple pole at $s = 1$.

[Hint: Write

$$\zeta(s) = \frac{1}{\Gamma(s)} \int_0^1 \frac{x^{s-1}}{e^x - 1} \, dx + \frac{1}{\Gamma(s)} \int_1^\infty \frac{x^{s-1}}{e^x - 1} \, dx.$$

The second integral defines an entire function, while

$$\int_0^1 \frac{x^{s-1}}{e^x - 1}\, dx = \sum_{m=0}^{\infty} \frac{B_m}{m!(s+m-1)},$$

where B_m denotes the mth Bernoulli number defined by

$$\frac{x}{e^x - 1} = \sum_{m=0}^{\infty} \frac{B_m}{m!} x^m.$$

Then $B_0 = 1$, and since $z/(e^z - 1)$ is holomorphic for $|z| < 2\pi$, we must have $\limsup_{m\to\infty} |B_m/m!|^{1/m} = 1/2\pi$.]

17. Let f be an indefinitely differentiable function on $\mathbb{R}$ that has compact support, or more generally, let f belong to the Schwartz space.[4] Consider

$$I(s) = \frac{1}{\Gamma(s)} \int_0^{\infty} f(x) x^{-1+s}\, dx.$$

(a) Observe that $I(s)$ is holomorphic for $\operatorname{Re}(s) > 0$. Prove that I has an analytic continuation as an entire function in the complex plane.

(b) Prove that $I(0) = f(0)$, and more generally

$$I(-n) = (-1)^n f^{(n)}(0) \qquad \text{for all } n \geq 0.$$

[Hint: To prove the analytic continuation, as well as the formulas in the second part, integrate by parts to show that $I(s) = \frac{(-1)^k}{\Gamma(s+k)} \int_0^{\infty} f^{(k)}(x) x^{s+k-1}\, dx$.]

4 Problems

1. This problem provides further estimates for ζ and ζ' near $\operatorname{Re}(s) = 1$.

(a) Use Proposition 2.5 and its corollary to prove

$$\zeta(s) = \sum_{1 \leq n < N} n^{-s} - \frac{N^{s-1}}{s-1} + \sum_{n \geq N} \delta_n(s)$$

for every integer $N \geq 2$, whenever $\operatorname{Re}(s) > 0$.

(b) Show that $|\zeta(1+it)| = O(\log |t|)$, as $|t| \to \infty$ by using the previous result with $N =$ greatest integer in $|t|$.

---

[4]The Schwartz space on $\mathbb{R}$ is denoted by $\mathcal{S}$ and consists of all indefinitely differentiable functions f, so that f and all its derivatives decay faster than any polynomials. In other words, $\sup_{x \in \mathbb{R}} |x|^m |f^{(\ell)}(x)| < \infty$ for all integers $m, \ell \geq 0$. This space appeared in the study of the Fourier transform in Book I.

(c) The second conclusion of Proposition 2.7 can be similarly refined.

(d) Show that if $t \neq 0$ and t is fixed, then the partial sums of the series $\sum_{n=1}^{\infty} 1/n^{1+it}$ are bounded, but the series does not converge.

2.* Prove that for $\text{Re}(s) > 0$

$$\zeta(s) = \frac{s}{s-1} - s \int_1^{\infty} \frac{\{x\}}{x^{s+1}}\, dx$$

where $\{x\}$ is the fractional part of x.

3.* If $Q(x) = \{x\} - 1/2$, then we can write the expression in the previous problem as

$$\zeta(s) = \frac{s}{s-1} - \frac{1}{2} - s \int_1^{\infty} \frac{Q(x)}{x^{s+1}}\, dx.$$

Let us construct $Q_k(x)$ recursively so that

$$\int_0^1 Q_k(x)\, dx = 0, \qquad \frac{dQ_{k+1}}{dx} = Q_k(x), \qquad Q_0(x) = Q(x) \qquad \text{and} \qquad Q_k(x+1) = Q_k(x).$$

Then we can write

$$\zeta(s) = \frac{s}{s-1} - \frac{1}{2} - s \int_1^{\infty} \left(\frac{d^k}{dx^k} Q_k(x) \right) x^{-s-1}\, dx\,,$$

and a k-fold integration by parts gives the analytic continuation for $\zeta(s)$ when $\text{Re}(s) > -k$.

4.* The functions Q_k in the previous problem are related to the Bernoulli polynomials $B_k(x)$ by the formula

$$Q_k(x) = \frac{B_{k+1}(x)}{(k+1)!} \qquad \text{for } 0 \leq x \leq 1.$$

Also, if k is a positive integer, then

$$2\zeta(2k) = (-1)^{k+1} \frac{(2\pi)^{2k}}{(2k)!} B_{2k},$$

where $B_k = B_k(0)$ are the Bernoulli numbers. For the definition of $B_k(x)$ and B_k see Chapter 3 in Book I.

7 The Zeta Function and Prime Number Theorem

Bernhard Riemann, whose extraordinary intuitive powers we have already mentioned, has especially renovated our knowledge of the distribution of prime numbers, also one of the most mysterious questions in mathematics. He has taught us to deduce results in that line from considerations borrowed from the integral calculus: more precisely, from the study of a certain quantity, a function of a variable s which may assume not only real, but also imaginary values. He proved some important properties of that function, but enunciated two or three as important ones without giving the proof. At the death of Riemann, a note was found among his papers, saying "These properties of $\zeta(s)$ (the function in question) are deduced from an expression of it which, however, I did not succeed in simplifying enough to publish it."

We still have not the slightest idea of what the expression could be. As to the properties he simply enunciated, some thirty years elapsed before I was able to prove all of them but one. The question concerning that last one remains unsolved as yet, though, by an immense labor pursued throughout this last half century, some highly interesting discoveries in that direction have been achieved. It seems more and more probable, but still not at all certain, that the "Riemann hypothesis" is true.

J. Hadamard, 1945

Euler found, through his product formula for the zeta function, a deep connection between analytical methods and arithmetic properties of numbers, in particular primes. An easy consequence of Euler's formula is that the sum of the reciprocals of all primes, $\sum_p 1/p$, diverges, a result that quantifies the fact that there are infinitely many prime numbers. The natural problem then becomes that of understanding how these primes are distributed. With this in mind, we consider the following function:

$$\pi(x) = \text{number of primes less than or equal to } x.$$

The erratic growth of the function $\pi(x)$ gives little hope of finding a simple formula for it. Instead, one is led to study the asymptotic behavior of $\pi(x)$ as x becomes large. About 60 years after Euler's discovery, Legendre and Gauss observed after numerous calculations that it was likely that

$$(1) \qquad\qquad \pi(x) \sim \frac{x}{\log x} \qquad \text{as } x \to \infty.$$

(The asymptotic relation $f(x) \sim g(x)$ as $x \to \infty$ means that $f(x)/g(x) \to 1$ as $x \to \infty$.) Another 60 years later, shortly before Riemann's work, Tchebychev proved by elementary methods (and in particular, without the zeta function) the weaker result that

$$(2) \qquad\qquad \pi(x) \approx \frac{x}{\log x} \qquad \text{as } x \to \infty.$$

Here, by definition, the symbol $\approx$ means that there are positive constants $A < B$ such that

$$A \frac{x}{\log x} \leq \pi(x) \leq B \frac{x}{\log x}$$

for all sufficiently large x.

In 1896, about 40 years after Tchebychev's result, Hadamard and de la Vallée Poussin gave a proof of the validity of the relation (1). Their result is known as the prime number theorem. The original proofs of this theorem, as well as the one we give below, use complex analysis. We should remark that since then other proofs have been found, some depending on complex analysis, and others more elementary in nature.

At the heart of the proof of the prime number theorem that we give below lies the fact that $\zeta(s)$ does not vanish on the line $\text{Re}(s) = 1$. In fact, it can be shown that these two propositions are equivalent.

1 Zeros of the zeta function

We have seen in Theorem 1.10, Chapter 8 in Book I, Euler's identity, which states that for $\text{Re}(s) > 1$ the zeta function can be expressed as an infinite product

$$\zeta(s) = \prod_p \frac{1}{1 - p^{-s}}.$$

The identity for $\mathrm{Re}(s) > 1$ follows by analytic continuation from the case $s > 1$, whose proof we now recall. The key observation is that $1/(1 - p^{-s})$ can be written as a convergent (geometric) power series

$$1 + \frac{1}{p^s} + \frac{1}{p^{2s}} + \cdots + \frac{1}{p^{Ms}} + \cdots,$$

and taking formally the product of these series over all primes p, yields the desired result. A precise argument goes as follows.

Suppose M and N are positive integers with $M > N$. Observe now that, by the fundamental theorem of arithmetic,[1] any positive integer $n \leq N$ can be written uniquely as a product of primes, and that each prime that occurs in the product must be less than or equal to N and repeated less than M times. Therefore

$$\sum_{n=1}^{N} \frac{1}{n^s} \leq \prod_{p \leq N} \left(1 + \frac{1}{p^s} + \frac{1}{p^{2s}} + \cdots + \frac{1}{p^{Ms}}\right)$$

$$\leq \prod_{p \leq N} \left(\frac{1}{1 - p^{-s}}\right)$$

$$\leq \prod_{p} \left(\frac{1}{1 - p^{-s}}\right).$$

Letting N tend to infinity in the series now yields

$$\sum_{n=1}^{\infty} \frac{1}{n^s} \leq \prod_{p} \left(\frac{1}{1 - p^{-s}}\right).$$

For the reverse inequality, we argue as follows. Again, by the fundamental theorem of arithmetic, we find that

$$\prod_{p \leq N} \left(1 + \frac{1}{p^s} + \frac{1}{p^{2s}} + \cdots + \frac{1}{p^{Ms}}\right) \leq \sum_{n=1}^{\infty} \frac{1}{n^s}.$$

Letting M tend to infinity gives

$$\prod_{p \leq N} \left(\frac{1}{1 - p^{-s}}\right) \leq \sum_{n=1}^{\infty} \frac{1}{n^s}.$$

---

[1] A proof of this elementary (but essential) fact is given in the first section of Chapter 8 in Book I.

Hence

$$\prod_p \left(\frac{1}{1-p^{-s}}\right) \leq \sum_{n=1}^{\infty} \frac{1}{n^s},$$

and the proof of the product formula for ζ is complete.

From the product formula we see, by Proposition 3.1 in Chapter 5, that $\zeta(s)$ does not vanish when $\mathrm{Re}(s) > 1$.

To obtain further information about the location of the zeros of ζ, we use the functional equation that provided the analytic continuation of ζ. We may write the fundamental relation $\xi(s) = \xi(1-s)$ in the form

$$\pi^{-s/2}\Gamma(s/2)\zeta(s) = \pi^{-(1-s)/2}\Gamma((1-s)/2)\zeta(1-s),$$

and therefore

$$\zeta(s) = \pi^{s-1/2}\frac{\Gamma((1-s)/2)}{\Gamma(s/2)}\zeta(1-s).$$

Now observe that for $\mathrm{Re}(s) < 0$ the following are true:

(i) $\zeta(1-s)$ has no zeros because $\mathrm{Re}(1-s) > 1$.

(ii) $\Gamma((1-s)/2)$ is zero free.

(iii) $1/\Gamma(s/2)$ has zeros at $s = -2, -4, -6, \ldots$.

Therefore, the only zeros of ζ in $\mathrm{Re}(s) < 0$ are located at the negative even integers $-2, -4, -6, \ldots$.

This proves the following theorem.

Theorem 1.1 *The only zeros of ζ outside the strip $0 \leq \mathrm{Re}(s) \leq 1$ are at the negative even integers, $-2, -4, -6, \ldots$.*

The region that remains to be studied is called the **critical strip**, $0 \leq \mathrm{Re}(s) \leq 1$. A key fact in the proof of the prime number theorem is that ζ has no zeros on the line $\mathrm{Re}(s) = 1$. As a simple consequence of this fact and the functional equation, it follows that ζ has no zeros on the line $\mathrm{Re}(s) = 0$.

In the seminal paper where Riemann introduced the analytic continuation of the ζ function and proved its functional equation, he applied these insights to the theory of prime numbers, and wrote down "explicit" formulas for determining the distribution of primes. While he did not succeed in fully proving and exploiting his assertions, he did initiate many important new ideas. His analysis led him to believe the truth of what has since been called the **Riemann hypothesis**:

> *The zeros of $\zeta(s)$ in the critical strip lie on the line $\mathrm{Re}(s) = 1/2$.*

He said about this: "It would certainly be desirable to have a rigorous demonstration of this proposition; nevertheless I have for the moment set this aside, after several quick but unsuccessful attempts, because it seemed unneeded for the immediate goal of my study." Although much of the theory and numerical results point to the validity of this hypothesis, a proof or a counter-example remains to be discovered. The Riemann hypothesis is today one of mathematics' most famous unresolved problems.

In particular, it is for this reason that the zeros of ζ located outside the critical strip are sometimes called the **trivial zeros** of the zeta function. See also Exercise 5 for an argument proving that ζ has no zeros on the real segment, $0 \leq \sigma \leq 1$, where $s = \sigma + it$.

In the rest of this section we shall restrict ourselves to proving the following theorem, together with related estimates on ζ, which we shall use in the proof of the prime number theorem.

Theorem 1.2 *The zeta function has no zeros on the line $\mathrm{Re}(s) = 1$.*

Of course, since we know that ζ has a pole at $s = 1$, there are no zeros in a neighborhood of this point, but what we need is the deeper property that

$$\zeta(1 + it) \neq 0 \quad \text{for all } t \in \mathbb{R}.$$

The next sequence of lemmas gathers the necessary ingredients for the proof of Theorem 1.2.

Lemma 1.3 *If $\mathrm{Re}(s) > 1$, then*

$$\log \zeta(s) = \sum_{p,m} \frac{p^{-ms}}{m} = \sum_{n=1}^{\infty} c_n n^{-s}$$

for some $c_n \geq 0$.

Proof. Suppose first that $s > 1$. Taking the logarithm of the Euler product formula, and using the power series expansion for the logarithm

$$\log\left(\frac{1}{1-x}\right) = \sum_{m=1}^{\infty} \frac{x^m}{m},$$

which holds for $0 \le x < 1$, we find that

$$\log \zeta(s) = \log \prod_{p} \frac{1}{1 - p^{-s}} = \sum_{p} \log\left(\frac{1}{1 - p^{-s}}\right) = \sum_{p,m} \frac{p^{-ms}}{m}.$$

Since the double sum converges absolutely, we need not specify the order of summation. See the Note at the end of this chapter. The formula then holds for all $\operatorname{Re}(s) > 1$ by analytic continuation. Note that, by Theorem 6.2 in Chapter 3, $\log \zeta(s)$ is well defined in the simply connected half-plane $\operatorname{Re}(s) > 1$, since ζ has no zeros there. Finally, it is clear that we have

$$\sum_{p,m} \frac{p^{-ms}}{m} = \sum_{n=1}^{\infty} c_n n^{-s},$$

where $c_n = 1/m$ if $n = p^m$ and $c_n = 0$ otherwise.

The proof of the theorem we shall give depends on a simple trick that is based on the following inequality.

Lemma 1.4 *If $\theta \in \mathbb{R}$, then $3 + 4\cos\theta + \cos 2\theta \ge 0$.*

This follows at once from the simple observation

$$3 + 4\cos\theta + \cos 2\theta = 2(1 + \cos\theta)^2.$$

Corollary 1.5 *If $\sigma > 1$ and t is real, then*

$$\log |\zeta^3(\sigma)\zeta^4(\sigma + it)\zeta(\sigma + 2it)| \ge 0.$$

Proof. Let $s = \sigma + it$ and note that

$$\operatorname{Re}(n^{-s}) = \operatorname{Re}(e^{-(\sigma + it)\log n}) = e^{-\sigma \log n} \cos(t \log n) = n^{-\sigma} \cos(t \log n).$$

Therefore,

$$\log |\zeta^3(\sigma)\zeta^4(\sigma + it)\zeta(\sigma + 2it)|$$
$$= 3\log |\zeta(\sigma)| + 4\log |\zeta(\sigma + it)| + \log |\zeta(\sigma + 2it)|$$
$$= 3\operatorname{Re}[\log \zeta(\sigma)] + 4\operatorname{Re}[\log \zeta(\sigma + it)] + \operatorname{Re}[\log \zeta(\sigma + 2it)]$$
$$= \sum c_n n^{-\sigma}(3 + 4\cos\theta_n + \cos 2\theta_n),$$

where $\theta_n = t \log n$. The positivity now follows from Lemma 1.4, and the fact that $c_n \ge 0$.

We can now finish the proof of our theorem.

Proof of Theorem 1.2. Suppose on the contrary that $\zeta(1 + it_0) = 0$ for some $t_0 \neq 0$. Since ζ is holomorphic at $1 + it_0$, it must vanish at least to order 1 at this point, hence

$$|\zeta(\sigma + it_0)|^4 \leq C(\sigma - 1)^4 \quad \text{as } \sigma \to 1,$$

for some constant $C > 0$. Also, we know that $s = 1$ is a simple pole for $\zeta(s)$, so that

$$|\zeta(\sigma)|^3 \leq C'(\sigma - 1)^{-3} \quad \text{as } \sigma \to 1,$$

for some constant $C' > 0$. Finally, since ζ is holomorphic at the points $\sigma + 2it_0$, the quantity $|\zeta(\sigma + 2it_0)|$ remains bounded as $\sigma \to 1$. Putting these facts together yields

$$|\zeta^3(\sigma)\zeta^4(\sigma + it)\zeta(\sigma + 2it)| \to 0 \quad \text{as } \sigma \to 1,$$

which contradicts Corollary 1.5, since the logarithm of real numbers between 0 and 1 is negative. This concludes the proof that ζ is zero free on the real line $\text{Re}(s) = 1$.

1.1 Estimates for $1/\zeta(s)$

The proof of the prime number theorem relies on detailed manipulations of the zeta function near the line $\text{Re}(s) = 1$; the basic object involved is the logarithmic derivative $\zeta'(s)/\zeta(s)$. For this reason, besides the non-vanishing of ζ on the line, we need to know about the growth of ζ' and $1/\zeta$. The former was dealt with in Proposition 2.7 of Chapter 6; we now treat the latter.

The proposition that follows is actually a quantitative version of Theorem 1.2.

Proposition 1.6 *For every $\epsilon > 0$, we have $1/|\zeta(s)| \leq c_\epsilon |t|^\epsilon$ when $s = \sigma + it$, $\sigma \geq 1$, and $|t| \geq 1$.*

Proof. From our previous observations, we clearly have that

$$|\zeta^3(\sigma)\zeta^4(\sigma + it)\zeta(\sigma + 2it)| \geq 1, \quad \text{whenever } \sigma \geq 1.$$

Using the estimate for ζ in Proposition 2.7 of Chapter 6, we find that

$$|\zeta^4(\sigma + it)| \geq c|\zeta^{-3}(\sigma)| \, |t|^{-\epsilon} \geq c'(\sigma - 1)^3 |t|^{-\epsilon},$$

for all $\sigma \geq 1$ and $|t| \geq 1$. Thus

(3) $|\zeta(\sigma + it)| \geq c'(\sigma - 1)^{3/4}|t|^{-\epsilon/4}$,   whenever $\sigma \geq 1$ and $|t| \geq 1$.

We now consider two separate cases, depending on whether the inequality $\sigma - 1 \geq A|t|^{-5\epsilon}$ holds, for some appropriate constant A (whose value we choose later).

If this inequality does hold, then (3) immediately provides

$$|\zeta(\sigma + it)| \geq A'|t|^{-4\epsilon},$$

and it suffices to replace 4ϵ by ϵ to conclude the proof of the desired estimate, in this case.

If, however, $\sigma - 1 < A|t|^{-5\epsilon}$, then we first select $\sigma' > \sigma$ with $\sigma' - 1 = A|t|^{-5\epsilon}$. The triangle inequality then implies

$$|\zeta(\sigma + it)| \geq |\zeta(\sigma' + it)| - |\zeta(\sigma' + it) - \zeta(\sigma + it)|,$$

and an application of the mean value theorem, together with the estimates for the derivative of ζ obtained in the previous chapter, give

$$|\zeta(\sigma' + it) - \zeta(\sigma + it)| \leq c''|\sigma' - \sigma|\,|t|^{\epsilon} \leq c''|\sigma' - 1|\,|t|^{\epsilon}.$$

These observations, together with an application of (3) where we set $\sigma = \sigma'$, show that

$$|\zeta(\sigma + it)| \geq c'(\sigma' - 1)^{3/4}|t|^{-\epsilon/4} - c''(\sigma' - 1)|t|^{\epsilon}.$$

Now choose $A = (c'/(2c''))^4$, and recall that $\sigma' - 1 = A|t|^{-5\epsilon}$. This gives precisely

$$c'(\sigma' - 1)^{3/4}|t|^{-\epsilon/4} = 2c''(\sigma' - 1)|t|^{\epsilon},$$

and therefore

$$|\zeta(\sigma + it)| \geq A''|t|^{-4\epsilon}.$$

On replacing 4ϵ by ϵ, the desired inequality is established, and the proof of the proposition is complete.

2 Reduction to the functions ψ and ψ_1

In his study of primes, Tchebychev introduced an auxiliary function whose behavior is to a large extent equivalent to the asymptotic distribution of primes, but which is easier to manipulate than $\pi(x)$. **Tchebychev's ψ-function** is defined by

$$\psi(x) = \sum_{p^m \leq x} \log p.$$

The sum is taken over those integers of the form p^m that are less than or equal to x. Here p is a prime number and m is a positive integer. There are two other formulations of ψ that we shall need. First, if we define

$$\Lambda(n) = \begin{cases} \log p & \text{if } n = p^m \text{ for some prime } p \text{ and some } m \geq 1, \\ 0 & \text{otherwise,} \end{cases}$$

then it is clear that

$$\psi(x) = \sum_{1 \leq n \leq x} \Lambda(n).$$

Also, it is immediate that

$$\psi(x) = \sum_{p \leq x} \left[\frac{\log x}{\log p} \right] \log p$$

where $[u]$ denotes the greatest integer $\leq u$, and the sum is taken over the primes less than x. This formula follows from the fact that if $p^m \leq x$, then $m \leq \log x / \log p$.

The fact that $\psi(x)$ contains enough information about $\pi(x)$ to prove our theorem is given a precise meaning in the statement of the next proposition. In particular, this reduces the prime number theorem to a corresponding asymptotic statement about ψ.

Proposition 2.1 *If* $\psi(x) \sim x$ *as* $x \to \infty$, *then* $\pi(x) \sim x/\log x$ *as* $x \to \infty$.

Proof. The argument here is elementary. By definition, it suffices to prove the following two inequalities:

$$(4) \qquad 1 \leq \liminf_{x \to \infty} \pi(x) \frac{\log x}{x} \quad \text{and} \quad \limsup_{x \to \infty} \pi(x) \frac{\log x}{x} \leq 1.$$

To do so, first note that crude estimates give

$$\psi(x) = \sum_{p \leq x} \left[\frac{\log x}{\log p} \right] \log p \leq \sum_{p \leq x} \frac{\log x}{\log p} \log p = \pi(x) \log x,$$

and dividing through by x yields

$$\frac{\psi(x)}{x} \leq \frac{\pi(x) \log x}{x}.$$

The asymptotic condition $\psi(x) \sim x$ implies the first inequality in (4). The proof of the second inequality is a little trickier. Fix $0 < \alpha < 1$, and note that

$$\psi(x) \geq \sum_{p \leq x} \log p \geq \sum_{x^\alpha < p \leq x} \log p \geq (\pi(x) - \pi(x^\alpha)) \log x^\alpha,$$

and therefore

$$\psi(x) + \alpha\pi(x^\alpha) \log x \geq \alpha\pi(x) \log x.$$

Dividing by x, noting that $\pi(x^\alpha) \leq x^\alpha$, $\alpha < 1$, and $\psi(x) \sim x$, gives

$$1 \geq \alpha \lim_{x \to \infty} \sup \pi(x) \frac{\log x}{x}.$$

Since $\alpha < 1$ was arbitrary, the proof is complete.

Remark. The converse of the proposition is also true: if $\pi(x) \sim x/\log x$ then $\psi(x) \sim x$. Since we shall not need this result, we leave the proof to the interested reader.

In fact, it will be more convenient to work with a close cousin of the ψ function. Define the function ψ_1 by

$$\psi_1(x) = \int_1^x \psi(u) \, du.$$

In the previous proposition we reduced the prime number theorem to the asymptotics of $\psi(x)$ as x tends to infinity. Next, we show that this follows from the asymptotics of ψ_1.

Proposition 2.2 *If $\psi_1(x) \sim x^2/2$ as $x \to \infty$, then $\psi(x) \sim x$ as $x \to \infty$, and therefore $\pi(x) \sim x/\log x$ as $x \to \infty$.*

Proof. By Proposition 2.1, it suffices to prove that $\psi(x) \sim x$ as $x \to \infty$. This will follow quite easily from the fact that if $\alpha < 1 < \beta$, then

$$\frac{1}{(1 - \alpha)x} \int_{\alpha x}^{x} \psi(u) \, du \leq \psi(x) \leq \frac{1}{(\beta - 1)x} \int_{x}^{\beta x} \psi(u) \, du.$$

The proof of this double inequality is immediate and relies simply on the fact that ψ is increasing. As a consequence, we find, for example, that

$$\psi(x) \leq \frac{1}{(\beta - 1)x}[\psi_1(\beta x) - \psi_1(x)],$$

and therefore

$$\frac{\psi(x)}{x} \leq \frac{1}{(\beta - 1)} \left[\frac{\psi_1(\beta x)}{(\beta x)^2} \beta^2 - \frac{\psi_1(x)}{x^2} \right].$$

In turn this implies

$$\limsup_{x \to \infty} \frac{\psi(x)}{x} \leq \frac{1}{\beta - 1} \left[\frac{1}{2} \beta^2 - \frac{1}{2} \right] = \frac{1}{2}(\beta + 1).$$

Since this result is true for all $\beta > 1$, we have proved that $\limsup_{x \to \infty} \psi(x)/x \leq 1$. A similar argument with $\alpha < 1$, then shows that $\liminf_{x \to \infty} \psi(x)/x \geq 1$, and the proof of the proposition is complete.

It is now time to relate ψ_1 (and therefore also ψ) and ζ. We proved in Lemma 1.3 that for $\text{Re}(s) > 1$

$$\log \zeta(s) = \sum_{m,p} \frac{p^{-ms}}{m}.$$

Differentiating this expression gives

$$\frac{\zeta'(s)}{\zeta(s)} = -\sum_{m,p} (\log p) p^{-ms} = -\sum_{n=1}^{\infty} \frac{\Lambda(n)}{n^s}.$$

We record this formula for $\text{Re}(s) > 1$ as

(5) $$-\frac{\zeta'(s)}{\zeta(s)} = \sum_{n=1}^{\infty} \frac{\Lambda(n)}{n^s}.$$

The asymptotic behavior $\psi_1(x) \sim x^2/2$ will be a consequence via (5) of the relationship between ψ_1 and ζ, which is expressed by the following noteworthy integral formula.

Proposition 2.3 *For all $c > 1$*

(6) $$\psi_1(x) = \frac{1}{2\pi i} \int_{c-i\infty}^{c+i\infty} \frac{x^{s+1}}{s(s+1)} \left(-\frac{\zeta'(s)}{\zeta(s)} \right) ds.$$

To make the proof of this formula clear, we isolate the necessary contour integrals in a lemma.

Lemma 2.4 *If $c > 0$, then*

$$\frac{1}{2\pi i}\int_{c-i\infty}^{c+i\infty}\frac{a^s}{s(s+1)}\,ds = \begin{cases} 0 & \text{if } 0 < a \leq 1, \\ 1 - 1/a & \text{if } 1 \leq a. \end{cases}$$

Here, the integral is over the vertical line $\mathrm{Re}(s) = c$.

Proof. First note that since $|a^s| = a^c$, the integral converges. We suppose first that $1 \leq a$, and write $a = e^\beta$ with $\beta = \log a \geq 0$. Let

$$f(s) = \frac{a^s}{s(s+1)} = \frac{e^{s\beta}}{s(s+1)}.$$

Then $\mathrm{res}_{s=0} f = 1$ and $\mathrm{res}_{s=-1} f = -1/a$. For $T > 0$, consider the path $\Gamma(T)$ shown on Figure 1.

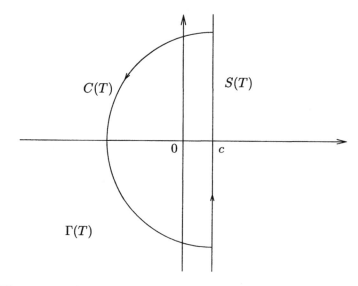

Figure 1. The contour in the proof of Lemma 2.4 when $a \geq 1$

The path $\Gamma(T)$ consists of the vertical segment $S(T)$ from $c - iT$ to $c + iT$, and of the half-circle $C(T)$ centered at c of radius T, lying to the left of the vertical segment. We equip $\Gamma(T)$ with the positive (counterclockwise) orientation, and note that we are dealing with a toy contour. If we choose T so large that 0 and -1 are contained in the interior of $\Gamma(T)$, then by the residue formula

$$\frac{1}{2\pi i}\int_{\Gamma(T)} f(s)\,ds = 1 - 1/a.$$

Since

$$\int_{\Gamma(T)} f(s)\, ds = \int_{S(T)} f(s)\, ds + \int_{C(T)} f(s)\, ds,$$

it suffices to prove that the integral over the half-circle goes to 0 as T tends to infinity. Note that if $s = \sigma + it \in C(T)$, then for all large T we have

$$|s(s+1)| \geq (1/2)T^2,$$

and since $\sigma \leq c$ we also have the estimate $|e^{\beta s}| \leq e^{\beta c}$. Therefore

$$\left| \int_{C(T)} f(s)\, ds \right| \leq \frac{C}{T^2} 2\pi T \to 0 \qquad \text{as } T \to \infty,$$

and the case when $a \geq 1$ is proved.

If $0 < a \leq 1$, consider an analogous contour but with the half-circle lying to the right of the line $\text{Re}(s) = c$. Noting that there are no poles in the interior of that contour, we can give an argument similar to the one given above to show that the integral over the half-circle also goes to 0 as T tends to infinity.

We are now ready to prove Proposition 2.3. First, observe that

$$\psi(u) = \sum_{n=1}^{\infty} \Lambda(n) f_n(u),$$

where $f_n(u) = 1$ if $n \leq u$ and $f_n(u) = 0$ otherwise. Therefore,

$$\psi_1(x) = \int_0^x \psi(u)\, du$$

$$= \sum_{n=1}^{\infty} \int_0^x \Lambda(n) f_n(u)\, du$$

$$= \sum_{n \leq x} \Lambda(n) \int_n^x du,$$

and hence

$$\psi_1(x) = \sum_{n \leq x} \Lambda(n)(x - n).$$

This fact, together with equation (5) and an application of Lemma 2.4 (with $a = x/n$), gives

$$\frac{1}{2\pi i} \int_{c-i\infty}^{c+i\infty} \frac{x^{s+1}}{s(s+1)} \left(-\frac{\zeta'(s)}{\zeta(s)} \right) ds = x \sum_{n=1}^{\infty} \Lambda(n) \frac{1}{2\pi i} \int_{c-i\infty}^{c+i\infty} \frac{(x/n)^s}{s(s+1)} ds$$

$$= x \sum_{n \leq x} \Lambda(n) \left(1 - \frac{n}{x} \right)$$

$$= \psi_1(x),$$

as was to be shown.

2.1 Proof of the asymptotics for ψ_1

In this section, we will show that

$$\psi_1(x) \sim x^2/2 \quad \text{as } x \to \infty,$$

and as a consequence, we will have proved the prime number theorem.

The key ingredients in the argument are:

- the formula in Proposition 2.3 connecting ψ_1 to ζ, namely

$$\psi_1(x) = \frac{1}{2\pi i} \int_{c-i\infty}^{c+i\infty} \frac{x^{s+1}}{s(s+1)} \left(-\frac{\zeta'(s)}{\zeta(s)} \right) ds$$

  for $c > 1$.

- the non-vanishing of the zeta function on $\text{Re}(s) = 1$,

$$\zeta(1+it) \neq 0 \quad \text{for all } t \in \mathbb{R},$$

  and the estimates for ζ near that line given in Proposition 2.7 of Chapter 6 together with Proposition 1.6 of this chapter.

Let us now discuss our strategy in more detail. In the integral above for $\psi_1(x)$ we want to change the line of integration $\text{Re}(s) = c$ with $c > 1$, to $\text{Re}(s) = 1$. If we could achieve that, the size of the factor x^{s+1} in the integrand would then be of order x^2 (which is close to what we want) instead of x^{c+1}, $c > 1$, which is much too large. However, there would still be two issues that must be dealt with. The first is the pole of $\zeta(s)$ at $s = 1$; it turns out that when it is taken into account, its contribution is exactly the main term $x^2/2$ of the asymptotic of $\psi_1(x)$. Second, what remains must be shown to be essentially smaller than this term, and so we must further refine the crude estimate of order x^2 when integrating on the line $\text{Re}(s) = 1$. We carry out our plan as follows.

Fix $c > 1$, say $c = 2$, and assume x is also fixed for the moment with $x \geq 2$. Let $F(s)$ denote the integrand

$$F(s) = \frac{x^{s+1}}{s(s+1)} \left(-\frac{\zeta'(s)}{\zeta(s)} \right).$$

First we deform the vertical line from $c - i\infty$ to $c + i\infty$ to the path $\gamma(T)$ shown in Figure 2. (The segments of $\gamma(T)$ on the line $\text{Re}(s) = 1$ consist of $T \leq t < \infty$, and $-\infty < t \leq -T$.) Here $T \geq 3$, and T will be chosen appropriately large later.

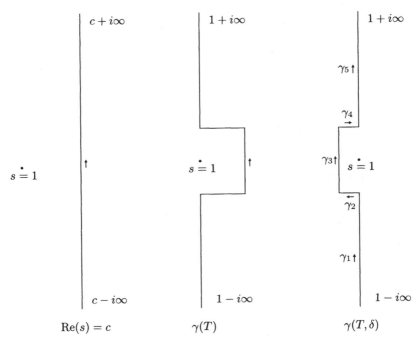

Figure 2. Three stages: the line $\text{Re}(s) = c$, the contours $\gamma(T)$ and $\gamma(T, \delta)$

The usual and familiar arguments using Cauchy's theorem allow us to see that

$$(7) \qquad \frac{1}{2\pi i} \int_{c-i\infty}^{c+i\infty} F(s) \, ds = \frac{1}{2\pi i} \int_{\gamma(T)} F(s) \, ds.$$

Indeed, we know on the basis of Proposition 2.7 in Chapter 6 and Proposition 1.6 that $|\zeta'(s)/\zeta(s)| \leq A|t|^{\eta}$ for any fixed $\eta > 0$, whenever $s = \sigma + it$, $\sigma \geq 1$, and $|t| \geq 1$. Thus $|F(s)| \leq A'|t|^{-2+\eta}$ in the two (infinite) rectangles bounded by the line $(c - i\infty, c + i\infty)$ and $\gamma(T)$. Since F is regular in that region, and its decrease at infinity is rapid enough, the assertion (7) is established.

Next, we pass from the contour $\gamma(T)$ to the contour $\gamma(T, \delta)$. (Again, see Figure 2.) For fixed T, we choose $\delta > 0$ small enough so that ζ has no zeros in the box

$$\{s = \sigma + it, \ 1 - \delta \leq \sigma \leq 1, \ |t| \leq T\}.$$

Such a choice can be made since ζ does not vanish on the line $\sigma = 1$.

Now $F(s)$ has a simple pole at $s = 1$. In fact, by Corollary 2.6 in Chapter 6, we know that $\zeta(s) = 1/(s - 1) + H(s)$, where $H(s)$ is regular near $s = 1$. Hence $-\zeta'(s)/\zeta(s) = 1/(s - 1) + h(s)$, where $h(s)$ is holomorphic near $s = 1$, and so the residue of $F(s)$ at $s = 1$ equals $x^2/2$. As a result

$$\frac{1}{2\pi i} \int_{\gamma(T)} F(s) \, ds = \frac{x^2}{2} + \frac{1}{2\pi i} \int_{\gamma(T,\delta)} F(s) \, ds.$$

We now decompose the contour $\gamma(T, \delta)$ as $\gamma_1 + \gamma_2 + \gamma_3 + \gamma_4 + \gamma_5$ and estimate each of the integrals $\int_{\gamma_j} F(s) \, ds$, $j = 1, 2, 3, 4, 5$, with the γ_j as in Figure 2.

First we contend that there exists T so large that

$$\left| \int_{\gamma_1} F(s) \, ds \right| \leq \frac{\epsilon}{2} x^2 \quad \text{and} \quad \left| \int_{\gamma_5} F(s) \, ds \right| \leq \frac{\epsilon}{2} x^2.$$

To see this, we first note that for $s \in \gamma_1$ one has

$$|x^{1+s}| = x^{1+\sigma} = x^2.$$

Then, by Proposition 1.6 we have, for example, that $|\zeta'(s)/\zeta(s)| \leq A|t|^{1/2}$, so

$$\left| \int_{\gamma_1} F(s) \, ds \right| \leq C x^2 \int_T^{\infty} \frac{|t|^{1/2}}{t^2} \, dt.$$

Since the integral converges, we can make the right-hand side $\leq \epsilon x^2/2$ upon taking T sufficiently large. The argument for the integral over γ_5 is the same.

Having now fixed T, we choose δ appropriately small. On γ_3, note that

$$|x^{1+s}| = x^{1+1-\delta} = x^{2-\delta},$$

from which we conclude that there exists a constant C_T (dependent on T) such that

$$\left|\int_{\gamma_3} F(s)\,ds\right| \leq C_T x^{2-\delta}.$$

Finally, on the small horizontal segment γ_2 (and similarly on γ_4), we can estimate the integral as follows:

$$\left|\int_{\gamma_2} F(s)\,ds\right| \leq C_T' \int_{1-\delta}^{1} x^{1+\sigma}\,d\sigma \leq C_T' \frac{x^2}{\log x}.$$

We conclude that there exist constants C_T and C_T' (possibly different from the ones above) such that

$$\left|\psi_1(x) - \frac{x^2}{2}\right| \leq \epsilon x^2 + C_T x^{2-\delta} + C_T' \frac{x^2}{\log x}.$$

Dividing through by $x^2/2$, we see that

$$\left|\frac{2\psi_1(x)}{x^2} - 1\right| \leq 2\epsilon + 2C_T x^{-\delta} + 2C_T' \frac{1}{\log x},$$

and therefore, for all large x we have

$$\left|\frac{2\psi_1(x)}{x^2} - 1\right| \leq 4\epsilon.$$

This concludes the proof that

$$\psi_1(x) \sim x^2/2 \quad \text{as } x \to \infty,$$

and thus, we have also completed the proof of the prime number theorem.

Note on interchanging double sums

We prove the following facts about the interchange of infinite sums: if $\{a_{k\ell}\}_{1 \leq k, \ell < \infty}$ is a sequence of complex numbers indexed by $\mathbb{N} \times \mathbb{N}$, such that

$$(8) \qquad \sum_{k=1}^{\infty}\left(\sum_{\ell=1}^{\infty} |a_{k\ell}|\right) < \infty,$$

then:

(i) The double sum $A = \sum_{k=1}^{\infty} \left(\sum_{\ell=1}^{\infty} a_{k\ell} \right)$ summed in this order converges, and we may in fact also interchange the order of summation, so that

$$A = \sum_{k=1}^{\infty} \sum_{\ell=1}^{\infty} a_{k\ell} = \sum_{\ell=1}^{\infty} \sum_{k=1}^{\infty} a_{k\ell}.$$

(ii) Given $\epsilon > 0$, there is a positive integer N so that for all $K, L > N$ we have
$$\left| A - \sum_{k=1}^{K} \sum_{\ell=1}^{L} a_{k\ell} \right| < \epsilon.$$

(iii) If $m \mapsto (k(m), \ell(m))$ is a bijection from $\mathbb{N}$ to $\mathbb{N} \times \mathbb{N}$, and if we write $c_m = a_{k(m)\ell(m)}$, then $A = \sum_{k=1}^{\infty} c_k$.

Statement (iii) says that any rearrangement of the sequence $\{a_{k\ell}\}$ can be summed without changing the limit. This is analogous to the case of absolutely convergent series, which can be summed in any desired order.

The condition (8) says that each sum $\sum_{\ell} a_{k\ell}$ converges absolutely, and moreover this convergence is "uniform" in k. An analogous situation arises for sequences of functions, where an important question is whether or not the interchange of limits

$$\lim_{x \to x_0} \lim_{n \to \infty} f_n(x) \overset{?}{=} \lim_{n \to \infty} \lim_{x \to x_0} f_n(x)$$

holds. It is a well-known fact that if the f_n's are continuous, and their convergence is uniform, then the above identity is true since the limit function is itself continuous. To take advantage of this fact, define $b_k = \sum_{\ell=1}^{\infty} |a_{k\ell}|$ and let $S = \{x_0, x_1, \dots\}$ be a countable set of points with $\lim_{n \to \infty} x_n = x_0$. Also, define functions on S as follows:

$$f_k(x_0) = \sum_{\ell=1}^{\infty} a_{k\ell} \qquad \text{for } k = 1, 2, \dots$$

$$f_k(x_n) = \sum_{\ell=1}^{n} a_{k\ell} \qquad \text{for } k = 1, 2, \dots \text{ and } n = 1, 2, \dots$$

$$g(x) = \sum_{k=1}^{\infty} f_k(x) \quad \text{for } x \in S.$$

By assumption (8), each f_k is continuous at x_0. Moreover $|f_k(x)| \le b_k$ and $\sum b_k < \infty$, so the series defining the function g is uniformly convergent on S, and therefore g is also continuous at x_0. As a consequence we find (i), since

$$\sum_{k=1}^{\infty} \sum_{\ell=1}^{\infty} a_{k\ell} = g(x_0) = \lim_{n \to \infty} g(x_n) = \lim_{n \to \infty} \sum_{k=1}^{\infty} \sum_{\ell=1}^{n} a_{k\ell}$$

$$= \lim_{n \to \infty} \sum_{\ell=1}^{n} \sum_{k=1}^{\infty} a_{k\ell} = \sum_{\ell=1}^{\infty} \sum_{k=1}^{\infty} a_{k\ell}.$$

For the second statement, first observe that

$$\left| A - \sum_{k=1}^{K} \sum_{\ell=1}^{L} a_{k\ell} \right| \le \sum_{k \le K} \sum_{\ell > L} |a_{k\ell}| + \sum_{k > K} \sum_{\ell=1}^{\infty} |a_{k\ell}|.$$

To estimate the second term, we use the fact that $\sum b_k$ converges, which implies $\sum_{k>K} \sum_{\ell=1}^{\infty} |a_{k\ell}| < \epsilon/2$ whenever $K > K_0$, for some K_0. For the first term above, note that $\sum_{k \leq K} \sum_{\ell > L} |a_{k\ell}| \leq \sum_{k=1}^{\infty} \sum_{\ell > L} |a_{k\ell}|$. But the argument above guarantees that we can interchange these last two sums; also $\sum_{\ell=1}^{\infty} \sum_{k=1}^{\infty} |a_{k\ell}| < \infty$, so that for all $L > L_0$ we have $\sum_{\ell > L} \sum_{k=1}^{\infty} |a_{k\ell}| < \epsilon/2$. Taking $N > \max(L_0, K_0)$ completes the proof of (ii).

The proof of (iii) is a direct consequence of (ii). Indeed, given any rectangle

$$R(K,L) = \{(k,\ell) \in \mathbb{N} \times \mathbb{N} : 1 \leq k \leq K \text{ and } 1 \leq \ell \leq L\},$$

there exists M such that the image of $[1, M]$ under the map $m \mapsto (k(m), \ell(m))$ contains $R(K, L)$.

When U denotes any open set in $\mathbb{R}^2$ that contains the origin, we define for $R > 0$ its dilate $U(R) = \{y \in \mathbb{R}^2 : y = Rx \text{ for some } x \in U\}$, and we can apply (ii) to see that

$$A = \lim_{R \to \infty} \sum_{(k,\ell) \in U(R)} a_{k\ell}.$$

In other words, under condition (8) the double sum $\sum_{k\ell} a_{k\ell}$ can be evaluated by summing over discs, squares, rectangles, ellipses, etc.

Finally, we leave the reader with the instructive task of finding a sequence of complex numbers $\{a_{k\ell}\}$ such that

$$\sum_k \sum_\ell a_{k\ell} \neq \sum_\ell \sum_k a_{k\ell}.$$

[Hint: Consider $\{a_{k\ell}\}$ as the entries of an infinite matrix with 0 above the diagonal, -1 on the diagonal, and $a_{k\ell} = 2^{\ell-k}$ if $k > \ell$.]

3 Exercises

1. Suppose that $\{a_n\}_{n=1}^{\infty}$ is a sequence of real numbers such that the partial sums

$$A_n = a_1 + \cdots + a_n$$

are bounded. Prove that the Dirichlet series

$$\sum_{n=1}^{\infty} \frac{a_n}{n^s}$$

converges for $\text{Re}(s) > 0$ and defines a holomorphic function in this half-plane.

[Hint: Use summation by parts to compare the original (non-absolutely convergent) series to the (absolutely convergent) series $\sum A_n(n^{-s} - (n+1)^{-s})$. An estimate for the term in parentheses is provided by the mean value theorem. To prove that the series is analytic, show that the partial sums converge uniformly on every compact subset of the half-plane $\text{Re}(s) > 0$.]

2. The following links the multiplication of Dirichlet series with the divisibility properties of their coefficients.

(a) Show that if $\{a_m\}$ and $\{b_k\}$ are two bounded sequences of complex numbers, then

$$\left(\sum_{m=1}^{\infty} \frac{a_m}{m^s}\right)\left(\sum_{k=1}^{\infty} \frac{b_k}{k^s}\right) = \sum_{n=1}^{\infty} \frac{c_n}{n^s} \qquad \text{where } c_n = \sum_{mk=n} a_m b_k.$$

The above series converge absolutely when $\text{Re}(s) > 1$.

(b) Prove as a consequence that one has

$$(\zeta(s))^2 = \sum_{n=1}^{\infty} \frac{d(n)}{n^s} \qquad \text{and} \qquad \zeta(s)\zeta(s-a) = \sum_{n=1}^{\infty} \frac{\sigma_a(n)}{n^s}$$

for $\text{Re}(s) > 1$ and $\text{Re}(s-a) > 1$, respectively. Here $d(n)$ equals the number of divisors of n, and $\sigma_a(n)$ is the sum of the ath powers of divisors of n. In particular, one has $\sigma_0(n) = d(n)$.

3. In line with the previous exercise, we consider the Dirichlet series for $1/\zeta$.

(a) Prove that for $\text{Re}(s) > 1$

$$\frac{1}{\zeta(s)} = \sum_{n=1}^{\infty} \frac{\mu(n)}{n^s},$$

where $\mu(n)$ is the **Möbius function** defined by

$$\mu(n) = \begin{cases} 1 & \text{if } n = 1, \\ (-1)^k & \text{if } n = p_1 \cdots p_k, \text{ and the } p_j \text{ are distinct primes}, \\ 0 & \text{otherwise}. \end{cases}$$

Note that $\mu(nm) = \mu(n)\mu(m)$ whenever n and m are relatively prime. [Hint: Use the Euler product formula for $\zeta(s)$.]

(b) Show that

$$\sum_{k|n} \mu(k) = \begin{cases} 1 & \text{if } n = 1, \\ 0 & \text{otherwise}. \end{cases}$$

4. Suppose $\{a_n\}_{n=1}^{\infty}$ is a sequence of complex numbers such that $a_n = a_m$ if $n \equiv m$ mod q for some positive integer q. Define the **Dirichlet L-series** associated to $\{a_n\}$ by

$$L(s) = \sum_{n=1}^{\infty} \frac{a_n}{n^s} \qquad \text{for } \text{Re}(s) > 1.$$

Also, with $a_0 = a_q$, let

$$Q(x) = \sum_{m=0}^{q-1} a_{q-m} e^{mx}.$$

Show, as in Exercises 15 and 16 of the previous chapter, that

$$L(s) = \frac{1}{\Gamma(s)} \int_0^\infty \frac{Q(x) x^{s-1}}{e^{qx} - 1} \, dx, \quad \text{for Re}(s) > 1.$$

Prove as a result that $L(s)$ is continuable into the complex plane, with the only possible singularity a pole at $s = 1$. In fact, $L(s)$ is regular at $s = 1$ if and only if $\sum_{m=0}^{q-1} a_m = 0$. Note the connection with the Dirichlet $L(s, \chi)$ series, taken up in Book I, Chapter 8, and that as a consequence, $L(s, \chi)$ is regular at $s = 1$ if and only if χ is a non-trivial character.

5. Consider the following function

$$\tilde{\zeta}(s) = 1 - \frac{1}{2^s} + \frac{1}{3^s} - \cdots = \sum_{n=1}^\infty \frac{(-1)^{n+1}}{n^s}.$$

(a) Prove that the series defining $\tilde{\zeta}(s)$ converges for $\text{Re}(s) > 0$ and defines a holomorphic function in that half-plane.

(b) Show that for $s > 1$ one has $\tilde{\zeta}(s) = (1 - 2^{1-s})\zeta(s)$.

(c) Conclude, since $\tilde{\zeta}$ is given as an alternating series, that ζ has no zeros on the segment $0 < \sigma < 1$. Extend this last assertion to $\sigma = 0$ by using the functional equation.

6. Show that for every $c > 0$

$$\lim_{N \to \infty} \frac{1}{2\pi i} \int_{c-iN}^{c+iN} a^s \frac{ds}{s} = \begin{cases} 1 & \text{if } a > 1, \\ 1/2 & \text{if } a = 1, \\ 0 & \text{if } 0 \le a < 1. \end{cases}$$

The integral is taken over the vertical segment from $c - iN$ to $c + iN$.

7. Show that the function

$$\xi(s) = \pi^{-s/2} \Gamma(s/2) \zeta(s)$$

is real when s is real, or when $\text{Re}(s) = 1/2$.

8. The function ζ has infinitely many zeros in the critical strip. This can be seen as follows.

(a) Let

$$F(s) = \xi(1/2 + s), \quad \text{where} \quad \xi(s) = \pi^{-s/2}\Gamma(s/2)\zeta(s).$$

Show that $F(s)$ is an even function of s, and as a result, there exists G so that $G(s^2) = F(s)$.

(b) Show that the function $(s - 1)\zeta(s)$ is an entire function of growth order 1, that is

$$|(s - 1)\zeta(s)| \le A_\epsilon e^{a_\epsilon |s|^{1+\epsilon}}.$$

As a consequence $G(s)$ is of growth order $1/2$.

(c) Deduce from the above that ζ has infinitely many zeros in the critical strip.

[Hint: To prove (a) and (b) use the functional equation for $\zeta(s)$. For (c), use a result of Hadamard, which states that an entire function with fractional order has infinitely many zeros (Exercise 14 in Chapter 5).]

9. Refine the estimates in Proposition 2.7 in Chapter 6 and Proposition 1.6 to show that

(a) $|\zeta(1 + it)| \le A \log |t|$,

(b) $|\zeta'(1 + it)| \le A(\log |t|)^2$,

(c) $1/|\zeta(1 + it)| \le A(\log |t|)^a$,

when $|t| \ge 2$ (with $a = 7$).

10. In the theory of primes, a better approximation to $\pi(x)$ (instead of $x/\log x$) turns out to be $\text{Li}(x)$ defined by

$$\text{Li}(x) = \int_2^x \frac{dt}{\log t}.$$

(a) Prove that

$$\text{Li}(x) = \frac{x}{\log x} + O\left(\frac{x}{(\log x)^2}\right) \quad \text{as } x \to \infty,$$

and that as a consequence

$$\pi(x) \sim \text{Li}(x) \quad \text{as } x \to \infty.$$

[Hint: Integrate by parts in the definition of $\text{Li}(x)$ and observe that it suffices to prove

$$\int_2^x \frac{dt}{(\log t)^2} = O\left(\frac{x}{(\log x)^2}\right).$$

To see this, split the integral from 2 to $\sqrt{x}$ and from $\sqrt{x}$ to x.]

(b) Refine the previous analysis by showing that for every integer $N > 0$ one has the following asymptotic expansion

$$\text{Li}(x) = \frac{x}{\log x} + \frac{x}{(\log x)^2} + 2\frac{x}{(\log x)^3} \cdots + (N-1)!\frac{x}{(\log x)^N} + O\left(\frac{x}{(\log x)^{N+1}}\right)$$

as $x \to \infty$.

11. Let

$$\varphi(x) = \sum_{p \leq x} \log p$$

where the sum is taken over all primes $\leq x$. Prove that the following are equivalent as $x \to \infty$:

(i) $\varphi(x) \sim x$,

(ii) $\pi(x) \sim x/\log x$,

(iii) $\psi(x) \sim x$,

(iv) $\psi_1(x) \sim x^2/2$.

12. If p_n denotes the nth prime, the prime number theorem implies that $p_n \sim n \log n$ as $n \to \infty$.

(a) Show that $\pi(x) \sim x/\log x$ implies that

$$\log \pi(x) + \log \log x \sim \log x.$$

(b) As a consequence, prove that $\log \pi(x) \sim \log x$, and take $x = p_n$ to conclude the proof.

4 Problems

1. Let $F(s) = \sum_{n=1}^{\infty} a_n/n^s$, where $|a_n| \leq M$ for all n.

(a) Then

$$\lim_{T \to \infty} \frac{1}{2T} \int_{-T}^{T} |F(\sigma + it)|^2 \, dt = \sum_{n=1}^{\infty} \frac{|a_n|^2}{n^{2\sigma}} \qquad \text{if } \sigma > 1.$$

How is this reminiscent of the Parseval-Plancherel theorem? See e.g. Chapter 3 in Book I.

(b) Show as a consequence the uniqueness of Dirichlet series: If $F(s) = \sum_{n=1}^{\infty} a_n n^{-s}$, where the coefficients are assumed to satisfy $|a_n| \leq cn^k$ for some k, and $F(s) \equiv 0$, then $a_n = 0$ for all n.

[Hint: For part (a) use the fact that

$$\frac{1}{2T} \int_{-T}^{T} (nm)^{-\sigma} n^{-it} m^{it} \, dt \to \begin{cases} n^{-2\sigma} & \text{if } n = m, \\ 0 & \text{if } n \neq m. \end{cases}$$]

2.* One of the "explicit formulas" in the theory of primes is as follows: if ψ_1 is the integrated Tchebychev function considered in Section 2, then

$$\psi_1(x) = \frac{x^2}{2} - \sum_{\rho} \frac{x^{\rho+1}}{\rho(\rho+1)} - E(x)$$

where the sum is taken over all zeros ρ of the zeta function in the critical strip. The error term is given by $E(x) = c_1 x + c_0 + \sum_{k=1}^{\infty} x^{1-2k}/(2k(2k-1))$, where $c_1 = \zeta'(0)/\zeta(0)$ and $c_0 = \zeta'(-1)/\zeta(-1)$. Note that $\sum_{\rho} 1/|\rho|^{1+\epsilon} < \infty$ for every $\epsilon > 0$, because $(1-s)\zeta(s)$ has order of growth 1. (See Exercise 8.) Also, obviously $E(x) = O(x)$ as $x \to \infty$.

3.* Using the previous problem one can show that

$$\pi(x) - \text{Li}(x) = O(x^{\alpha+\epsilon}) \qquad \text{as } x \to \infty$$

for every $\epsilon > 0$, where α is fixed and $1/2 \leq \alpha < 1$ if and only if $\zeta(s)$ has no zeros in the strip $\alpha < \text{Re}(s) < 1$. The case $\alpha = 1/2$ corresponds to the Riemann hypothesis.

4.* One can combine ideas from the prime number theorem with the proof of Dirichlet's theorem about primes in arithmetic progression (given in Book I) to prove the following. Let q and ℓ be relatively prime integers. We consider the primes belonging to the arithmetic progression $\{qk + \ell\}_{k=1}^{\infty}$, and let $\pi_{q,\ell}(x)$ denote the number of such primes $\leq x$. Then one has

$$\pi_{q,\ell}(x) \sim \frac{x}{\varphi(q) \log x} \qquad \text{as } x \to \infty,$$

where $\varphi(q)$ denotes the number of positive integers less than q and relatively prime to q.

8 Conformal Mappings

> The results I found for polygons can be extended un-
> der very general assumptions. I have undertaken this
> research because it is a step towards a deeper un-
> derstanding of the mapping problem, for which not
> much has happened since Riemann's inaugural disser-
> tation; this, even though the theory of mappings, with
> its close connection with the fundamental theorems of
> Riemann's function theory, deserves in the highest de-
> gree to be developed further.
>
> *E. B. Christoffel,* 1870

The problems and ideas we present in this chapter are more geometric in nature than the ones we have seen so far. In fact, here we will
be primarily interested in mapping properties of holomorphic functions.
In particular, most of our results will be "global," as opposed to the
more "local" analytical results proved in the first three chapters. The
motivation behind much of our presentation lies in the following simple
question:

> Given two open sets U and V in $\mathbb{C}$, does there exist a holo-
> morphic bijection between them?

By a holomorphic bijection we simply mean a function that is both
holomorphic and bijective. (It will turn out that the inverse map is then
automatically holomorphic.) A solution to this problem would permit
a transfer of questions about analytic functions from one open set with
little geometric structure to another with possibly more useful properties.
The prime example consists in taking $V = \mathbb{D}$ the unit disc, where many
ideas have been developed to study analytic functions.[1] In fact, since
the disc seems to be the most fruitful choice for V we are led to a variant
of the above question:

> Given an open subset Ω of $\mathbb{C}$, what conditions on Ω guarantee
> that there exists a holomorphic bijection from Ω to $\mathbb{D}$?

---

[1] For the corresponding problem when $V = \mathbb{C}$, the solution is trivial: only $U = \mathbb{C}$ is
possible. See Exercise 14 in Chapter 3.

In some instances when a bijection exists it can be given by explicit formulas, and we turn to this aspect of the theory first. For example, the upper half-plane can be mapped by a holomorphic bijection to the disc, and this is given by a fractional linear transformation. From there, one can construct many other examples, by composing simple maps already encountered earlier, such as rational functions, trigonometric functions, logarithms, etc. As an application, we discuss the consequence of these constructions to the solution of the Dirichlet problem for the Laplacian in some particular domains.

Next, we pass from the specific examples to prove the first general result of the chapter, namely the Schwarz lemma, with an immediate application to the determination of all holomorphic bijections ("automorphisms" of the disc to itself). These are again given by fractional linear transformations.

Then comes the heart of the matter: the Riemann mapping theorem, which states that Ω can be mapped to the unit disc whenever it is simply connected and not all of $\mathbb{C}$. This is a remarkable theorem, since little is assumed about Ω, not even regularity of its boundary $\partial\Omega$. (After all, the boundary of the disc is smooth.) In particular, the interiors of triangles, squares, and in fact any polygon can be mapped via a bijective holomorphic function to the disc. A precise description of the mapping in the case of polygons, called the Schwarz-Christoffel formula, will be taken up in the last section of the chapter. It is interesting to note that the mapping functions for rectangles are given by "elliptic integrals," and these lead to doubly-periodic functions. The latter are the subject of the next chapter.

1 Conformal equivalence and examples

We fix some terminology that we shall use in the rest of this chapter. A bijective holomorphic function $f : U \to V$ is called a **conformal map** or **biholomorphism**. Given such a mapping f, we say that U and V are **conformally equivalent** or simply **biholomorphic**. An important fact is that the inverse of f is then automatically holomorphic.

Proposition 1.1 *If* $f : U \to V$ *is holomorphic and injective, then* $f'(z) \neq 0$ *for all* $z \in U$. *In particular, the inverse of* f *defined on its range is holomorphic, and thus the inverse of a conformal map is also holomorphic.*

Proof. We argue by contradiction, and suppose that $f'(z_0) = 0$ for some $z_0 \in U$. Then

$$f(z) - f(z_0) = a(z - z_0)^k + G(z) \qquad \text{for all } z \text{ near } z_0,$$

with $a \neq 0$, $k \geq 2$ and G vanishing to order $k+1$ at z_0. For sufficiently small w, we write

$$f(z) - f(z_0) - w = F(z) + G(z), \qquad \text{where } F(z) = a(z - z_0)^k - w.$$

Since $|G(z)| < |F(z)|$ on a small circle centered at z_0, and F has at least two zeros inside that circle, Rouché's theorem implies that $f(z) - f(z_0) - w$ has at least two zeros there. Since $f'(z) \neq 0$ for all $z \neq z_0$ but sufficiently close to z_0 it follows that the roots of $f(z) - f(z_0) - w$ are distinct, hence f is not injective, a contradiction.

Now let $g = f^{-1}$ denote the inverse of f on its range, which we can assume is V. Suppose $w_0 \in V$ and w is close to w_0. Write $w = f(z)$ and $w_0 = f(z_0)$. If $w \neq w_0$, we have

$$\frac{g(w) - g(w_0)}{w - w_0} = \frac{1}{\frac{w - w_0}{g(w) - g(w_0)}} = \frac{1}{\frac{f(z) - f(z_0)}{z - z_0}}.$$

Since $f'(z_0) \neq 0$, we may let $z \to z_0$ and conclude that g is holomorphic at w_0 with $g'(w_0) = 1/f'(g(w_0))$.

From this proposition we conclude that two open sets U and V are conformally equivalent if and only if there exist holomorphic functions $f : U \to V$ and $g : V \to U$ such that $g(f(z)) = z$ and $f(g(w)) = w$ for all $z \in U$ and $w \in V$.

We point out that the terminology adopted here is not universal. Some authors call a holomorphic map $f : U \to V$ conformal if $f'(z) \neq 0$ for all $z \in U$. This definition is clearly less restrictive than ours; for example, $f(z) = z^2$ on the punctured disc $\mathbb{C} - \{0\}$ satisfies $f'(z) \neq 0$, but is not injective. However, the condition $f'(z) \neq 0$ is tantamount to f being a local bijection (Exercise 1). There is a geometric consequence of the condition $f'(z) \neq 0$ and it is at the root of this discrepency of terminology in the definitions. A holomorphic map that satisfies this condition preserves angles. Loosely speaking, if two curves γ and η intersect at z_0, and α is the oriented angle between the tangent vectors to these curves, then the image curves $f \circ \gamma$ and $f \circ \eta$ intersect at $f(z_0)$, and their tangent vectors form the same angle α. Problem 2 develops this idea.

We begin our study of conformal mappings by looking at a number of specific examples. The first gives the conformal equivalence between the unit disc and the upper half-plane, which plays an important role in many problems.

1.1 The disc and upper half-plane

The upper half-plane, which we denote by $\mathbb{H}$, consists of those complex numbers with positive imaginary part; that is,

$$\mathbb{H} = \{z \in \mathbb{C} : \text{Im}(z) > 0\}.$$

A remarkable fact, which at first seems surprising, is that the unbounded set $\mathbb{H}$ is conformally equivalent to the unit disc. Moreover, an explicit formula giving this equivalence exists. Indeed, let

$$F(z) = \frac{i - z}{i + z} \quad \text{and} \quad G(w) = i\frac{1 - w}{1 + w}.$$

Theorem 1.2 *The map* $F : \mathbb{H} \to \mathbb{D}$ *is a conformal map with inverse* $G : \mathbb{D} \to \mathbb{H}$.

Proof. First we observe that both maps are holomorphic in their respective domains. Then we note that any point in the upper half-plane is closer to i than to $-i$, so $|F(z)| < 1$ and F maps $\mathbb{H}$ into $\mathbb{D}$. To prove that G maps into the upper half-plane, we must compute $\text{Im}(G(w))$ for $w \in \mathbb{D}$. To this end we let $w = u + iv$, and note that

$$\begin{aligned}
\text{Im}(G(w)) &= \text{Re}\left(\frac{1 - u - iv}{1 + u + iv}\right) \\
&= \text{Re}\left(\frac{(1 - u - iv)(1 + u - iv)}{(1 + u)^2 + v^2}\right) \\
&= \frac{1 - u^2 - v^2}{(1 + u)^2 + v^2} > 0
\end{aligned}$$

since $|w| < 1$. Therefore G maps the unit disc to the upper half-plane. Finally,

$$F(G(w)) = \frac{i - i\frac{1-w}{1+w}}{i + i\frac{1-w}{1+w}} = \frac{1 + w - 1 + w}{1 + w + 1 - w} = w,$$

and similarly $G(F(z)) = z$. This proves the theorem.

An interesting aspect of these functions is their behavior on the boundaries of our open sets.[2] Observe that F is holomorphic everywhere on $\mathbb{C}$

---

[2]The boundary behavior of conformal maps is a recurrent theme that plays an important role in this chapter.

except at $z = -i$, and in particular it is continuous everywhere on the boundary of $\mathbb{H}$, namely the real line. If we take $z = x$ real, then the distance from x to i is the same as the distance from x to $-i$, therefore $|F(x)| = 1$. Thus F maps $\mathbb{R}$ onto the boundary of $\mathbb{D}$. We get more information by writing

$$F(x) = \frac{i-x}{i+x} = \frac{1-x^2}{1+x^2} + i\frac{2x}{1+x^2},$$

and parametrizing the real line by $x = \tan t$ with $t \in (-\pi/2, \pi/2)$. Since

$$\sin 2a = \frac{2\tan a}{1+\tan^2 a} \quad \text{and} \quad \cos 2a = \frac{1-\tan^2 a}{1+\tan^2 a},$$

we have $F(x) = \cos 2t + i\sin 2t = e^{i2t}$. Hence the image of the real line is the arc consisting of the circle omitting the point -1. Moreover, as x travels from $-\infty$ to ∞, $F(x)$ travels along that arc starting from -1 and first going through that part of the circle that lies in the lower half-plane.

The point -1 on the circle corresponds to the "point at infinity" of the upper half-plane.

Remark. Mappings of the form

$$z \longmapsto \frac{az+b}{cz+d},$$

where a, b, c, and d are complex numbers, and where the denominator is assumed not to be a multiple of the numerator, are usually referred to as **fractional linear transformations**. Other instances occur as the automorphisms of the disc and of the upper half-plane in Theorems 2.1 and 2.4.

1.2 Further examples

We gather here several illustrations of conformal mappings. In certain cases we discuss the behavior of the map on the boundary of the relevant domain. Some of the mappings are pictured in Figure 1.

EXAMPLE 1. Translations and dilations provide the first simple examples. Indeed, if $h \in \mathbb{C}$, the translation $z \mapsto z + h$ is a conformal map from $\mathbb{C}$ to itself whose inverse is $w \mapsto w - h$. If h is real, then this translation is also a conformal map from the upper half-plane to itself.

For any non-zero complex number c, the map $f : z \mapsto cz$ is a conformal map from the complex plane to itself, whose inverse is simply $g : w \mapsto c^{-1}w$. If c has modulus 1, so that $c = e^{i\varphi}$ for some real φ, then f is a **rotation** by φ. If $c > 0$ then f corresponds to a dilation. Finally, if $c < 0$ the map f consists of a dilation by $|c|$ followed by a rotation of π.

EXAMPLE 2. If n is a positive integer, then the map $z \mapsto z^n$ is conformal from the sector $S = \{z \in \mathbb{C} : 0 < \arg(z) < \pi/n\}$ to the upper half-plane. The inverse of this map is simply $w \mapsto w^{1/n}$, defined in terms of the principal branch of the logarithm.

More generally, if $0 < \alpha < 2$ the map $f(z) = z^\alpha$ takes the upper half-plane to the sector $S = \{w \in \mathbb{C} : 0 < \arg(w) < \alpha\pi\}$. Indeed, if we choose the branch of the logarithm obtained by deleting the positive real axis, and $z = re^{i\theta}$ with $r > 0$ and $0 < \theta < \pi$, then

$$f(z) = z^\alpha = |z|^\alpha e^{i\alpha\theta}.$$

Therefore f maps $\mathbb{H}$ into S. Moreover, a simple verification shows that the inverse of f is given by $g(w) = w^{1/\alpha}$, where the branch of the logarithm is chosen so that $0 < \arg w < \alpha\pi$.

By composing the map just discussed with the translations and rotations in the previous example, we may map the upper half-plane conformally to any (infinite) sector in $\mathbb{C}$.

Let us note the boundary behavior of f. If x travels from $-\infty$ to 0 on the real line, then $f(x)$ travels from $\infty e^{i\alpha\pi}$ to 0 on the half-line determined by $\arg z = \alpha\pi$. As x goes from 0 to ∞ on the real line, the image $f(x)$ goes from 0 to ∞ on the real line as well.

EXAMPLE 3.   The map $f(z) = (1+z)/(1-z)$ takes the upper half-disc $\{z = x + iy : |z| < 1 \text{ and } y > 0\}$ conformally to the first quadrant $\{w = u + iv : u > 0 \text{ and } v > 0\}$. Indeed, if $z = x + iy$ we have

$$f(z) = \frac{1 - (x^2 + y^2)}{(1-x)^2 + y^2} + i\frac{2y}{(1-x)^2 + y^2},$$

so f maps the half-disc in the upper half-plane into the first quadrant. The inverse map, given by $g(w) = (w-1)/(w+1)$, is clearly holomorphic in the first quadrant. Moreover, $|w+1| > |w-1|$ for all w in the first quadrant because the distance from w to -1 is greater than the distance from w to 1; thus g maps into the unit disc. Finally, an easy calculation shows that the imaginary part of $g(w)$ is positive whenever w is in the first quadrant. So g transforms the first quadrant into the desired half-disc and we conclude that f is conformal because g is the inverse of f.

To examine the action of f on the boundary, note that if $z = e^{i\theta}$ belongs to the upper half-circle, then

$$f(z) = \frac{1 + e^{i\theta}}{1 - e^{i\theta}} = \frac{e^{-i\theta/2} + e^{i\theta/2}}{e^{-i\theta/2} - e^{i\theta/2}} = \frac{i}{\tan(\theta/2)}.$$

As θ travels from 0 to π we see that $f(e^{i\theta})$ travels along the imaginary axis from infinity to 0. Moreover, if $z = x$ is real, then

$$f(z) = \frac{1 + x}{1 - x}$$

is also real; and one sees from this, that f is actually a bijection from $(-1, 1)$ to the positive real axis, with $f(x)$ increasing from 0 to infinity as x travels from -1 to 1. Note also that $f(0) = 1$.

EXAMPLE 4. The map $z \mapsto \log z$, defined as the branch of the logarithm obtained by deleting the negative imaginary axis, takes the upper half-plane to the strip $\{w = u + iv : u \in \mathbb{R}, \ 0 < v < \pi\}$. This is immediate from the fact that if $z = re^{i\theta}$ with $-\pi/2 < \theta < 3\pi/2$, then by definition,

$$\log z = \log r + i\theta.$$

The inverse map is then $w \mapsto e^w$.

As x travels from $-\infty$ to 0, the point $f(x)$ travels from $\infty + i\pi$ to $-\infty + i\pi$ on the line $\{x + i\pi : -\infty < x < \infty\}$. When x travels from 0 to ∞ on the real line, its image $f(x)$ then goes from $-\infty$ to ∞ along the reals.

EXAMPLE 5.   With the previous example in mind, we see that $z \mapsto \log z$ also defines a conformal map from the half-disc $\{z = x + iy : |z| < 1, \ y > 0\}$ to the half-strip $\{w = u + iv : u < 0, \ 0 < v < \pi\}$. As x travels from 0 to 1 on the real line, then $\log x$ goes from $-\infty$ to 0. When x goes from 1 to -1 on the half-circle in the upper half-plane, then the point $\log x$ travels from 0 to πi on the vertical segment of the strip. Finally, as x goes from -1 to 0, the point $\log x$ goes from πi to $-\infty + i\pi$ on the top half-line of the strip.

EXAMPLE 6.   The map $f(z) = e^{iz}$ takes the half-strip $\{z = x + iy : -\pi/2 < x < \pi/2, \ y > 0\}$ conformally to the half-disc $\{w = u + iv : |w| < 1, \ u > 0\}$. This is immediate from the fact that if $z = x + iy$, then

$$e^{iz} = e^{-y}e^{ix}.$$

If x goes from $\pi/2 + i\infty$ to $\pi/2$, then $f(x)$ goes from 0 to i, and as x goes from $\pi/2$ to $-\pi/2$, then $f(x)$ travels from i to $-i$ on the half-circle. Finally, as x goes from $-\pi/2$ to $-\pi/2 + i\infty$, we see that $f(x)$ travels from $-i$ back to 0.

The mapping f is closely related to the inverse of the map in Example 5.

EXAMPLE 7. The function $f(z) = -\frac{1}{2}(z + 1/z)$ is a conformal map from the half-disc $\{z = x + iy : |z| < 1, \ y > 0\}$ to the upper half-plane (Exercise 5).

The boundary behavior of f is as follows. If x travels from 0 to 1, then $f(x)$ goes from ∞ to 1 on the real axis. If $z = e^{i\theta}$, then $f(z) = \cos\theta$ and as x travels from 1 to -1 along the unit half-circle in the upper half-plane, the $f(x)$ goes from 1 to -1 on the real segment. Finally, when x goes from -1 to 0, $f(x)$ goes from -1 to $-\infty$ along the real axis.

EXAMPLE 8. The map $f(z) = \sin z$ takes the half-strip $\{w = x + iy : -\pi/2 < x < \pi/2 \ y > 0\}$ conformally onto the upper half-plane. To see this, note that if $\zeta = e^{iz}$, then

$$\sin z = \frac{e^{iz} - e^{-iz}}{2i} = \frac{-1}{2}\left(i\zeta + \frac{1}{i\zeta}\right),$$

and therefore f is obtained first by applying the map in Example 6, then multiplying by i (that is, rotating by $\pi/2$), and finally applying the map in Example 7.

As x travels from $-\pi/2 + i\infty$ to $-\pi/2$, the point $f(x)$ goes from $-\infty$ to -1. When x is real, between $-\pi/2$ and $\pi/2$, then $f(x)$ is also real between -1 and 1. Finally, if x goes from $\pi/2$ to $\pi/2 + i\infty$, then $f(x)$ travels from 1 to ∞ on the real axis.

1.3 The Dirichlet problem in a strip

The Dirichlet problem in the open set Ω consists of solving

$$(1) \qquad \begin{cases} \Delta u = 0 & \text{in } \Omega, \\ u = f & \text{on } \partial\Omega, \end{cases}$$

where Δ denotes the Laplacian $\partial^2/\partial x^2 + \partial^2/\partial y^2$, and f is a given function on the boundary of Ω. In other words, we wish to find a harmonic function in Ω with prescribed boundary values f. This problem was already considered in Book I in the cases where Ω is the unit disc or the

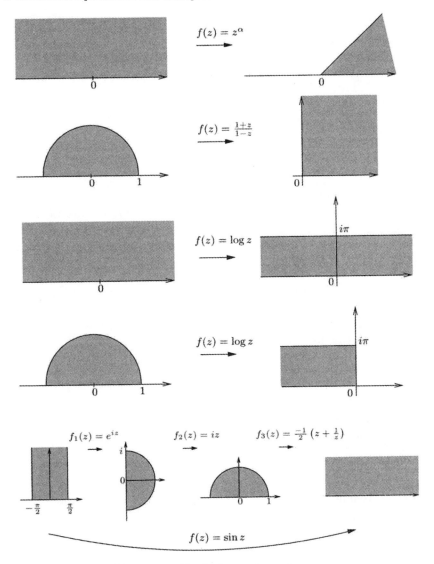

Figure 1. Explicit conformal maps

upper half-plane, where it arose in the solution of the steady-state heat equation. In these specific examples, explicit solutions were obtained in terms of convolutions with the Poisson kernels.

Our goal here is to connect the Dirichlet problem with the conformal maps discussed so far. We begin by providing a formula for a solution to the problem (1) in the special case where Ω is a strip. In fact, this example was studied in Problem 3 of Chapter 5, Book I, where the problem was solved using the Fourier transform. Here, we recover this solution using only conformal mappings and the known solution in the disc.

The first important fact that we use is that the composition of a harmonic function with a holomorphic function is still harmonic.

Lemma 1.3 *Let V and U be open sets in $\mathbb{C}$ and $F : V \to U$ a holomorphic function. If $u : U \to \mathbb{C}$ is a harmonic function, then $u \circ F$ is harmonic on V.*

Proof. The thrust of the lemma is purely local, so we may assume that U is an open disc. We let G be a holomorphic function in U whose real part is u (such a G exists by Exercise 12 in Chapter 2, and is determined up to an additive constant). Let $H = G \circ F$ and note that $u \circ F$ is the real part of H. Hence $u \circ F$ is harmonic because H is holomorphic.

For an alternate (computational) proof of this lemma, see Exercise 6.

With this result in hand, we may now consider the problem (1) when Ω consists of the horizontal strip

$$\Omega = \{x + iy : x \in \mathbb{R},\ 0 < y < 1\},$$

whose boundary is the union of the two horizontal lines $\mathbb{R}$ and $i + \mathbb{R}$. We express the boundary data as two functions f_0 and f_1 defined on $\mathbb{R}$, and ask for a solution $u(x, y)$ in Ω of $\triangle u = 0$ that satisfies

$$u(x, 0) = f_0(x) \quad \text{and} \quad u(x, 1) = f_1(x).$$

We shall assume that f_0 and f_1 are continuous and vanish at infinity, that is, that $\lim_{|x| \to \infty} f_j(x) = 0$ for $j = 0, 1$.

The method we shall follow consists of relocating the problem from the strip to the unit disc via a conformal map. In the disc the solution $\tilde{u}$ is then expressed in terms of a convolution with the Poisson kernel. Finally, $\tilde{u}$ is moved back to the strip using the inverse of the previous conformal map, thereby giving our final answer to the problem.

To achieve our goal, we introduce the mappings $F : \mathbb{D} \to \Omega$ and $G : \Omega \to \mathbb{D}$, that are defined by

$$F(w) = \frac{1}{\pi} \log \left(i \frac{1 - w}{1 + w} \right) \quad \text{and} \quad G(z) = \frac{i - e^{\pi z}}{i + e^{\pi z}}.$$

These two functions, which are obtained from composing mappings from examples in the previous sections, are conformal and inverses to one

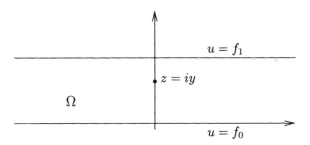

Figure 2. The Dirichlet problem in a strip

another. Tracing through the boundary behavior of F, we find that it maps the lower half-circle to the line $i + \mathbb{R}$, and the upper half-circle to $\mathbb{R}$. More precisely, as φ travels from $-\pi$ to 0, then $F(e^{i\varphi})$ goes from $i + \infty$ to $i - \infty$, and as φ travels from 0 to π, then $F(e^{i\varphi})$ goes from $-\infty$ to ∞ on the real line.

With the behavior of F on the circle in mind, we define

$$\tilde{f}_1(\varphi) = f_1(F(e^{i\varphi}) - i) \quad \text{whenever } -\pi < \varphi < 0,$$

and

$$\tilde{f}_0(\varphi) = f_0(F(e^{i\varphi})) \quad \text{whenever } 0 < \varphi < \pi.$$

Then, since f_0 and f_1 vanish at infinity, the function $\tilde{f}$ that is equal to $\tilde{f}_1$ on the lower semi-circle, $\tilde{f}_0$ on the upper semi-circle, and 0 at the points $\varphi = \pm\pi, 0$, is continuous on the whole circle. The solution to the Dirichlet problem in the unit disc with boundary data $\tilde{f}$ is given by the Poisson integral[3]

$$\tilde{u}(w) = \frac{1}{2\pi} \int_{-\pi}^{\pi} P_r(\theta - \varphi)\tilde{f}(\varphi)\, d\varphi$$

$$= \frac{1}{2\pi} \int_{-\pi}^{0} P_r(\theta - \varphi)\tilde{f}_1(\varphi)\, d\varphi + \frac{1}{2\pi} \int_{0}^{\pi} P_r(\theta - \varphi)\tilde{f}_0(\varphi)\, d\varphi,$$

where $w = re^{i\theta}$, and

$$P_r(\theta) = \frac{1 - r^2}{1 - 2r\cos\theta + r^2}$$

---

[3]We refer the reader to Chapter 2 in Book I for a detailed discussion of the Dirichlet problem in the disc and the Poisson integral formula. Also, the Poisson integral formula is deduced in Exercise 12 of Chapter 2 and Problem 2 in Chapter 3 of this book.

is the Poisson kernel. Lemma 1.3 guarantees that the function u, defined by

$$u(z) = \tilde{u}(G(z)),$$

is harmonic in the strip. Moreover, our construction also insures that u has the correct boundary values.

A formula for u in terms of f_0 and f_1 is first obtained at the points $z = iy$ with $0 < y < 1$. The appropriate change of variables (see Exercise 7) shows that if $re^{i\theta} = G(iy)$, then

$$\frac{1}{2\pi} \int_0^\pi P_r(\theta - \varphi) \tilde{f}_0(\varphi) \, d\varphi = \frac{\sin \pi y}{2} \int_{-\infty}^\infty \frac{f_0(t)}{\cosh \pi t - \cos \pi y} \, dt.$$

A similar calculation also establishes

$$\frac{1}{2\pi} \int_0^\pi P_r(\theta - \varphi) \tilde{f}_1(\varphi) \, d\varphi = \frac{\sin \pi y}{2} \int_{-\infty}^\infty \frac{f_1(t)}{\cosh \pi t + \cos \pi y} \, dt.$$

Adding these last two integrals provides a formula for $u(0, y)$. In general, we recall from Exercise 13 in Chapter 5 of Book I, that a solution to the Dirichlet problem in the strip vanishing at infinity is unique. Consequently, a translation of the boundary condition by x results in a translation of the solution by x as well. We may therefore apply the same argument to $f_0(x + t)$ and $f_1(x + t)$ (with x fixed), and a final change of variables shows that

$$u(x, y) = \frac{\sin \pi y}{2} \left(\int_{-\infty}^\infty \frac{f_0(x - t)}{\cosh \pi t - \cos \pi y} \, dt + \int_{-\infty}^\infty \frac{f_1(x - t)}{\cosh \pi t + \cos \pi y} \, dt \right),$$

which gives a solution to the Dirichlet problem in the strip. In particular, we find that the solution is given in terms of convolutions with the functions f_0 and f_1. Also, note that at the mid-point of the strip $(y = 1/2)$, the solution is given by integration with respect to the function $1/\cosh \pi t$; this function happens to be its own Fourier transform, as we saw in Example 3, Chapter 3.

Remarks about the Dirichlet problem

The example above leads us to envisage the solution of the more general Dirichlet problem for Ω (a suitable region), if we know a conformal map F from the disc $\mathbb{D}$ to Ω. That is, suppose we wish to solve (1), where f is an assigned continuous function and $\partial\Omega$ is the boundary of Ω. Assuming we have a conformal map F from $\mathbb{D}$ to Ω (that extends to a continuous bijection of the boundary of the disc to the boundary of Ω), then $\tilde{f} = f \circ F$ is defined on the circle, and we can solve the Dirichlet problem for the disc with boundary data $\tilde{f}$. The solution is given by the Poisson integral formula

$$\tilde{u}(re^{i\theta}) = \frac{1}{2\pi} \int_0^{2\pi} P_r(\theta - \varphi)\tilde{f}(e^{i\varphi})\,d\varphi,$$

where P_r is the Poisson kernel. Then, one can expect that the solution of the original problem is given by $u = \tilde{u} \circ F^{-1}$.

Success with this approach requires that we are able to resolve affirmatively two questions:

- Does there exist a conformal map $\Phi = F^{-1}$ from Ω to $\mathbb{D}$?

- If so, does this map extend to a continuous bijection from the boundary of Ω to the boundary of $\mathbb{D}$?

The first question, that of existence, is settled by the Riemann mapping theorem, which we prove in the next section. It is completely general (assuming only that Ω is a proper subset of $\mathbb{C}$ that is simply connected), and necessitates no regularity of the boundary of Ω. A positive answer to the second question requires some regularity of $\partial\Omega$. A particular case, when Ω is the interior of a polygon, is treated below in Section 4.3. (See Exercise 18 and Problem 6 for more general assertions.)

It is interesting to note that in Riemann's original approach to the mapping problem, the chain of implications was reversed: his idea was that the existence of the conformal map Φ from Ω to $\mathbb{D}$ is a consequence of the solvability of the Dirichlet problem in Ω. He argued as follows. Suppose we wish to find such a Φ, with the property that a given point $z_0 \in \Omega$ is mapped to 0. Then Φ must be of the form

$$\Phi(z) = (z - z_0)G(z),$$

where G is holomorphic and non-vanishing in Ω. Hence we can take

$$\Phi(z) = (z - z_0)e^{H(z)},$$

for suitable H. Now if $u(z)$ is the harmonic function given by $u = \mathrm{Re}(H)$, then the fact that $|\Phi(z)| = 1$ on $\partial\Omega$ means that u must satisfy the boundary condition $u(z) = \log(1/|z - z_0|)$ for $z \in \partial\Omega$. So if we can find such a solution u of the Dirichlet problem,[4] we can construct H, and from this the mapping function Φ.

---

[4]The harmonic function $u(z)$ is also known as the Green's function with source z_0 for the region Ω.

However, there are several shortcomings to this method. First, one has to verify that Φ is a bijection. In addition, to succeed, this method requires some regularity of the boundary of Ω. Moreover, one is still faced with the question of solving the Dirichlet problem for Ω. At this stage Riemann proposed using the "Dirichlet principle." But applying this idea involves difficulties that must be overcome.[5]

Nevertheless, using different methods, one can prove the existence of the mapping in the general case. This approach is carried out below in Section 3.

2 The Schwarz lemma; automorphisms of the disc and upper half-plane

The statement and proof of the Schwarz lemma are both simple, but the applications of this result are far-reaching. We recall that a rotation is a map of the form $z \mapsto cz$ with $|c| = 1$, namely $c = e^{i\theta}$, where $\theta \in \mathbb{R}$ is called the angle of rotation and is well-defined up to an integer multiple of 2π.

Lemma 2.1 *Let $f : \mathbb{D} \to \mathbb{D}$ be holomorphic with $f(0) = 0$. Then*

(i) $|f(z)| \leq |z|$ *for all $z \in \mathbb{D}$.*

(ii) *If for some $z_0 \neq 0$ we have $|f(z_0)| = |z_0|$, then f is a rotation.*

(iii) $|f'(0)| \leq 1$, *and if equality holds, then f is a rotation.*

Proof. We first expand f in a power series centered at 0 and convergent in all of $\mathbb{D}$

$$f(z) = a_0 + a_1 z + a_2 z^2 + \cdots .$$

Since $f(0) = 0$ we have $a_0 = 0$, and therefore $f(z)/z$ is holomorphic in $\mathbb{D}$ (since it has a removable singularity at 0). If $|z| = r < 1$, then since $|f(z)| \leq 1$ we have

$$\left| \frac{f(z)}{z} \right| \leq \frac{1}{r},$$

and by the maximum modulus principle, we can conclude that this is true whenever $|z| \leq r$. Letting $r \to 1$ gives the first result.

For (ii), we see that $f(z)/z$ attains its maximum in the interior of $\mathbb{D}$ and must therefore be constant, say $f(z) = cz$. Evaluating this expression

---

[5] An implementation of Dirichlet's principle in the present two-dimensional situation is taken up in Book III.

at z_0 and taking absolute values, we find that $|c| = 1$. Therefore, there exists $\theta \in \mathbb{R}$ such that $c = e^{i\theta}$, and that explains why f is a rotation.

Finally, observe that if $g(z) = f(z)/z$, then $|g(z)| \leq 1$ throughout $\mathbb{D}$, and moreover

$$g(0) = \lim_{z \to 0} \frac{f(z) - f(0)}{z} = f'(0).$$

Hence, if $|f'(0)| = 1$, then $|g(0)| = 1$, and by the maximum principle g is constant, which implies $f(z) = cz$ with $|c| = 1$.

Our first application of this lemma is to the determination of the automorphisms of the disc.

2.1 Automorphisms of the disc

A conformal map from an open set Ω to *itself* is called an **automorphism** of Ω. The set of all automorphisms of Ω is denoted by $\mathrm{Aut}(\Omega)$, and carries the structure of a group. The group operation is composition of maps, the identity element is the map $z \mapsto z$, and the inverses are simply the inverse functions. It is clear that if f and g are automorphisms of Ω, then $f \circ g$ is also an automorphism, and in fact, its inverse is given by

$$(f \circ g)^{-1} = g^{-1} \circ f^{-1}.$$

As mentioned above, the identity map is always an automorphism. We can give other more interesting automorphisms of the unit disc. Obviously, any rotation by an angle $\theta \in \mathbb{R}$, that is, $r_\theta : z \mapsto e^{i\theta}z$, is an automorphism of the unit disc whose inverse is the rotation by the angle $-\theta$, that is, $r_{-\theta} : z \mapsto e^{-i\theta}z$. More interesting, are the automorphisms of the form

$$\psi_\alpha(z) = \frac{\alpha - z}{1 - \overline{\alpha}z}, \qquad \text{where } \alpha \in \mathbb{C} \text{ with } |\alpha| < 1.$$

These mappings, which where introduced in Exercise 7 of Chapter 1, appear in a number of problems in complex analysis because of their many useful properties. The proof that they are automorphisms of $\mathbb{D}$ is quite simple. First, observe that since $|\alpha| < 1$, the map ψ_α is holomorphic in the unit disc. If $|z| = 1$ then $z = e^{i\theta}$ and

$$\psi_\alpha(e^{i\theta}) = \frac{\alpha - e^{i\theta}}{e^{i\theta}(e^{-i\theta} - \overline{\alpha})} = e^{-i\theta}\frac{w}{\overline{w}},$$

where $w = \alpha - e^{i\theta}$, therefore $|\psi_\alpha(z)| = 1$. By the maximum modulus principle, we conclude that $|\psi_\alpha(z)| < 1$ for all $z \in \mathbb{D}$. Finally we make the following very simple observation:

$$(\psi_\alpha \circ \psi_\alpha)(z) = \frac{\alpha - \frac{\alpha - z}{1 - \bar{\alpha}z}}{1 - \bar{\alpha}\frac{\alpha - z}{1 - \bar{\alpha}z}}$$

$$= \frac{\alpha - |\alpha|^2 z - \alpha + z}{1 - \bar{\alpha}z - |\alpha|^2 + \bar{\alpha}z}$$

$$= \frac{(1 - |\alpha|^2)z}{1 - |\alpha|^2}$$

$$= z,$$

from which we conclude that ψ_α is its own inverse! Another important property of ψ_α is that it vanishes at $z = \alpha$; moreover it interchanges 0 and α, namely

$$\psi_\alpha(0) = \alpha \quad \text{and} \quad \psi_\alpha(\alpha) = 0.$$

The next theorem says that the rotations combined with the maps ψ_α exhaust all the automorphisms of the disc.

Theorem 2.2 *If f is an automorphism of the disc, then there exist $\theta \in \mathbb{R}$ and $\alpha \in \mathbb{D}$ such that*

$$f(z) = e^{i\theta}\frac{\alpha - z}{1 - \bar{\alpha}z}.$$

Proof. Since f is an automorphism of the disc, there exists a unique complex number $\alpha \in \mathbb{D}$ such that $f(\alpha) = 0$. Now we consider the automorphism g defined by $g = f \circ \psi_\alpha$. Then $g(0) = 0$, and the Schwarz lemma gives

(2) $$|g(z)| \leq |z| \quad \text{for all } z \in \mathbb{D}.$$

Moreover, $g^{-1}(0) = 0$, so applying the Schwarz lemma to g^{-1}, we find that

$$|g^{-1}(w)| \leq |w| \quad \text{for all } w \in \mathbb{D}.$$

Using this last inequality for $w = g(z)$ for each $z \in \mathbb{D}$ gives

(3) $$|z| \leq |g(z)| \quad \text{for all } z \in \mathbb{D}.$$

Combining (2) and (3) we find that $|g(z)| = |z|$ for all $z \in \mathbb{D}$, and by the Schwarz lemma we conclude that $g(z) = e^{i\theta}z$ for some $\theta \in \mathbb{R}$. Replacing z by $\psi_\alpha(z)$ and using the fact that $(\psi_\alpha \circ \psi_\alpha)(z) = z$, we deduce that $f(z) = e^{i\theta}\psi_\alpha(z)$, as claimed.

Setting $\alpha = 0$ in the theorem yields the following result.

Corollary 2.3 *The only automorphisms of the unit disc that fix the origin are the rotations.*

Note that by the use of the mappings ψ_α, we can see that the group of automorphisms of the disc acts **transitively**, in the sense that given any pair of points α and β in the disc, there is an automorphism ψ mapping α to β. One such ψ is given by $\psi = \psi_\beta \circ \psi_\alpha$.

The explicit formulas for the automorphisms of $\mathbb{D}$ give a good description of the group $\mathrm{Aut}(\mathbb{D})$. In fact, this group of automorphisms is "almost" isomorphic to a group of 2×2 matrices with complex entries often denoted by $\mathrm{SU}(1,1)$. This group consists of all 2×2 matrices that preserve the hermitian form on $\mathbb{C}^2 \times \mathbb{C}^2$ defined by

$$\langle Z, W \rangle = z_1 \overline{w}_1 - z_2 \overline{w}_2,$$

where $Z = (z_1, z_2)$ and $W = (w_1, w_2)$. For more information about this subject, we refer the reader to Problem 4.

2.2 Automorphisms of the upper half-plane

Our knowledge of the automorphisms of $\mathbb{D}$ together with the conformal map $F : \mathbb{H} \to \mathbb{D}$ found in Section 1.1 allow us to determine the group of automorphisms of $\mathbb{H}$ which we denote by $\mathrm{Aut}(\mathbb{H})$.

Consider the map

$$\Gamma : \mathrm{Aut}(\mathbb{D}) \to \mathrm{Aut}(\mathbb{H})$$

given by "conjugation by F":

$$\Gamma(\varphi) = F^{-1} \circ \varphi \circ F.$$

It is clear that $\Gamma(\varphi)$ is an automorphism of $\mathbb{H}$ whenever φ is an automorphism of $\mathbb{D}$, and Γ is a bijection whose inverse is given by $\Gamma^{-1}(\psi) = F \circ \psi \circ F^{-1}$. In fact, we prove more, namely that Γ preserves the operations on the corresponding groups of automorphisms. Indeed, suppose that $\varphi_1, \varphi_2 \in \mathrm{Aut}(\mathbb{D})$. Since $F \circ F^{-1}$ is the identity on $\mathbb{D}$ we find that

$$\begin{aligned}
\Gamma(\varphi_1 \circ \varphi_2) &= F^{-1} \circ \varphi_1 \circ \varphi_2 \circ F \\
&= F^{-1} \circ \varphi_1 \circ F \circ F^{-1} \circ \varphi_2 \circ F \\
&= \Gamma(\varphi_1) \circ \Gamma(\varphi_2).
\end{aligned}$$

The conclusion is that the two groups $\mathrm{Aut}(\mathbb{D})$ and $\mathrm{Aut}(\mathbb{H})$ are the same, since Γ defines an isomorphism between them. We are still left with the

task of giving a description of elements of Aut($\mathbb{H}$). A series of calculations, which consist of pulling back the automorphisms of the disc to the
upper half-plane via F, can be used to verify that Aut($\mathbb{H}$) consists of all
maps

$$z \mapsto \frac{az + b}{cz + d},$$

where a, b, c, and d are real numbers with $ad - bc = 1$. Again, a matrix
group is lurking in the background. Let $\mathrm{SL}_2(\mathbb{R})$ denote the group of all
2×2 matrices with real entries and determinant 1, namely

$$\mathrm{SL}_2(\mathbb{R}) = \left\{ M = \begin{pmatrix} a & b \\ c & d \end{pmatrix} : \ a, b, c, d \in \mathbb{R} \text{ and } \det(M) = ad - bc = 1 \right\}.$$

This group is called the **special li 1ear group**.

Given a matrix $M \in \mathrm{SL}_2(\mathbb{R})$ we define the mapping f_M by

$$f_M(z) = \frac{az + b}{cz + d}.$$

Theorem 2.4 *Every automorphism of* $\mathbb{H}$ *takes the form* f_M *for some*
$M \in \mathrm{SL}_2(\mathbb{R})$*. Conversely, every map of this form is an automorphism of*
$\mathbb{H}$*.*

The proof consists of a sequence of steps. For brevity, we denote the
group $\mathrm{SL}_2(\mathbb{R})$ by $\mathcal{G}$.

Step 1. If $M \in \mathcal{G}$, then f_M maps $\mathbb{H}$ to itself. This is clear from the
observation that

$$(4) \quad \mathrm{Im}(f_M(z)) = \frac{(ad - bc)\mathrm{Im}(z)}{|cz + d|^2} = \frac{\mathrm{Im}(z)}{|cz + d|^2} > 0 \quad \text{whenever } z \in \mathbb{H}.$$

Step 2. If M and M' are two matrices in $\mathcal{G}$, then $f_M \circ f_{M'} = f_{MM'}$.
This follows from a straightforward calculation, which we omit. As a
consequence, we can prove the first half of the theorem. Each f_M is an
automorphism because it has a holomorphic inverse $(f_M)^{-1}$, which is
simply $f_{M^{-1}}$. Indeed, if I is the identity matrix, then

$$(f_M \circ f_{M^{-1}})(z) = f_{MM^{-1}}(z) = f_I(z) = z.$$

Step 3. Given any two points z and w in $\mathbb{H}$, there exists $M \in \mathcal{G}$ such
that $f_M(z) = w$, and therefore $\mathcal{G}$ acts transitively on $\mathbb{H}$. To prove this,

it suffices to show that we can map any $z \in \mathbb{H}$ to i. Setting $d = 0$ in equation (4) above gives

$$\text{Im}(f_M(z)) = \frac{\text{Im}(z)}{|cz|^2}$$

and we may choose a real number c so that $\text{Im}(f_M(z)) = 1$. Next we choose the matrix

$$M_1 = \begin{pmatrix} 0 & -c^{-1} \\ c & 0 \end{pmatrix}$$

so that $f_{M_1}(z)$ has imaginary part equal to 1. Then we translate by a matrix of the form

$$M_2 = \begin{pmatrix} 1 & b \\ 0 & 1 \end{pmatrix} \quad \text{with } b \in \mathbb{R},$$

to bring $f_{M_1}(z)$ to i. Finally, the map f_M with $M = M_2 M_1$ takes z to i.

Step 4. If θ is real, then the matrix

$$M_\theta = \begin{pmatrix} \cos\theta & -\sin\theta \\ \sin\theta & \cos\theta \end{pmatrix}$$

belongs to $\mathcal{G}$, and if $F : \mathbb{H} \to \mathbb{D}$ denotes the standard conformal map, then $F \circ f_{M_\theta} \circ F^{-1}$ corresponds to the rotation of angle -2θ in the disc. This follows from the fact that $F \circ f_{M_\theta} = e^{-2i\theta} F(z)$, which is easily verified.

Step 5. We can now complete the proof of the theorem. We suppose f is an automorphism of $\mathbb{H}$ with $f(\beta) = i$, and consider a matrix $N \in \mathcal{G}$ such that $f_N(i) = \beta$. Then $g = f \circ f_N$ satisfies $g(i) = i$, and therefore $F \circ g \circ F^{-1}$ is an automorphism of the disc that fixes the origin. So $F \circ g \circ F^{-1}$ is a rotation, and by Step 4 there exists $\theta \in \mathbb{R}$ such that

$$F \circ g \circ F^{-1} = F \circ f_{M_\theta} \circ F^{-1}.$$

Hence $g = f_{M_\theta}$, and we conclude that $f = f_{M_\theta N^{-1}}$ which is of the desired form.

A final observation is that the group $\text{Aut}(\mathbb{H})$ is not quite isomorphic with $SL_2(\mathbb{R})$. The reason for this is because the two matrices M and $-M$ give rise to the same function $f_M = f_{-M}$. Therefore, if we identify the two matrices M and $-M$, then we obtain a new group $PSL_2(\mathbb{R})$ called the **projective special linear group**; this group is isomorphic with $\text{Aut}(\mathbb{H})$.

3 The Riemann mapping theorem

3.1 Necessary conditions and statement of the theorem

We now come to the promised cornerstone of this chapter. The basic problem is to determine conditions on an open set Ω that guarantee the existence of a conformal map $F : \Omega \to \mathbb{D}$.

A series of simple observations allow us to find necessary conditions on Ω. First, if $\Omega = \mathbb{C}$ there can be no conformal map $F : \Omega \to \mathbb{D}$, since by Liouville's theorem F would have to be a constant. Therefore, a necessary condition is to assume that $\Omega \neq \mathbb{C}$. Since $\mathbb{D}$ is connected, we must also impose the requirement that Ω be connected. There is still one more condition that is forced upon us: since $\mathbb{D}$ is simply connected, the same must be true of Ω (see Exercise 3). It is remarkable that these conditions on Ω are also sufficient to guarantee the existence of a biholomorpism from Ω to $\mathbb{D}$.

For brevity, we shall call a subset Ω of $\mathbb{C}$ **proper** if it is non-empty and not the whole of $\mathbb{C}$.

Theorem 3.1 (Riemann mapping theorem) *Suppose Ω is proper and simply connected. If $z_0 \in \Omega$, then there exists a unique conformal map $F : \Omega \to \mathbb{D}$ such that*

$$F(z_0) = 0 \quad and \quad F'(z_0) > 0.$$

Corollary 3.2 *Any two proper simply connected open subsets in $\mathbb{C}$ are conformally equivalent.*

Clearly, the corollary follows from the theorem, since we can use as an intermediate step the unit disc. Also, the uniqueness statement in the theorem is straightforward, since if F and G are conformal maps from Ω to $\mathbb{D}$ that satisfy these two conditions, then $H = F \circ G^{-1}$ is an automorphism of the disc that fixes the origin. Therefore $H(z) = e^{i\theta} z$, and since $H'(0) > 0$, we must have $e^{i\theta} = 1$, from which we conclude that $F = G$.

The rest of this section is devoted to the proof of the existence of the conformal map F. The idea of the proof is as follows. We consider all injective holomorphic maps $f : \Omega \to \mathbb{D}$ with $f(z_0) = 0$. From these we wish to choose an f so that its image fills out all of $\mathbb{D}$, and this can be achieved by making $f'(z_0)$ as large as possible. In doing this, we shall need to be able to extract f as a limit from a given sequence of functions. We turn to this point first.

3.2 Montel's theorem

Let Ω be an open subset of $\mathbb{C}$. A family $\mathcal{F}$ of holomorphic functions on Ω is said to be **normal** if every sequence in $\mathcal{F}$ has a subsequence that converges uniformly on every compact subset of Ω (the limit need not be in $\mathcal{F}$).

The proof that a family of functions is normal is, in practice, the consequence of two related properties, uniform boundedness and equicontinuity. These we shall now define.

The family $\mathcal{F}$ is said to be **uniformly bounded on compact subsets of Ω** if for each compact set $K \subset \Omega$ there exists $B > 0$, such that

$$|f(z)| \leq B \quad \text{for all } z \in K \text{ and } f \in \mathcal{F}.$$

Also, the family $\mathcal{F}$ is **equicontinuous** on a compact set K if for every $\epsilon > 0$ there exists $\delta > 0$ such that whenever $z, w \in K$ and $|z - w| < \delta$, then

$$|f(z) - f(w)| < \epsilon \quad \text{for all } f \in \mathcal{F}.$$

Equicontinuity is a strong condition, which requires uniform continuity, uniformly in the family. For instance, any family of differentiable functions on $[0, 1]$ whose derivatives are uniformly bounded is equicontinuous. This follows directly from the mean value theorem. On the other hand, note that the family $\{f_n\}$ on $[0, 1]$ given by $f_n(x) = x^n$ is *not* equicontinuous since for any fixed $0 < x_0 < 1$ we have $|f_n(1) - f_n(x_0)| \to 1$ as n tends to infinity.

The theorem that follows puts together these new concepts and is an important ingredient in the proof of the Riemann mapping theorem.

Theorem 3.3 *Suppose $\mathcal{F}$ is a family of holomorphic functions on Ω that is uniformly bounded on compact subsets of Ω. Then:*

(i) *$\mathcal{F}$ is equicontinuous on every compact subset of Ω.*

(ii) *$\mathcal{F}$ is a normal family.*

The theorem really consists of two separate parts. The first part says that $\mathcal{F}$ is equicontinuous under the assumption that $\mathcal{F}$ is a family of *holomorphic* functions that is uniformly bounded on compact subsets of Ω. The proof follows from an application of the Cauchy integral formula and hence relies on the fact that $\mathcal{F}$ consists of holomorphic functions. This conclusion is in sharp contrast with the real situation as illustrated by the family of functions given by $f_n(x) = \sin(nx)$ on $(0, 1)$, which is uniformly bounded. However, this family is not equicontinuous and has no convergent subsequence on any compact subinterval of $(0, 1)$.

The second part of the theorem is not complex-analytic in nature. Indeed, the fact that $\mathcal{F}$ is a normal family follows from assuming only that $\mathcal{F}$ is uniformly bounded and equicontinuous on compact subsets of Ω. This result is sometimes known as the Arzela-Ascoli theorem and its proof consists primarily of a diagonalization argument.

We are required to prove convergence on arbitrary compact subsets of Ω, therefore it is useful to introduce the following notion. A sequence $\{K_\ell\}_{\ell=1}^\infty$ of compact subsets of Ω is called an **exhaustion** if

(a) K_ℓ is contained in the interior of $K_{\ell+1}$ for all $\ell = 1, 2, \ldots$.

(b) Any compact set $K \subset \Omega$ is contained in K_ℓ for some ℓ. In particular
$$\Omega = \bigcup_{\ell=1}^\infty K_\ell.$$

Lemma 3.4 *Any open set Ω in the complex plane has an exhaustion.*

Proof. If Ω is bounded, we let K_ℓ denote the set of all points in Ω at distance $\geq 1/\ell$ from the boundary of Ω. If Ω is not bounded, let K_ℓ denote the same set as above except that we also require $|z| \leq \ell$ for all $z \in K_\ell$.

We may now begin the proof of Montel's theorem. Let K be a compact subset of Ω and choose $r > 0$ so small that $D_{3r}(z)$ is contained in Ω for all $z \in K$. It suffices to choose r so that $3r$ is less than the distance from K to the boundary of Ω. Let $z, w \in K$ with $|z - w| < r$, and let γ denote the boundary circle of the disc $D_{2r}(w)$. Then, by Cauchy's integral formula, we have

$$f(z) - f(w) = \frac{1}{2\pi i} \int_\gamma f(\zeta) \left[\frac{1}{\zeta - z} - \frac{1}{\zeta - w} \right] d\zeta.$$

Observe that

$$\left| \frac{1}{\zeta - z} - \frac{1}{\zeta - w} \right| = \frac{|z - w|}{|\zeta - z| \, |\zeta - w|} \leq \frac{|z - w|}{r^2}$$

since $\zeta \in \gamma$ and $|z - w| < r$. Therefore

$$|f(z) - f(w)| \leq \frac{1}{2\pi} \frac{2\pi r}{r^2} B|z - w|,$$

where B denotes the uniform bound for the family $\mathcal{F}$ in the compact set consisting of all points in Ω at a distance $\leq 2r$ from K. Therefore $|f(z) - f(w)| < C|z - w|$, and this estimate is true for all $z, w \in K$ with $|z - w| < r$ and $f \in \mathcal{F}$; thus this family is equicontinuous, as was to be shown.

To prove the second part of the theorem, we argue as follows. Let $\{f_n\}_{n=1}^\infty$ be a sequence in $\mathcal{F}$ and K a compact subset of Ω. Choose a sequence of points $\{w_j\}_{j=1}^\infty$ that is dense in Ω. Since $\{f_n\}$ is uniformly bounded, there exists a subsequence $\{f_{n,1}\} = \{f_{1,1}, f_{2,1}, f_{3,1}, \ldots\}$ of $\{f_n\}$ such that $f_{n,1}(w_1)$ converges.

From $\{f_{n,1}\}$ we can extract a subsequence $\{f_{n,2}\} = \{f_{1,2}, f_{2,2}, f_{3,2}, \ldots\}$ so that $f_{n,2}(w_2)$ converges. We may continue this process, and extract a subsequence $\{f_{n,j}\}$ of $\{f_{n,j-1}\}$ such that $f_{n,j}(w_j)$ converges.

Finally, let $g_n = f_{n,n}$ and consider the diagonal subsequence $\{g_n\}$. By construction, $g_n(w_j)$ converges for each j, and we claim that equicontinuity implies that g_n converges uniformly on K. Given $\epsilon > 0$, choose δ as in the definition of equicontinuity, and note that for some J, the set K is contained in the union of the discs $D_\delta(w_1), \ldots, D_\delta(w_J)$. Pick N so large that if $n, m > N$, then

$$|g_m(w_j) - g_n(w_j)| < \epsilon \quad \text{for all } j = 1, \ldots, J.$$

So if $z \in K$, then $z \in D_\delta(w_j)$ for some $1 \leq j \leq J$. Therefore,

$$|g_n(z) - g_m(z)| \leq |g_n(z) - g_n(w_j)| + |g_n(w_j) - g_m(w_j)| +$$

$$+ |g_m(w_j) - g_m(z)| < 3\epsilon$$

whenever $n, m > N$. Hence $\{g_n\}$ converges uniformly on K.

Finally, we need one more diagonalization argument to obtain a subsequence that converges uniformly on *every* compact subset of Ω. Let $K_1 \subset K_2 \subset \cdots \subset K_\ell \subset \cdots$ be an exhaustion of Ω, and suppose $\{g_{n,1}\}$ is a subsequence of the original sequence $\{f_n\}$ that converges uniformly on K_1. Extract from $\{g_{n,1}\}$ a subsequence $\{g_{n,2}\}$ that converges uniformly on K_2, and so on. Then, $\{g_{n,n}\}$ is a subsequence of $\{f_n\}$ that converges uniformly on every K_ℓ and since the K_ℓ exhaust Ω, the sequence $\{g_{n,n}\}$ converges uniformly on any compact subset of Ω, as was to be shown.

We need one further result before we can give the proof of the Riemann mapping theorem.

Proposition 3.5 *If Ω is a connected open subset of $\mathbb{C}$ and $\{f_n\}$ a sequence of injective holomorphic functions on Ω that converges uniformly*

*on every compact subset of Ω to a holomorphic function f, then f is
either injective or constant.*

Proof. We argue by contradiction and suppose that f is not injective,
so there exist distinct complex numbers z_1 and z_2 in Ω such that $f(z_1) = f(z_2)$. Define a new sequence by $g_n(z) = f_n(z) - f_n(z_1)$, so that g_n has
no other zero besides z_1, and the sequence $\{g_n\}$ converges uniformly on
compact subsets of Ω to $g(z) = f(z) - f(z_1)$. If g is not identically zero,
then z_2 is an isolated zero for g (because Ω is connected); therefore

$$1 = \frac{1}{2\pi i} \int_\gamma \frac{g'(\zeta)}{g(\zeta)}\, d\zeta,$$

where γ is a small circle centered at z_2 chosen so that g does not vanish
on γ or at any point of its interior besides z_2. Therefore, $1/g_n$ converges
uniformly to $1/g$ on γ, and since $g_n' \to g'$ uniformly on γ we have

$$\frac{1}{2\pi i} \int_\gamma \frac{g_n'(\zeta)}{g_n(\zeta)}\, d\zeta \to \frac{1}{2\pi i} \int_\gamma \frac{g'(\zeta)}{g(\zeta)}\, d\zeta.$$

But this is a contradiction since g_n has no zeros inside γ, and hence

$$\frac{1}{2\pi i} \int_\gamma \frac{g_n'(\zeta)}{g_n(\zeta)}\, d\zeta = 0 \quad \text{for all } n.$$

3.3 Proof of the Riemann mapping theorem

Once we have established the technical results above, the rest of the
proof of the Riemann mapping theorem is very elegant. It consists of
three steps, which we isolate.

Step 1. Suppose that Ω is a simply connected proper open subset of
$\mathbb{C}$. We claim that Ω is conformally equivalent to an open subset of the
unit disc that contains the origin. Indeed, choose a complex number α
that does not belong to Ω, (recall that Ω is proper), and observe that
$z - \alpha$ never vanishes on the simply connected set Ω. Therefore, we can
define a holomorphic function

$$f(z) = \log(z - \alpha)$$

with the desired properties of the logarithm. As a consequence one has,
$e^{f(z)} = z - \alpha$, which proves in particular that f is injective. Pick a point
$w \in \Omega$, and observe that

$$f(z) \neq f(w) + 2\pi i \quad \text{for all } z \in \Omega$$

for otherwise, we exponentiate this relation to find that $z = w$, hence $f(z) = f(w)$, a contradiction. In fact, we claim that $f(z)$ stays strictly away from $f(w) + 2\pi i$, in the sense that there exists a disc centered at $f(w) + 2\pi i$ that contains no points of the image $f(\Omega)$. Otherwise, there exists a sequence $\{z_n\}$ in Ω such that $f(z_n) \to f(w) + 2\pi i$. We exponentiate this relation, and, since the exponential function is continuous, we must have $z_n \to w$. But this implies $f(z_n) \to f(w)$, which is a contradiction. Finally, consider the map

$$F(z) = \frac{1}{f(z) - (f(w) + 2\pi i)}.$$

Since f is injective, so is F, hence $F : \Omega \to F(\Omega)$ is a conformal map. Moreover, by our analysis, $F(\Omega)$ is bounded. We may therefore translate and rescale the function F in order to obtain a conformal map from Ω to an open subset of $\mathbb{D}$ that contains the origin.

Step 2. By the first step, we may assume that Ω is an open subset of $\mathbb{D}$ with $0 \in \Omega$. Consider the family $\mathcal{F}$ of all injective holomorphic functions on Ω that map into the unit disc and fix the origin:

$$\mathcal{F} = \{f : \Omega \to \mathbb{D} \text{ holomorphic, injective and } f(0) = 0\}.$$

First, note that $\mathcal{F}$ is non-empty since it contains the identity. Also, this family is uniformly bounded by construction, since all functions are required to map into the unit disc.

Now, we turn to the question of finding a function $f \in \mathcal{F}$ that maximizes $|f'(0)|$. First, observe that the quantities $|f'(0)|$ are uniformly bounded as f ranges in $\mathcal{F}$. This follows from the Cauchy inequality (Corollary 4.3 in Chapter 2) for f' applied to a small disc centered at the origin.

Next, we let

$$s = \sup_{f \in \mathcal{F}} |f'(0)|,$$

and we choose a sequence $\{f_n\} \subset \mathcal{F}$ such that $|f_n'(0)| \to s$ as $n \to \infty$. By Montel's theorem (Theorem 3.3), this sequence has a subsequence that converges uniformly on compact sets to a holomorphic function f on Ω. Since $s \geq 1$ (because $z \mapsto z$ belongs to $\mathcal{F}$), f is non-constant, hence injective, by Proposition 3.5. Also, by continuity we have $|f(z)| \leq 1$ for all $z \in \Omega$ and from the maximum modulus principle we see that $|f(z)| < 1$. Since we clearly have $f(0) = 0$, we conclude that $f \in \mathcal{F}$ with $|f'(0)| = s$.

Step 3. In this last step, we demonstrate that f is a conformal map from Ω to $\mathbb{D}$. Since f is already injective, it suffices to prove that f is also surjective. If this were not true, we could construct a function in $\mathcal{F}$ with derivative at 0 greater than s. Indeed, suppose there exists $\alpha \in \mathbb{D}$ such that $f(z) \neq \alpha$, and consider the automorphism ψ_α of the disc that interchanges 0 and α, namely

$$\psi_\alpha(z) = \frac{\alpha - z}{1 - \overline{\alpha}z}.$$

Since Ω is simply connected, so is $U = (\psi_\alpha \circ f)(\Omega)$, and moreover, U does not contain the origin. It is therefore possible to define a square root function on U by

$$g(w) = e^{\frac{1}{2} \log w}.$$

Next, consider the function

$$F = \psi_{g(\alpha)} \circ g \circ \psi_\alpha \circ f.$$

We claim that $F \in \mathcal{F}$. Clearly F is holomorphic and it maps 0 to 0. Also F maps into the unit disc since this is true of each of the functions in the composition. Finally, F is injective. This is clearly true for the automorphisms ψ_α and $\psi_{g(\alpha)}$; it is also true for the square root g and the function f, since the latter is injective by assumption. If h denotes the square function $h(w) = w^2$, then we must have

$$f = \psi_\alpha^{-1} \circ h \circ \psi_{g(\alpha)}^{-1} \circ F = \Phi \circ F.$$

But Φ maps $\mathbb{D}$ into $\mathbb{D}$ with $\Phi(0) = 0$, and is not injective because F is and h is not. By the last part of the Schwarz lemma, we conclude that $|\Phi'(0)| < 1$. The proof is complete once we observe that

$$f'(0) = \Phi'(0)F'(0),$$

and thus

$$|f'(0)| < |F'(0)|,$$

contradicting the maximality of $|f'(0)|$ in $\mathcal{F}$.

Finally, we multiply f by a complex number of absolute value 1 so that $f'(0) > 0$, which ends the proof.

For a variant of this proof, see Problem 7.

Remark. It is worthwhile to point out that the only places where the hypothesis of simple-connectivity entered in the proof were in the uses of the logarithm and the square root. Thus it would have sufficed to have assumed (in addition to the hypothesis that Ω is proper) that Ω is **holomorphically simply connected** in the sense that for any holomorphic function f in Ω and any closed curve γ in Ω, we have $\int_{\gamma} f(z)\,dz = 0$. Further discussion of this point, and various equivalent properties of simple-connectivity, are given in Appendix B.

4 Conformal mappings onto polygons

The Riemann mapping theorem guarantees the existence of a conformal map from any proper, simply connected open set to the disc, or equivalently to the upper half-plane, but this theorem gives little insight as to the exact form of this map. In Section 1 we gave various explicit formulas in the case of regions that have symmetries, but it is of course unreasonable to ask for an explicit formula in the general case. There is, however, another class of open sets for which there are nice formulas, namely the polygons. Our aim in this last section is to give a proof of the Schwarz-Christoffel formula, which describes the nature of conformal maps from the disc (or upper half-plane) to polygons.

4.1 Some examples

We begin by studying some motivating examples. The first two correspond to easy (but infinite and degenerate) cases.

EXAMPLE 1. First, we investigate the conformal map from the upper half-plane to the sector $\{z : 0 < \arg z < \alpha\pi\}$, with $0 < \alpha < 2$, given in Section 1 by $f(z) = z^{\alpha}$. Anticipating the Schwarz-Christoffel formula below, we write

$$z^{\alpha} = f(z) = \int_{0}^{z} f'(\zeta)\,d\zeta = \alpha \int_{0}^{z} \zeta^{-\beta}\,d\zeta$$

with $\alpha + \beta = 1$, and where the integral is taken along any path in the upper half-plane. In fact, by continuity and Cauchy's theorem, we may take the path of integration to lie in the closure of the upper half-plane. Although the behavior of f follows immediately from the original definition, we study it in terms of the integral expression above, since this provides insight for the general case treated later.

   Note first that $\zeta^{-\beta}$ is integrable near 0 since $\beta < 1$, therefore $f(0) = 0$. Observe that when z is real and positive ($z = x$), then $f'(x) = \alpha x^{\alpha-1}$ is positive; also it is *not* finitely integrable at ∞. Therefore, as x travels from 0 to ∞, we see that $f(x)$ increases from 0 to ∞, thus f maps $[0, \infty)$ to $[0, \infty)$. On the other hand, when $z = x$ is negative, then

$$f'(z) = \alpha |x|^{\alpha-1} e^{i\pi(\alpha-1)} = -\alpha |x|^{\alpha-1} e^{i\pi\alpha},$$

so f maps the segment $(-\infty, 0]$ to $(e^{i\pi\alpha}\infty, 0]$. The situation is illustrated in Figure 3 where the infinite segment A is mapped to A' and the segment B is mapped to B', with the direction of travel indicated in Figure 3.

Figure 3. The conformal map z^α

EXAMPLE 2. Next, we consider for $z \in \mathbb{H}$,

$$f(z) = \int_0^z \frac{d\zeta}{(1-\zeta^2)^{1/2}},$$

where the integral is taken from 0 to z along any path in the closed upper half-plane. We choose the branch for $(1 - \zeta^2)^{1/2}$ that makes it holomorphic in the upper half-plane and positive when $-1 < \zeta < 1$. As a result

$$(1-\zeta^2)^{-1/2} = i(\zeta^2 - 1)^{-1/2} \quad \text{when } \zeta > 1.$$

We observe that f maps the real line to the boundary of the half-strip pictured in Figure 4.

In fact, since $f(\pm 1) = \pm\pi/2$, and $f'(x) > 0$ if $-1 < x < 1$, we see that f maps the segment B to B'. Moreover

$$f(x) = \frac{\pi}{2} + \int_1^x f'(x)\,dx \quad \text{when } x > 1, \quad \text{and} \quad \int_1^\infty \frac{dx}{(x^2-1)^{1/2}} = \infty.$$

Thus, as x travels along the segment C, the image traverses the infinite segment C'. Similarly segment A is mapped to A'.

Note the connection of this example with Example 8 in Section 1.2. In fact, one can show that the function $f(z)$ is the inverse to the function

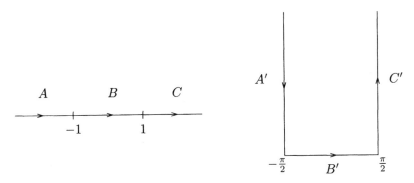

Figure 4. Mapping of the boundary in Example 2

$\sin z$, and hence f takes $\mathbb{H}$ conformally to the interior of the half-strip bounded by the segments $A', B',$ and C'.

EXAMPLE 3. Here we take

$$f(z) = \int_0^z \frac{d\zeta}{[(1 - \zeta^2)(1 - k^2\zeta^2)]^{1/2}}, \quad z \in \mathbb{H},$$

where k is a fixed real number with $0 < k < 1$ (the branch of $[(1 - \zeta^2)(1 - k^2\zeta^2)]^{1/2}$ in the upper half-plane is chosen to be the one that is positive when ζ is real and $-1 < \zeta < 1$). Integrals of this kind are called **elliptic integrals**, because variants of these arise in the calculation of the arc-length of an ellipse. We shall observe that f maps the real axis onto the rectangle shown in Figure 5(b), where K and K' are determined by

$$K = \int_0^1 \frac{dx}{[(1 - x^2)(1 - k^2x^2)]^{1/2}}, \quad K' = \int_1^{1/k} \frac{dx}{[(x^2 - 1)(1 - k^2x^2)]^{1/2}}.$$

We divide the real axis into four "segments," with division points $-1/k,\ -1,\ 1,$ and $1/k$ (see Figure 5(a)). The segments are $[-1/k, -1]$, $[-1, 1]$, $[1, 1/k]$, and $[1/k, -1/k]$, the last consisting of the union of the two half-segments $[1/k, \infty)$ and $(-\infty, -1/k]$. It is clear from the definitions that $f(\pm 1) = \pm K$, and since $f'(x) > 0$, when $-1 < x < 1$, it follows that f maps the segment $[-1, 1]$ to $[-K, K]$. Moreover, since

$$f(z) = K + \int_1^x \frac{d\zeta}{[(1 - \zeta^2)(1 - k^2\zeta^2)]^{1/2}} \quad \text{if } 1 < x < 1/k,$$

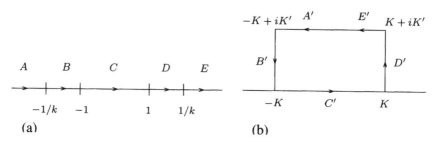

Figure 5. Mapping of the boundary in Example 3

we see that f maps the segment $[1, 1/k]$ to $[K, K + iK']$, where K' was defined above. Similarly, f maps $[-1/k, -1]$ to $[-K + iK', -K]$. Next, when $x > 1/k$ we have

$$f'(x) = -\frac{1}{[(x^2 - 1)(k^2 x^2 - 1)]^{1/2}},$$

and therefore,

$$f(x) = K + iK' - \int_{1/k}^{x} \frac{dx}{[(x^2 - 1)(k^2 x^2 - 1)]^{1/2}}.$$

However,

$$\int_{1/k}^{\infty} \frac{dx}{[(x^2 - 1)(k^2 x^2 - 1)]^{1/2}} = \int_{0}^{1} \frac{dx}{[(1 - x^2)(1 - k^2 x^2)]^{1/2}},$$

as can be seen by making the change of variables $x = 1/ku$ in the integral on the left. Thus f maps the segment $[1/k, \infty)$ to the segment $[K + iK', iK')$. Similarly f maps $(-\infty, -1/k]$ to $[-K + iK', iK')$. Altogether, then, f maps the real axis to the above rectangle, with the point at infinity corresponding to the mid-point of the upper side of the rectangle.

The results obtained so far lead naturally to two problems.

The first, which we pursue next, consists of a generalization of the above examples. More precisely we define the Schwarz-Christoffel integral and prove that it maps the real line to a polygonal line.

Second, we note that in the examples above little was inferred about the behavior of f in $\mathbb{H}$ itself. In particular, we have not shown that f maps $\mathbb{H}$ conformally to the interior of the corresponding polygon. After a careful study of the boundary behavior of conformal maps, we prove a theorem that guarantees that the conformal map from the upper half-plane to a simply connected region bounded by a polygonal line is essentially given by a Schwarz-Christoffel integral.

4.2 The Schwarz-Christoffel integral

With the examples of the previous section in mind, we define the general **Schwarz-Christoffel integral** by

$$(5) \qquad S(z) = \int_0^z \frac{d\zeta}{(\zeta - A_1)^{\beta_1} \cdots (\zeta - A_n)^{\beta_n}}.$$

Here $A_1 < A_2 < \cdots < A_n$ are n distinct points on the real axis arranged in increasing order. The exponents β_k will be assumed to satisfy the conditions $\beta_k < 1$ for each k and $1 < \sum_{k=1}^n \beta_k$.[6]

The integrand in (5) is defined as follows: $(z - A_k)^{\beta_k}$ is that branch (defined in the complex plane slit along the infinite ray $\{A_k + iy : y \leq 0\}$) which is positive when $z = x$ is real and $x > A_k$. As a result

$$(z - A_k)^{\beta_k} = \begin{cases} (x - A_k)^{\beta_k} & \text{if } x \text{ is real and } x > A_k, \\ |x - A_k|^{\beta_k} e^{i\pi\beta_k} & \text{if } x \text{ is real and } x < A_k. \end{cases}$$

The complex plane slit along the union of the rays $\cup_{k=1}^n \{A_k + iy : y \leq 0\}$ is simply connected (see Exercise 19), so the integral that defines $S(z)$ is holomorphic in this open set. Since the requirement $\beta_k < 1$ implies that the singularities $(\zeta - A_k)^{-\beta_k}$ are integrable near A_k, the function S is continuous up to the real line, including the points A_k, with $k = 1, \ldots, n$. Finally, this continuity condition implies that the integral can be taken along any path in the complex plane that avoids the union of the *open* slits $\cup_{k=1}^n \{A_k + iy : y < 0\}$.

Now

$$\left| \prod_{k=1}^n (\zeta - A_k)^{-\beta_k} \right| \leq c|\zeta|^{-\sum \beta_k}$$

for $|\zeta|$ large, so the assumption $\sum \beta_k > 1$ guarantees the convergence of the integral (5) at infinity. This fact and Cauchy's theorem imply that $\lim_{r \to \infty} S(re^{i\theta})$ exists and is independent of the angle θ, $0 \leq \theta \leq \pi$. We call this limit a_∞, and we let $a_k = S(A_k)$ for $k = 1, \ldots, n$.

---

[6]Note that the case $\sum \beta_k \leq 1$, which occurs in Examples 1 and 2 above is excluded. However, a modification of the proposition that follows can be made to take these cases into account; but then $S(z)$ is no longer bounded in the upper half-plane.

Proposition 4.1 *Suppose $S(z)$ is given by (5).*

(i) *If $\sum_{k=1}^{n} \beta_k = 2$, and $\mathfrak{p}$ denotes the polygon whose vertices are given (in order) by $a_1, \ldots, a_n$, then S maps the real axis onto $\mathfrak{p} - \{a_\infty\}$. The point a_∞ lies on the segment $[a_n, a_1]$ and is the image of the point at infinity. Moreover, the (interior) angle at the vertex a_k is $\alpha_k \pi$ where $\alpha_k = 1 - \beta_k$.*

(ii) *There is a similar conclusion when $1 < \sum_{k=1}^{n} \beta_k < 2$, except now the image of the extended line is the polygon of $n+1$ sides with vertices $a_1, a_2, \ldots, a_n, a_\infty$. The angle at the vertex a_∞ is $\alpha_\infty \pi$ with $\alpha_\infty = 1 - \beta_\infty$, where $\beta_\infty = 2 - \sum_{k=1}^{n} \beta_k$.*

Figure 6 illustrates the proposition. The idea of the proof is already captured in Example 1 above.

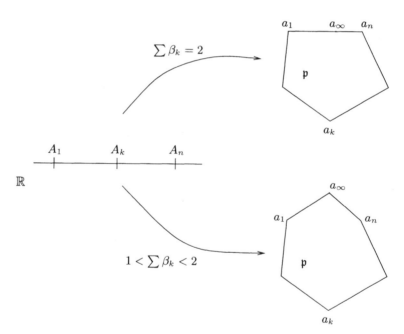

Figure 6. Action of the integral $S(z)$

Proof. We assume that $\sum_{k=1}^{n} \beta_k = 2$. If $A_k < x < A_{k+1}$ when $1 \le k \le n-1$, then

$$S'(x) = \prod_{j \le k} (x - A_j)^{-\beta_j} \prod_{j > k} (x - A_j)^{-\beta_j}.$$

Hence

$$\arg S'(x) = \arg\left(\prod_{j>k}(x - A_j)^{-\beta_j}\right) = \arg\prod_{j>k} e^{-i\pi\beta_j} = -\pi\sum_{j>k}\beta_j,$$

which of course is constant when x traverses the interval (A_k, A_{k+1}). Since

$$S(x) = S(A_k) + \int_{A_k}^{x} S'(y)\,dy,$$

we see that as x varies from A_k to A_{k+1}, $S(x)$ varies from $S(A_k) = a_k$ to $S(A_{k+1}) = a_{k+1}$ along the straight line segment[7] $[a_k, a_{k+1}]$, and this makes an angle of $-\pi\sum_{j>k}\beta_j$ with the real axis. Similarly, when $A_n < x$ then $S'(x)$ is positive, while if $x < A_1$, the argument of $S'(x)$ is $-\pi\sum_{k=1}^{n}\beta_k = -2\pi$, and so $S'(x)$ is again positive. Thus as x varies on $[A_n, +\infty)$, $S(x)$ varies along a straight line (parallel to the x-axis) between a_n and a_∞; similarly $S(x)$ varies along a straight line (parallel to that axis) between a_∞ and a_1 as x varies in $(-\infty, A_1]$. Moreover, the union of $[a_n, a_\infty)$ and $(a_\infty, a_1]$ is the segment $[a_n, a_1]$ with the point a_∞ removed.

Now the increase of the angle of $[a_{k+1}, a_k]$ over that of $[a_{k-1}, a_k]$ is $\pi\beta_k$, which means that the angle at the vertex a_k is $\pi\alpha_k$. The proof when $1 < \sum_{k=1}^{n}\beta_k < 2$ is similar, and is left to the reader.

As elegant as this proposition is, it does not settle the problem of finding a conformal map from the half-plane to a given region P that is bounded by a polygon. There are two reasons for this.

1. It is not true for general n and generic choices of $A_1, \ldots, A_n$, that the polygon (which is the image of the real axis under S) is simple, that is, it does not cross itself. Nor is it true in general that the mapping S is conformal on the upper half-plane.

2. Neither does the proposition show that starting with a simply connected region P (whose boundary is a polygonal line $\mathfrak{p}$) the mapping S is, for certain choices of $A_1, \ldots, A_n$ and simple modifications, a conformal map from $\mathbb{H}$ to P. That however is the case, and is the result whose proof we now turn to.

---

[7]We denote the closed straight line segment between two complex numbers z and w by $[z, w]$, that is, $[z, w] = \{(1-t)z + tw : t \in [0,1]\}$. If we restrict $0 < t < 1$, then (z, w) denotes the open line segment between z and w. Similarly for the half-open segments $[z, w)$ and $(z, w]$ obtained by restricting $0 \le t < 1$ and $0 < t \le 1$, respectively.

4.3 Boundary behavior

In what follows we shall consider a **polygonal region** P, namely a bounded, simply connected open set whose boundary is a polygonal line $\mathfrak{p}$. In this context, we always assume that the polygonal line is closed, and we sometimes refer to $\mathfrak{p}$ as a **polygon**.

To study conformal maps from the half-plane $\mathbb{H}$ to P, we consider first the conformal maps from the disc $\mathbb{D}$ to P, and their boundary behavior.

Theorem 4.2 *If $F : \mathbb{D} \to P$ is a conformal map, then F extends to a continuous bijection from the closure $\overline{\mathbb{D}}$ of the disc to the closure $\overline{P}$ of the polygonal region. In particular, F gives rise to a bijection from the boundary of the disc to the boundary polygon $\mathfrak{p}$.*

The main point consists in showing that if z_0 belongs to the unit circle, then $\lim_{z \to z_0} F(z)$ exists. To prove this, we need a preliminary result, which uses the fact that if $f : U \to f(U)$ is conformal, then

$$\text{Area}(f(U)) = \iint_U |f'(z)|^2 \, dx \, dy.$$

This assertion follows from the definition, $\text{Area}(f(U)) = \iint_{f(U)} dx \, dy$, and the fact that the determinant of the Jacobian in the change of variables $w = f(z)$ is simply $|f'(z)|^2$, an observation we made in equation (4), Section 2.2, Chapter 1.

Lemma 4.3 *For each $0 < r < 1/2$, let C_r denote the circle centered at z_0 of radius r. Suppose that for all sufficiently small r we are given two points z_r and z'_r in the unit disc that also lie on C_r. If we let $\rho(r) = |f(z_r) - f(z'_r)|$, then there exists a sequence $\{r_n\}$ of radii that tends to zero, and such that $\lim_{n \to \infty} \rho(r_n) = 0$.*

Proof. If not, there exist $0 < c$ and $0 < R < 1/2$ such that $c \leq \rho(r)$ for all $0 < r \leq R$. Observe that

$$f(z_r) - f(z'_r) = \int_\alpha f'(\zeta) \, d\zeta,$$

where the integral is taken over the arc α on C_r that joins z_r and z'_r in $\mathbb{D}$. If we parametrize this arc by $z_0 + re^{i\theta}$ with $\theta_1(r) \leq \theta \leq \theta_2(r)$, then

$$\rho(r) \leq \int_{\theta_1(r)}^{\theta_2(r)} |f'(z)| r \, d\theta.$$

We now apply the Cauchy-Schwarz inequality to see that

$$\rho(r) \leq \left(\int_{\theta_1(r)}^{\theta_2(r)} |f'(z)|^2 r \, d\theta \right)^{1/2} \left(\int_{\theta_1(r)}^{\theta_2(r)} r \, d\theta \right)^{1/2}.$$

Squaring both sides and dividing by r yields

$$\frac{\rho(r)^2}{r} \leq 2\pi \int_{\theta_1(r)}^{\theta_2(r)} |f'(z)|^2 r \, d\theta.$$

We may now integrate both sides from 0 to R, and since $c \leq \rho(r)$ on that region we obtain

$$c^2 \int_0^R \frac{dr}{r} \leq 2\pi \int_0^R \int_{\theta_1(r)}^{\theta_2(r)} |f'(z)|^2 r \, d\theta dr \leq 2\pi \int\int_{\mathbb{D}} |f'(z)|^2 \, dx dy.$$

Now the left-hand side is infinite because $1/r$ is not integrable near the origin, and the right-hand side is bounded because the area of the polygonal region is bounded, so this yields the desired contradiction and concludes the proof of the lemma.

Lemma 4.4 *Let z_0 be a point on the unit circle. Then $F(z)$ tends to a limit as z approaches z_0 within the unit disc.*

Proof. If not, there are two sequences $\{z_1, z_2, \ldots\}$ and $\{z_1', z_2', \ldots\}$ in the unit disc that converge to z_0 and are so that $F(z_n)$ and $F(z_n')$ converge to two distinct points ζ and ζ' in the closure of P. Since F is conformal, the points ζ and ζ' must lie on the boundary $\mathfrak{p}$ of P. We may therefore choose two disjoint discs D and D' centered at ζ and ζ', respectively, that are at a distance $d > 0$ from each other. For all large n, $F(z_n) \in D$ and $F(z_n') \in D'$. Therefore, there exist two continuous curves[8] Λ and Λ' in $D \cap P$ and $D' \cap P$, respectively, with $F(z_n) \in \Lambda$ and $F(z_n') \in \Lambda'$ for all large n, and with the end-points of Λ and Λ' equal to ζ and ζ', respectively.

Define $\lambda = F^{-1}(\Lambda)$ and $\lambda' = F^{-1}(\Lambda')$. Then λ and λ' are two continuous curves in $\mathbb{D}$. Moreover, both λ and λ' contain infinitely many points in each sequence $\{z_n\}$ and $\{z_n'\}$. Recall that these sequences converge to z_0. By continuity, the circle C_r centered at z_0 and of radius r will intersect λ and λ' for all small r, say at some points $z_r \in \lambda$ and $z_r' \in \lambda'$. This

---

[8]By a continuous curve, we mean the image of a continuous (not necessarily piecewise-smooth) function from a closed interval $[a, b]$ to $\mathbb{C}$.

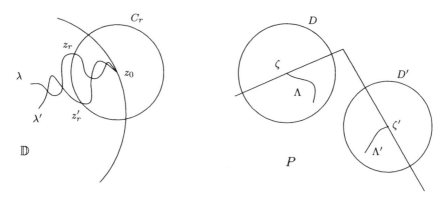

Figure 7. Illustration for the proof of Lemma 4.4

contradicts the previous lemma, because $|F(z_r) - F(z_r')| > d$. Therefore $F(z)$ converges to a limit on $\mathfrak{p}$ as z approaches z_0 from within the unit disc, and the proof is complete.

Lemma 4.5 *The conformal map F extends to a continuous function from the closure of the disc to the closure of the polygon.*

Proof. By the previous lemma, the limit

$$\lim_{z \to z_0} F(z)$$

exists, and we define $F(z_0)$ to be the value of this limit. There remains to prove that F is continuous on the closure of the unit disc. Given ϵ, there exists δ such that whenever $z \in \mathbb{D}$ and $|z - z_0| < \delta$, then $|F(z) - F(z_0)| < \epsilon$. Now if z belongs to the boundary of $\mathbb{D}$ and $|z - z_0| < \delta$, then we may choose w such that $|F(z) - F(w)| < \epsilon$ and $|w - z_0| < \delta$. Therefore

$$|F(z) - F(z_0)| \leq |F(z) - F(w)| + |F(w) - F(z_0)| < 2\epsilon,$$

and the lemma is established.

We may now complete the proof of the theorem. We have shown that F extends to a continuous function from $\overline{\mathbb{D}}$ to $\overline{P}$. The previous argument can be applied to the inverse G of F. Indeed, the key geometric property of the unit disc that we used was that if z_0 belongs to the boundary of $\mathbb{D}$, and C is any small circle centered at z_0, then $C \cap \mathbb{D}$ consists of an arc. Clearly, this property also holds at every boundary point of the polygonal region P. Therefore, G also extends to a continuous function from $\overline{P}$ to $\overline{\mathbb{D}}$. It suffices to now prove that the extensions of F and G are inverses of each other. If $z \in \partial \mathbb{D}$ and $\{z_k\}$ is a sequence in the disc that converges to z, then $G(F(z_k)) = z_k$, so after taking the limit and using the fact that F is continuous, we conclude that $G(F(z)) = z$ for all $z \in \overline{\mathbb{D}}$. Similarly, $F(G(w)) = w$ for all $w \in \overline{P}$, and the theorem is proved.

The circle of ideas used in this proof can be used to prove more general theorems on the boundary continuity of conformal maps. See Exercise 18 and Problem 6 below.

4.4 The mapping formula

Suppose P is a polygonal region bounded by a polygon $\mathfrak{p}$ whose vertices are ordered consecutively $a_1, a_2, \ldots, a_n$, and with $n \geq 3$. We denote by $\pi \alpha_k$ the interior angle at a_k, and define the exterior angle $\pi \beta_k$ by $\alpha_k + \beta_k = 1$. A simple geometric argument provides $\sum_{k=1}^{n} \beta_k = 2$.

We shall consider conformal mappings of the half-plane $\mathbb{H}$ to P, and make use of the results of the previous section regarding conformal maps from the disc $\mathbb{D}$ to P. The standard correspondences $w = (i - z)/(i + z)$, $z = i(1 - w)/(1 + w)$ allows us to go back and forth between $z \in \mathbb{H}$ and $w \in \mathbb{D}$. Notice that the boundary point $w = -1$ of the circle corresponds to the point at infinity on the line, and so the conformal map of $\mathbb{H}$ to $\mathbb{D}$ extends to a continuous bijection of the boundary of $\mathbb{H}$, which for the purpose of this discussion includes the point at infinity.

Let F be a conformal map from $\mathbb{H}$ to P. (Its existence is guaranteed by the Riemann mapping theorem and the previous discussion.) We assume first that none of the vertices of $\mathfrak{p}$ correspond to the point at infinity. Therefore, there are real numbers $A_1, A_2, \ldots, A_n$ so that $F(A_k) = a_k$ for all k. Since F is continuous and injective, and the vertices are numbered consecutively, we may conclude that the A_k's are in either increasing or decreasing order. After relabeling the vertices a_k and the points A_k, we may assume that $A_1 < A_2 < \cdots < A_n$. These points divide the real line into $n - 1$ segments $[A_k, A_{k+1}]$, $1 \leq k \leq n - 1$, and the segment that consists of the join of the two half-segments $(-\infty, A_1] \cup [A_n, \infty)$. These are mapped bijectively onto the corresponding sides of the polygon, that is, the segments $[a_k, a_{k+1}]$, $1 \leq k \leq n - 1$, and $[a_n, a_1]$ (see Figure 8).

Theorem 4.6 *There exist complex numbers c_1 and c_2 so that the conformal map F of $\mathbb{H}$ to P is given by*

$$F(z) = c_1 S(z) + c_2$$

where S is the Schwarz-Christoffel integral introduced in Section 4.2.

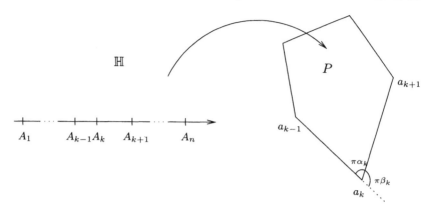

Figure 8. The mapping F

Proof. We first consider z in the upper half-plane lying above the two adjacent segments $[A_{k-1}, A_k]$ and $[A_k, A_{k+1}]$, where $1 < k < n$. We note that F maps these two segments to two segments that intersect at $a_k = F(A_k)$ at an angle $\pi\alpha_k$.

By choosing a branch of the logarithm we can in turn define

$$h_k(z) = (F(z) - a_k)^{1/\alpha_k}$$

for all z in the half-strip in the upper half-plane bounded by the lines $\operatorname{Re}(z) = A_{k-1}$ and $\operatorname{Re}(z) = A_{k+1}$. Since F continues to the boundary of $\mathbb{H}$, the map h_k is actually continuous up to the segment (A_{k-1}, A_{k+1}) on the real line. By construction h_k will map the segment $[A_{k-1}, A_{k+1}]$ to a (straight) segment L_k in the complex plane, with A_k mapped to 0. We may therefore apply the Schwarz reflection principle to see that h_k is analytically continuable to a holomorphic function in the two-way infinite strip $A_{k-1} < \operatorname{Re}(z) < A_{k+1}$ (see Figure 9). We claim that h_k' never vanishes in that strip. First, if z belongs to the open upper half-strip, then

$$\frac{F'(z)}{F(z) - F(A_k)} = \alpha_k \frac{h_k'(z)}{h_k(z)},$$

and since F is conformal, we have $F'(z) \neq 0$ so $h_k'(z) \neq 0$ (Proposition 1.1). By reflection, this also holds in the lower half-strip, and it remains to investigate points on the segment (A_{k-1}, A_{k+1}). If $A_{k-1} < x < A_{k+1}$, we note that the image under h_k of a small half-disc centered at x and contained in $\mathbb{H}$ lies on one side of the straight line segment

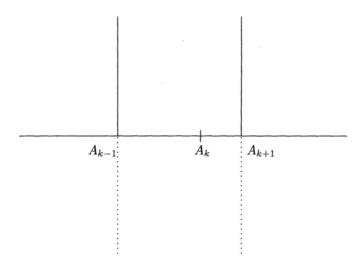

Figure 9. Schwarz reflection

L_k. Since h_k is injective up to L_k (because F is) the symmetry in the Schwarz reflection principle guarantees that h_k is injective in the whole disc centered at x, whence $h'_k(x) \neq 0$, whence $h'_k(z) \neq 0$ for all z in the strip $A_{k-1} < \operatorname{Re}(z) < A_{k+1}$.

Now because $F' = \alpha_k h_k^{-\beta_k} h'_k$ and $F'' = -\beta_k \alpha_k h_k^{-\beta_k - 1}(h'_k)^2 + \alpha_k h_k^{-\beta_k} h''_k$, the fact that $h'_k(z) \neq 0$ implies that

$$\frac{F''(z)}{F'(z)} = \frac{-\beta_k}{z - A_k} + E_k(z),$$

where E_k is holomorphic in the strip $A_{k-1} < \operatorname{Re}(z) < A_{k+1}$. A similar result holds for $k = 1$ and $k = n$, namely

$$\frac{F''(z)}{F'(z)} = -\frac{\beta_1}{z - A_1} + E_1,$$

where E_1 is holomorphic in the strip $-\infty < \operatorname{Re}(z) < A_2$, and

$$\frac{F''(z)}{F'(z)} = -\frac{\beta_n}{z - A_n} + E_n,$$

where E_n is holomorphic in the strip $A_{n-1} < \operatorname{Re}(z) < \infty$. Finally, another application of the reflection principle shows that F is continuable in the exterior of a disc $|z| \leq R$, for large R (say $R > \max_{1 \leq k \leq n} |A_k|$). Indeed, we may continue F across the union of the segments

$(-\infty, A_1) \cup (A_n, \infty)$ since their image under F is a straight line segment and Schwarz reflection applies. The fact that F maps the upper half-plane to a bounded region shows that the analytic continuation of F outside a large disc is also bounded, and hence holomorphic at infinity. Thus F''/F' is holomorphic at infinity and we claim that it goes to 0 as $|z| \to \infty$. Indeed, we may expand F at $z = \infty$ as

$$F(z) = c_0 + \frac{c_1}{z} + \frac{c_2}{z^2} + \cdots .$$

This after differentiation shows that F''/F' decays like $1/z$ as $|z|$ becomes large, and proves our claim.

Altogether then, because the various strips overlap and cover the entire complex plane,

$$\frac{F''(z)}{F'(z)} + \sum_{k=1}^{n} \frac{\beta_k}{z - A_k}$$

is holomorphic in the entire plane and vanishes at infinity; thus, by Liouville's theorem it is zero. Hence

$$\frac{F''(z)}{F'(z)} = -\sum_{k=1}^{n} \frac{\beta_k}{z - A_k}.$$

From this we contend that $F'(z) = c(z - A_1)^{-\beta_1} \cdots (z - A_n)^{-\beta_n}$. Indeed, denoting this product by $Q(z)$, we have

$$\frac{Q'(z)}{Q(z)} = -\sum_{k=1}^{n} \frac{\beta_k}{z - A_k}.$$

Therefore

$$\frac{d}{dz} \left(\frac{F'(z)}{Q(z)} \right) = 0,$$

which proves the contention. A final integration yields the theorem.

We may now withdraw the hypothesis we made at the beginning that F did not map the point at infinity to a vertex of P, and obtain a formula for that case as well.

Theorem 4.7 *If F is a conformal map from the upper half-plane to the polygonal region P and maps the points $A_1, \ldots, A_{n-1}, \infty$ to the vertices of p, then there exist constants C_1 and C_2 such that*

$$F(z) = C_1 \int_0^z \frac{d\zeta}{(\zeta - A_1)^{\beta_1} \cdots (\zeta - A_{n-1})^{\beta_{n-1}}} + C_2.$$

In other words, the formula is obtained by deleting the last term in the Schwarz-Christoffel integral (5).

Proof. After a preliminary translation, we may assume that $A_j \neq 0$ for $j = 1, \ldots, n-1$. Choose a point $A_n^* > 0$ on the real line, and consider the fractional linear map defined by

$$\Phi(z) = A_n^* - \frac{1}{z}.$$

Then Φ is an automorphism of the upper half-plane. Let $A_k^* = \Phi(A_k)$ for $k = 1, \ldots, n-1$, and note that $A_n^* = \Phi(\infty)$. Then

$$(F \circ \Phi^{-1})(A_k^*) = a_k \quad \text{for all } k = 1, 2, \ldots, n.$$

We can now apply the Schwarz-Christoffel formula just proved to find that

$$(F \circ \Phi^{-1})(z') = C_1 \int_0^{z'} \frac{d\zeta}{(\zeta - A_1^*)^{\beta_1} \cdots (\zeta - A_n^*)^{\beta_n}} + C_2.$$

The change of variables $\zeta = \Phi(w)$ satisfies $d\zeta = dw/w^2$, and since we can write $2 = \beta_1 + \cdots + \beta_n$, we obtain

$$(F \circ \Phi^{-1})(z') = C_1 \int_0^{\Phi^{-1}(z')} \frac{dw}{(w(A_n^* - A_1^*) - 1)^{\beta_1} \cdots (w(A_n^* - A_{n-1}^*) - 1)^{\beta_{n-1}}} + C_2'$$

$$= C_1' \int_0^{\Phi^{-1}(z')} \frac{dw}{(w - 1/(A_n^* - A_1^*))^{\beta_1} \cdots (w - 1/(A_n^* - A_{n-1}^*))^{\beta_{n-1}}} + C_2'.$$

Finally, we note that $1/(A_n^* - A_k^*) = A_k$ and set $\Phi^{-1}(z') = z$ in the above equation to conclude that

$$F(z) = C_1' \int_0^z \frac{dw}{(w - A_1)^{\beta_1} \cdots (w - A_{n-1})^{\beta_{n-1}}} + C_2',$$

as was to be shown.

4.5 Return to elliptic integrals

We consider again the elliptic integral

$$I(z) = \int_0^z \frac{d\zeta}{[(1 - \zeta^2)(1 - k^2\zeta^2)]^{1/2}} \quad \text{with } 0 < k < 1,$$

which arose in Example 3 of Section 4.1. We saw that it mapped the real axis to the rectangle R with vertices $-K$, K, $K + iK'$, and $-K + iK'$. We will now see that this mapping is a conformal mapping of $\mathbb{H}$ to the interior of R.

According to Theorem 4.6 there is a conformal map F to the rectangle, that maps four points on the real axis to the vertices of R. By preceding this map with a suitable automorphism of $\mathbb{H}$ we may assume that F maps -1, 0, 1 to $-K$, 0, K, respectively. Indeed, by using a preliminary automorphism, we may assume that $-K$, 0, K are the images of points A_1, 0, A_2 with $A_1 < 0 < A_2$; then we can further take $A_1 = -1$ and $A_2 = 1$. See Exercise 15.

Next, let ℓ be chosen with $0 < \ell < 1$, so that $1/\ell$ is the point on the real line mapped by F to the vertex $K + iK'$, which is the vertex next in order after $-K$ and K. We claim that $F(-1/\ell)$ is the vertex $-K + iK'$. Indeed, if $F^*(z) = -F(-\bar{z})$, then by the symmetry of R, F^* is also a conformal map of $\mathbb{H}$ to R; moreover $F^*(0) = 0$, and $F^*(\pm 1) = \pm K$. Thus $F^{-1} \circ F^*$ is an automorphism of $\mathbb{H}$ that fixes the points -1, 0, and 1. Hence $F^{-1} \circ F^*$ is the identity (see Exercise 15), and $F = F^*$, from which it follows that

$$F(-1/\ell) = -\overline{F}(1/\ell) = -K + iK'.$$

Therefore, by Theorem 4.6

$$F(z) = c_1 \int_0^z \frac{d\zeta}{[(1 - \zeta^2)(1 - \ell^2 \zeta^2)]^{1/2}} + c_2.$$

Setting $z = 0$ gives $c_2 = 0$, and letting $z = 1$, $z = 1/\ell$, yields

$$K(k) = c_1 K(\ell) \quad \text{and} \quad K'(k) = c_1 K'(\ell),$$

where

$$K(k) = \int_0^1 \frac{dx}{[(1 - x^2)(1 - k^2 x^2)]^{1/2}},$$

$$K'(k) = \int_1^{1/k} \frac{dx}{[(x^2 - 1)(1 - k^2 x^2)]^{1/2}}.$$

Now $K(k)$ is clearly strictly increasing as k varies in $(0, 1)$. Moreover, a change of variables (Exercise 24) establishes the identity

$$K'(k) = K(\tilde{k}) \quad \text{where } \tilde{k}^2 = 1 - k^2 \text{ and } \tilde{k} > 0,$$

and this shows that $K'(k)$ is strictly decreasing. Hence $K(k)/K'(k)$ is strictly increasing. Since $K(k)/K'(k) = K(\ell)/K'(\ell)$, we must have $k = \ell$, and finally $c_1 = 1$. This shows that $I(z) = F(z)$, and hence I is conformal, as was to be proved.

A final observation is of significance. A basic insight into elliptic integrals is obtained by passing to their inverse functions. We therefore consider $z \mapsto \operatorname{sn}(z)$, the inverse map of $z \mapsto I(z)$.[9] It transforms the closed rectangle into the closed upper half-plane. Now consider the series of rectangles $R = R_0, R_1, R_2, \ldots$ gotten by reflecting successively along the lower sides (Figure 10).

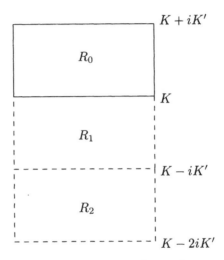

Figure 10. Reflections of $R = R_0$

With $\operatorname{sn}(z)$ defined in R_0, we can by the reflection principle extend it to R_1 by setting $\operatorname{sn}(z) = \overline{\operatorname{sn}(\overline{z})}$ whenever $z \in R_1$ (note that then $\overline{z} \in R_0$). Next we can extend $\operatorname{sn}(z)$ to R_2 by setting $\operatorname{sn}(z) = \overline{\operatorname{sn}(-iK' + \overline{z})}$ if $z \in R_2$ and noting that if $z \in R_2$, then $-iK' + \overline{z} \in R_1$. Combining these reflections and continuing this way we see that we can extend $\operatorname{sn}(z)$ in the entire strip $-K < \operatorname{Re}(z) < K$, so that $\operatorname{sn}(z) = \operatorname{sn}(z + 2iK')$.

Similarly, by reflecting in a series of horizontal rectangles, and combining these with the previous reflections, we see that $\operatorname{sn}(z)$ can be continued to the complex plane and also satisfies $\operatorname{sn}(z) = \operatorname{sn}(z + 4K)$. Thus $\operatorname{sn}(z)$ is *doubly periodic* (with periods $4K$ and $2iK'$). A further examination

---

[9]The notation $\operatorname{sn}(z)$ in somewhat different form is due to Jacobi, and was adopted because of the analogy with $\sin z$.

shows that the only singularities $\text{sn}(z)$ are poles. Functions of this type, called "elliptic functions," are the subject of the next chapter.

5 Exercises

1. A holomorphic mapping $f : U \to V$ is a **local bijection** on U if for every $z \in U$ there exists an open disc $D \subset U$ centered at z, so that $f : D \to f(D)$ is a bijection.

Prove that a holomorphic map $f : U \to V$ is a local bijection on U if and only if $f'(z) \neq 0$ for all $z \in U$.

[Hint: Use Rouché's theorem as in the proof of Proposition 1.1.]

2. Supppose $F(z)$ is holomorphic near $z = z_0$ and $F(z_0) = F'(z_0) = 0$, while $F''(z_0) \neq 0$. Show that there are two curves Γ_1 and Γ_2 that pass through z_0, are orthogonal at z_0, and so that F restricted to Γ_1 is real and has a minimum at z_0, while F restricted to Γ_2 is also real but has a maximum at z_0.

[Hint: Write $F(z) = (g(z))^2$ for z near z_0, and consider the mapping $z \mapsto g(z)$ and its inverse.]

3. Suppose U and V are conformally equivalent. Prove that if U is simply connected, then so is V. Note that this conclusion remains valid if we merely assume that there exists a continuous bijection between U and V.

4. Does there exist a holomorphic surjection from the unit disc to $\mathbb{C}$?

[Hint: Move the upper half-plane "down" and then square it to get $\mathbb{C}$.]

5. Prove that $f(z) = -\frac{1}{2}(z + 1/z)$ is a conformal map from the half-disc $\{z = x + iy : |z| < 1, \ y > 0\}$ to the upper half-plane.

[Hint: The equation $f(z) = w$ reduces to the quadratic equation $z^2 + 2wz + 1 = 0$, which has two distinct roots in $\mathbb{C}$ whenever $w \neq \pm 1$. This is certainly the case if $w \in \mathbb{H}$.]

6. Give another proof of Lemma 1.3 by showing directly that the Laplacian of $u \circ F$ is zero.

[Hint: The real and imaginary parts of F satisfy the Cauchy-Riemann equations.]

7. Provide all the details in the proof of the formula for the solution of the Dirichlet problem in a strip discussed in Section 1.3. Recall that it suffices to compute the solution at the points $z = iy$ with $0 < y < 1$.

(a) Show that if $re^{i\theta} = G(iy)$, then

$$re^{i\theta} = i\frac{\cos \pi y}{1 + \sin \pi y}.$$

This leads to two separate cases: either $0 < y \leq 1/2$ and $\theta = \pi/2$, or $1/2 \leq$

$y < 1$ and $\theta = -\pi/2$. In either case, show that

$$r^2 = \frac{1 - \sin \pi y}{1 + \sin \pi y} \quad \text{and} \quad P_r(\theta - \varphi) = \frac{\sin \pi y}{1 - \cos \pi y \sin \varphi}.$$

(b) In the integral $\frac{1}{2\pi} \int_0^\pi P_r(\theta - \varphi) \tilde{f}_0(\varphi) \, d\varphi$ make the change of variables $t = F(e^{i\varphi})$. Observe that

$$e^{i\varphi} = \frac{i - e^{\pi t}}{i + e^{\pi t}},$$

and then take the imaginary part and differentiate both sides to establish the two identities

$$\sin \varphi = \frac{1}{\cosh \pi t} \quad \text{and} \quad \frac{d\varphi}{dt} = \frac{\pi}{\cosh \pi t}.$$

Hence deduce that

$$\frac{1}{2\pi} \int_0^\pi P_r(\theta - \varphi) \tilde{f}_0(\varphi) \, d\varphi = \frac{1}{2\pi} \int_0^\pi \frac{\sin \pi y}{1 - \cos \pi y \sin \varphi} \tilde{f}_0(\varphi) \, d\varphi$$

$$= \frac{\sin \pi y}{2} \int_{-\infty}^\infty \frac{f_0(t)}{\cosh \pi t - \cos \pi y} \, dt.$$

(c) Use a similar argument to prove the formula for the integral $\frac{1}{2\pi} \int_{-\pi}^0 P_r(\theta - \varphi) \tilde{f}_1(\varphi) \, d\varphi$.

8. Find a harmonic function u in the open first quadrant that extends continuously up to the boundary except at the points 0 and 1, and that takes on the following boundary values: $u(x, y) = 1$ on the half-lines $\{y = 0, \ x > 1\}$ and $\{x = 0, \ y > 0\}$, and $u(x, y) = 0$ on the segment $\{0 < x < 1, y = 0\}$.

[Hint: Find conformal maps $F_1, F_2, \ldots, F_5$ indicated in Figure 11. Note that $\frac{1}{\pi} \arg(z)$ is harmonic on the upper half-plane, equals 0 on the positive real axis, and 1 on the negative real axis.]

9. Prove that the function u defined by

$$u(x, y) = \operatorname{Re}\left(\frac{i + z}{i - z}\right) \quad \text{and} \quad u(0, 1) = 0$$

is harmonic in the unit disc and vanishes on its boundary. Note that u is not bounded in $\mathbb{D}$.

10. Let $F : \mathbb{H} \to \mathbb{C}$ be a holomorphic function that satisfies

$$|F(z)| \leq 1 \quad \text{and} \quad F(i) = 0.$$

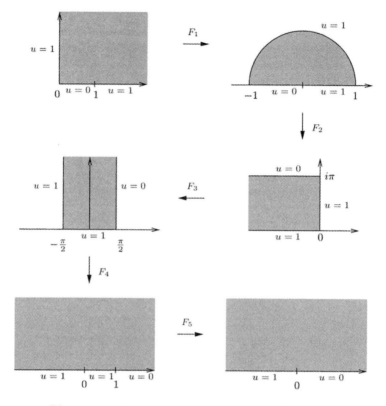

Figure 11. Successive conformal maps in Exercise 8

Prove that

$$|F(z)| \le \left|\frac{z-i}{z+i}\right| \qquad \text{for all } z \in \mathbb{H}.$$

11. Show that if $f : D(0, R) \to \mathbb{C}$ is holomorphic, with $|f(z)| \le M$ for some $M > 0$, then

$$\left|\frac{f(z) - f(0)}{M^2 - \overline{f(0)}f(z)}\right| \le \frac{|z|}{MR}.$$

[Hint: Use the Schwarz lemma.]

12. A complex number $w \in \mathbb{D}$ is a **fixed point** for the map $f : \mathbb{D} \to \mathbb{D}$ if $f(w) = w$.

(a) Prove that if $f : \mathbb{D} \to \mathbb{D}$ is analytic and has two distinct fixed points, then f is the identity, that is, $f(z) = z$ for all $z \in \mathbb{D}$.

(b) Must every holomorphic function $f : \mathbb{D} \to \mathbb{D}$ have a fixed point? [Hint: Consider the upper half-plane.]

13. The **pseudo-hyperbolic distance** between two points $z, w \in \mathbb{D}$ is defined by

$$\rho(z, w) = \left| \frac{z - w}{1 - \overline{w}z} \right|.$$

(a) Prove that if $f : \mathbb{D} \to \mathbb{D}$ is holomorphic, then

$$\rho(f(z), f(w)) \leq \rho(z, w) \qquad \text{for all } z, w \in \mathbb{D}.$$

Moreover, prove that if f is an automorphism of $\mathbb{D}$ then f preserves the pseudo-hyperbolic distance

$$\rho(f(z), f(w)) = \rho(z, w) \qquad \text{for all } z, w \in \mathbb{D}.$$

[Hint: Consider the automorphism $\psi_\alpha(z) = (z - \alpha)/(1 - \overline{\alpha}z)$ and apply the Schwarz lemma to $\psi_{f(w)} \circ f \circ \psi_w^{-1}$.]

(b) Prove that

$$\frac{|f'(z)|}{1 - |f(z)|^2} \leq \frac{1}{1 - |z|^2} \qquad \text{for all } z \in \mathbb{D}.$$

This result is called the Schwarz-Pick lemma. See Problem 3 for an important application of this lemma.

14. Prove that all conformal mappings from the upper half-plane $\mathbb{H}$ to the unit disc $\mathbb{D}$ take the form

$$e^{i\theta} \frac{z - \beta}{z - \overline{\beta}}, \qquad \theta \in \mathbb{R} \text{ and } \beta \in \mathbb{H}.$$

15. Here are two properties enjoyed by automorphisms of the upper half-plane.

(a) Suppose Φ is an automorphism of $\mathbb{H}$ that fixes three distinct points on the real axis. Then Φ is the identity.

(b) Suppose (x_1, x_2, x_3) and (y_1, y_2, y_3) are two pairs of three distinct points on the real axis with

$$x_1 < x_2 < x_3 \qquad \text{and} \qquad y_1 < y_2 < y_3.$$

Prove that there exists (a unique) automorphism Φ of $\mathbb{H}$ so that $\Phi(x_j) = y_j$, $j = 1, 2, 3$. The same conclusion holds if $y_3 < y_1 < y_2$ or $y_2 < y_3 < y_1$.

16. Let

$$f(z) = \frac{i - z}{i + z} \quad \text{and} \quad f^{-1}(w) = i\frac{1 - w}{1 + w}.$$

(a) Given $\theta \in \mathbb{R}$, find real numbers a, b, c, d such that $ad - bc = 1$, and so that for any $z \in \mathbb{H}$

$$\frac{az + b}{cz + d} = f^{-1}\left(e^{i\theta}f(z)\right).$$

(b) Given $\alpha \in \mathbb{D}$ find real numbers a, b, c, d so that $ad - bc = 1$, and so that for any $z \in \mathbb{H}$

$$\frac{az + b}{cz + d} = f^{-1}(\psi_\alpha(f(z))),$$

with ψ_α defined in Section 2.1.

(c) Prove that if g is an automorphism of the unit disc, then there exist real numbers a, b, c, d such that $ad - bc = 1$ and so that for any $z \in \mathbb{H}$

$$\frac{az + b}{cz + d} = f^{-1} \circ g \circ f(z).$$

[Hint: Use parts (a) and (b).]

17. If $\psi_\alpha(z) = (\alpha - z)/(1 - \bar{\alpha}z)$ for $|\alpha| < 1$, prove that

$$\frac{1}{\pi} \iint_{\mathbb{D}} |\psi_\alpha'|^2 \, dx dy = 1 \quad \text{and} \quad \frac{1}{\pi} \iint_{\mathbb{D}} |\psi_\alpha'| \, dx dy = \frac{1 - |\alpha|^2}{|\alpha|^2} \log \frac{1}{1 - |\alpha|^2},$$

where in the case $\alpha = 0$ the expression on the right is understood as the limit as $|\alpha| \to 0$.

[Hint: The first integral can be evaluated without a calculation. For the second, use polar coordinates, and for each fixed r use contour integration to evaluate the integral in θ.]

18. Suppose that Ω is a simply connected domain that is bounded by a piecewise-smooth closed curve γ (in the terminology of Chapter 1). Then any conformal map F of $\mathbb{D}$ to Ω extends to a continuous bijection of $\overline{\mathbb{D}}$ to $\overline{\Omega}$. The proof is simply a generalization of the argument used in Theorem 4.2.

19. Prove that the complex plane slit along the union of the rays $\cup_{k=1}^n \{A_k + iy : y \leq 0\}$ is simply connected.

[Hint: Given a curve, first "raise" it so that it is completely contained in the upper half-plane.]

20. Other examples of elliptic integrals providing conformal maps from the upper half-plane to rectangles are given below.

(a) The function

$$\int_0^z \frac{d\zeta}{\sqrt{\zeta(\zeta-1)(\zeta-\lambda)}}, \qquad \text{with } \lambda \in \mathbb{R} \text{ and } \lambda \neq 1$$

maps the upper half-plane conformally to a rectangle, one of whose vertices is the image of the point at infinity.

(b) In the case $\lambda = -1$, the image of

$$\int_0^z \frac{d\zeta}{\sqrt{\zeta(\zeta^2-1)}}$$

is a square whose side lengths are $\frac{\Gamma^2(1/4)}{2\sqrt{2\pi}}$.

21. We consider conformal mappings to triangles.

(a) Show that

$$\int_0^z z^{-\beta_1}(1-z)^{-\beta_2}\,dz,$$

with $0 < \beta_1 < 1$, $0 < \beta_2 < 1$, and $1 < \beta_1 + \beta_2 < 2$, maps $\mathbb{H}$ to a triangle whose vertices are the images of 0, 1, and ∞, and with angles $\alpha_1\pi$, $\alpha_2\pi$, and $\alpha_3\pi$, where $\alpha_j + \beta_j = 1$ and $\beta_1 + \beta_2 + \beta_3 = 2$.

(b) What happens when $\beta_1 + \beta_2 = 1$?

(c) What happens when $0 < \beta_1 + \beta_2 < 1$?

(d) In (a), the length of the side of the triangle opposite angle $\alpha_j\pi$ is $\frac{\sin(\alpha_j\pi)}{\pi}\Gamma(\alpha_1)\Gamma(\alpha_2)\Gamma(\alpha_3)$.

22. If P is a simply connected region bounded by a polygon with vertices $a_1,\ldots,a_n$ and angles $\alpha_1\pi,\ldots,\alpha_n\pi$, and F is a conformal map of the disc $\mathbb{D}$ to P, then there exist complex numbers $B_1,\ldots,B_n$ on the unit circle, and constants c_1 and c_2 so that

$$F(z) = c_1 \int_1^z \frac{d\zeta}{(\zeta-B_1)^{\beta_1}\cdots(\zeta-B_n)^{\beta_n}} + c_2.$$

[Hint: This follows from the standard correspondence between $\mathbb{H}$ and $\mathbb{D}$ and an argument similar to that used in the proof of Theorem 4.7.]

23. If

$$F(z) = \int_1^z \frac{d\zeta}{(1-\zeta^n)^{2/n}},$$

then F maps the unit disc conformally onto the interior of a regular polygon with n sides and perimeter

$$2^{\frac{n-2}{n}} \int_0^\pi (\sin\theta)^{-2/n}\, d\theta.$$

24. The elliptic integrals K and K' defined for $0 < k < 1$ by

$$K(k) = \int_0^1 \frac{dx}{((1-x^2)(1-k^2x^2))^{1/2}} \quad \text{and} \quad K'(k) = \int_1^{1/k} \frac{dx}{((x^2-1)(1-k^2x^2))^{1/2}}$$

satisfy various interesting identities. For instance:

(a) Show that if $\tilde{k}^2 = 1 - k^2$ and $0 < \tilde{k} < 1$, then

$$K'(k) = K(\tilde{k}).$$

[Hint: Change variables $x = (1 - \tilde{k}^2 y^2)^{-1/2}$ in the integral defining $K'(k)$.]

(b) Prove that if $\tilde{k}^2 = 1 - k^2$, and $0 < \tilde{k} < 1$, then

$$K(k) = \frac{2}{1+\tilde{k}} K\left(\frac{1-\tilde{k}}{1+\tilde{k}}\right).$$

[Hint: Change variables $x = 2t/(1 + \tilde{k} + (1 - \tilde{k})t^2)$.]

(c) Show that for $0 < k < 1$ one has

$$K(k) = \frac{\pi}{2} F(1/2, 1/2, 1; k^2),$$

where F the hypergeometric series. [Hint: This follows from the integral representation for F given in Exercise 9, Chapter 6.]

6 Problems

1. Let f be a complex-valued C^1 function defined in the neighborhood of a point z_0. There are several notions closely related to conformality at z_0. We say that f is **isogonal** at z_0 if whenever $\gamma(t)$ and $\eta(t)$ are two smooth curves with $\gamma(0) = \eta(0) = z_0$, that make an angle θ there ($|\theta| < \pi$), then $f(\gamma(t))$ and $f(\eta(t))$ make an angle of θ' at $t = 0$ with $|\theta'| = |\theta|$ for all θ. Also, f is said to be **isotropic** if it magnifies lengths by some factor for all directions emanating from z_0, that is, if the limit

$$\lim_{r \to 0} \frac{|f(z_0 + re^{i\theta}) - f(z_0)|}{r}$$

exists, is non-zero, and independent of θ.

Then f is isogonal at z_0 if and only if it is isotropic at z_0; moreover, f is isogonal at z_0 if and only if either $f'(z_0)$ exists and is non-zero, or the same holds for f replaced by $\bar{f}$.

2. The angle between two non-zero complex numbers z and w (taken in that order) is simply the oriented angle, in $(-\pi, \pi]$, that is formed between the two vectors in $\mathbb{R}^2$ corresponding to the points z and w. This oriented angle, say α, is uniquely determined by the two quantities

$$\frac{(z, w)}{|z|\,|w|} \quad \text{and} \quad \frac{(z, -iw)}{|z|\,|w|}$$

which are simply the cosine and sine of α, respectively. Here, the notation $(\cdot, \cdot)$ corresponds to the usual Euclidian inner product in $\mathbb{R}^2$, which in terms of complex numbers takes the form $(z, w) = \mathrm{Re}(z\bar{w})$.

In particular, we may now consider two smooth curves $\gamma : [a, b] \to \mathbb{C}$ and $\eta : [a, b] \to \mathbb{C}$, that intersect at z_0, say $\gamma(t_0) = \eta(t_0) = z_0$, for some $t_0 \in (a, b)$. If the quantities $\gamma'(t_0)$ and $\eta'(t_0)$ are non-zero, then they represent the tangents to the curves γ and η at the point z_0, and we say that the two curves intersect at z_0 at the angle formed by the two vectors $\gamma'(t_0)$ and $\eta'(t_0)$.

A holomorphic function f defined near z_0 is said to **preserve angles** at z_0 if for any two smooth curves γ and η intersecting at z_0, the angle formed between the curves γ and η at z_0 equals the angle formed between the curves $f \circ \gamma$ and $f \circ \eta$ at $f(z_0)$. (See Figure 12 for an illustration.) In particular, we assume that the tangents to the curves γ, η, $f \circ \gamma$, and $f \circ \eta$ at the point z_0 and $f(z_0)$ are all non-zero.

Figure 12. Preservation of angles at z_0

(a) Prove that if $f : \Omega \to \mathbb{C}$ is holomorphic, and $f'(z_0) \neq 0$, then f preserves angles at z_0. [Hint: Observe that

$$(f'(z_0)\gamma'(t_0), f'(z_0)\eta'(t_0)) = |f'(z_0)|^2 (\gamma'(t_0), \eta'(t_0)).]$$

(b) Conversely, prove the following: suppose $f : \Omega \to \mathbb{C}$ is a complex-valued function, that is real-differentiable at $z_0 \in \Omega$, and $J_f(z_0) \neq 0$. If f preserves angles at z_0, then f is holomorphic at z_0 with $f'(z_0) \neq 0$.

3.[*] The Schwarz-Pick lemma (see Exercise 13) is the infinitesimal version of an important observation in complex analysis and geometry.

For complex numbers $w \in \mathbb{C}$ and $z \in \mathbb{D}$ we define the **hyperbolic length** of w at z by

$$\|w\|_z = \frac{|w|}{1 - |z|^2},$$

where $|w|$ and $|z|$ denote the usual absolute values. This length is sometimes referred to as the **Poincaré metric**, and as a Riemann metric it is written as

$$ds^2 = \frac{|dz|^2}{(1 - |z|^2)^2}.$$

The idea is to think of w as a vector lying in the tangent space at z. Observe that for a fixed w, its hyperbolic length grows to infinity as z approaches the boundary of the disc. We pass from the infinitesimal hyperbolic length of tangent vectors to the global hyperbolic distance between two points by integration.

(a) Given two complex numbers z_1 and z_2 in the disc, we define the **hyperbolic distance** between them by

$$d(z_1, z_2) = \inf_{\gamma} \int_0^1 \|\gamma'(t)\|_{\gamma(t)} \, dt,$$

where the infimum is taken over all smooth curves $\gamma : [0, 1] \to \mathbb{D}$ joining z_1 and z_2. Use the Schwarz-Pick lemma to prove that if $f : \mathbb{D} \to \mathbb{D}$ is holomorphic, then

$$d(f(z_1), f(z_2)) \le d(z_1, z_2) \qquad \text{for any } z_1, z_2 \in \mathbb{D}.$$

In other words, holomorphic functions are distance-decreasing in the hyperbolic metric.

(b) Prove that automorphisms of the unit disc preserve the hyperbolic distance, namely

$$d(\varphi(z_1), \varphi(z_2)) = d(z_1, z_2), \qquad \text{for any } z_1, z_2 \in \mathbb{D}$$

and any automorphism φ. Conversely, if $\varphi : \mathbb{D} \to \mathbb{D}$ preserves the hyperbolic distance, then either φ or $\overline{\varphi}$ is an automorphism of $\mathbb{D}$.

(c) Given two points $z_1, z_2 \in \mathbb{D}$, show that there exists an automorphism φ such that $\varphi(z_1) = 0$ and $\varphi(z_2) = s$ for some s on the segment $[0, 1)$ on the real line.

(d) Prove that the hyperbolic distance between 0 and $s \in [0, 1)$ is

$$d(0, s) = \frac{1}{2} \log \frac{1 + s}{1 - s}.$$

(e) Find a formula for the hyperbolic distance between any two points in the unit disc.

4. * Consider the group of matrices of the form

$$M = \begin{pmatrix} a & b \\ c & d \end{pmatrix},$$

that satisfy the following conditions:

(i) $a, b, c,$ and $d \in \mathbb{C}$,

(ii) the determinant of M is equal to 1,

(iii) the matrix M preserves the following hermitian form on $\mathbb{C}^2 \times \mathbb{C}^2$:

$$\langle Z, W \rangle = z_1 \overline{w}_1 - z_2 \overline{w}_2,$$

where $Z = (z_1, z_2)$ and $W = (w_1, w_2)$. In other words, for all $Z, W \in \mathbb{C}^2$

$$\langle MZ, MW \rangle = \langle Z, W \rangle.$$

This group of matrices is denoted by $\mathrm{SU}(1, 1)$.

(a) Prove that all matrices in $\mathrm{SU}(1, 1)$ are of the form

$$\begin{pmatrix} a & b \\ \overline{b} & \overline{a} \end{pmatrix},$$

where $|a|^2 - |b|^2 = 1$. To do so, consider the matrix

$$J = \begin{pmatrix} 1 & 0 \\ 0 & -1 \end{pmatrix},$$

and observe that $\langle Z, W \rangle = {}^t W J Z$, where ${}^t W$ denotes the conjugate transpose of W.

(b) To every matrix in $\mathrm{SU}(1, 1)$ we can associate a fractional linear transformation

$$\frac{az + b}{cz + d}.$$

Prove that the group $\mathrm{SU}(1, 1)/\{\pm 1\}$ is isomorphic to the group of automorphisms of the disc. [Hint: Use the following association.]

$$e^{2i\theta} \frac{z - \alpha}{1 - \overline{\alpha}z} \longrightarrow \begin{pmatrix} \dfrac{e^{i\theta}}{\sqrt{1 - |\alpha|^2}} & -\dfrac{\alpha e^{i\theta}}{\sqrt{1 - |\alpha|^2}} \\ -\dfrac{\overline{\alpha}e^{-i\theta}}{\sqrt{1 - |\alpha|^2}} & \dfrac{e^{-i\theta}}{\sqrt{1 - |\alpha|^2}} \end{pmatrix}.$$

5. The following result is relevant to Problem 4 in Chapter 10 which treats modular functions.

(a) Suppose that $F : \mathbb{H} \to \mathbb{C}$ is holomorphic and bounded. Also, suppose that $F(z)$ vanishes when $z = ir_n$, $n = 1, 2, 3, \ldots$, where $\{r_n\}$ is a bounded sequence of positive numbers. Prove that if $\sum_{n=1}^{\infty} r_n = \infty$, then $F = 0$.

(b) If $\sum r_n < \infty$, it is possible to construct a bounded function on the upper half-plane with zeros precisely at the points ir_n.

For related results in the unit disc, see Problems 1 and 2 in Chapter 5.]

6.[*] The results of Exercise 18 extend to the case when γ is assumed merely to be closed, simple, and continuous. The proof, however, requires further ideas.

7.[*] Applying ideas of Carathéodory, Koebe gave a proof of the Riemann mapping theorem by constructing (more explicitly) a sequence of functions that converges to the desired conformal map.

Starting with a Koebe domain, that is, a simply connected domain $\mathcal{K}_0 \subset \mathbb{D}$ that is not all of $\mathbb{D}$, and which contains the origin, the strategy is to find an injective function f_0 such that $f_0(\mathcal{K}_0) = \mathcal{K}_1$ is a Koebe domain "larger" than $\mathcal{K}_0$. Then, one iterates this process, finally obtaining functions $F_n = f_n \circ \cdots \circ f_0 : \mathcal{K}_0 \to \mathbb{D}$ such that $F_n(\mathcal{K}_0) = \mathcal{K}_{n+1}$ and $\lim F_n = F$ is a conformal map from $\mathcal{K}_0$ to $\mathbb{D}$.

The **inner radius** of a region $\mathcal{K} \subset \mathbb{D}$ that contains the origin is defined by $r_\mathcal{K} = \sup\{\rho \geq 0 : D(0, \rho) \subset \mathcal{K}\}$. Also, a holomorphic injection $f : \mathcal{K} \to \mathbb{D}$ is said to be an **expansion** if $f(0) = 0$ and $|f(z)| > |z|$ for all $z \in \mathcal{K} - \{0\}$.

(a) Prove that if f is an expansion, then $r_{f(\mathcal{K})} \geq r_\mathcal{K}$ and $|f'(0)| > 1$. [Hint: Write $f(z) = zg(z)$ and use the maximum principle to prove that $|f'(0)| = |g(0)| > 1$.]

Suppose we begin with a Koebe domain $\mathcal{K}_0$ and a sequence of expansions $\{f_0, f_1, \ldots, f_n, \ldots\}$, so that $\mathcal{K}_{n+1} = f_n(\mathcal{K}_n)$ are also Koebe domains. We then define holomorphic maps $F_n : \mathcal{K}_0 \to \mathbb{D}$ by $F_n = f_n \circ \cdots \circ f_0$.

(b) Prove that for each n, the function F_n is an expansion. Moreover, $F_n'(0) = \prod_{k=0}^{n} f_k'(0)$, and conclude that $\lim_{n\to\infty} |f_n'(0)| = 1$. [Hint: Prove that the sequence $\{|F_n'(0)|\}$ has a limit by showing that it is bounded above and monotone increasing. Use the Schwarz lemma.]

(c) Show that if the sequence is osculating, that is, $r_{\mathcal{K}_n} \to 1$ as $n \to \infty$, then $\{F_n\}$ converges uniformly on compact subsets of $\mathcal{K}_0$ to a conformal map $F : \mathcal{K}_0 \to \mathbb{D}$. [Hint: If $r_{F(\mathcal{K}_0)} \geq 1$ then F is surjective.]

To construct the desired osculating sequence we shall use the automorphisms $\psi_\alpha = (\alpha - z)/(1 - \bar{\alpha}z)$.

(d) Given a Koebe domain $\mathcal{K}$, choose a point $\alpha \in \mathbb{D}$ on the boundary of $\mathcal{K}$ such that $|\alpha| = r_\mathcal{K}$, and also choose $\beta \in \mathbb{D}$ such that $\beta^2 = \alpha$. Let S denote the square root of ψ_α on $\mathcal{K}$ such that $S(0) = 0$. Why is such a function well defined? Prove that the function $f : \mathcal{K} \to \mathbb{D}$ defined by $f(z) = \psi_\beta \circ S \circ \psi_\alpha$

is an expansion. Moreover, show that $|f'(0)| = (1 + r_{\mathcal{K}})/2\sqrt{r_{\mathcal{K}}}$. [Hint: To prove that $|f(z)| > |z|$ on $\mathcal{K} - \{0\}$ apply the Schwarz lemma to the inverse function, namely $\psi_\alpha \circ g \circ \psi_\beta$ where $g(z) = z^2$.]

(e) Use part (d) to construct the desired sequence.

8.* Let f be an injective holomorphic function in the unit disc, with $f(0) = 0$ and $f'(0) = 1$. If we write $f(z) = z + a_2 z^2 + a_3 z^3 \cdots$, then Problem 1 in Chapter 3 shows that $|a_2| \leq 2$. Bieberbach conjectured that in fact $|a_n| \leq n$ for all $n \geq 2$; this was proved by deBranges. This problem outlines an argument to prove the conjecture under the additional assumption that the coefficients a_n are real.

(a) Let $z = re^{i\theta}$ with $0 < r < 1$, and show that if $v(r, \theta)$ denotes the imaginary part of $f(re^{i\theta})$, then

$$a_n r^n = \frac{2}{\pi} \int_0^\pi v(r, \theta) \sin n\theta \, d\theta.$$

(b) Show that for $0 \leq \theta \leq \pi$ and $n = 1, 2, \ldots$ we have $|\sin n\theta| \leq n \sin \theta$.

(c) Use the fact that $a_n \in \mathbb{R}$ to show that $f(\mathbb{D})$ is symmetric with respect to the real axis, and use this fact to show that f maps the upper half-disc into either the upper or lower part of $f(\mathbb{D})$.

(d) Show that for r small,

$$v(r, \theta) = r \sin \theta [1 + O(r)],$$

and use the previous part to conclude that $v(r, \theta) \sin \theta \geq 0$ for all $0 < r < 1$ and $0 \leq \theta \leq \pi$.

(e) Prove that $|a_n r^n| \leq nr$, and let $r \to 1$ to conclude that $|a_n| \leq n$.

(f) Check that the function $f(z) = z/(1-z)^2$ satisfies all the hypotheses and that $|a_n| = n$ for all n.

9.* Gauss found a connection between elliptic integrals and the familiar operations of forming arithmetic and geometric means.

We start with any pair (a, b) of numbers that satisfy $a \geq b > 0$, and form the arithmetic and geometric means of a and b, that is,

$$a_1 = \frac{a+b}{2} \quad \text{and} \quad b_1 = (ab)^{1/2}.$$

We then repeat these operations with a and b replaced by a_1 and b_1. Iterating this process provides two sequences $\{a_n\}$ and $\{b_n\}$ where a_{n+1} and b_{n+1} are the arithmetic and geometric means of a_n and b_n, respectively.

(a) Prove that the two sequences $\{a_n\}$ and $\{b_n\}$ have a common limit. This limit, which we denote by $M(a, b)$, is called the **arithmetic-geometric mean** of a and b. [Hint: Show that $a \geq a_1 \geq a_2 \geq \cdots \geq a_n \geq b_n \geq \cdots \geq b_1 \geq b$ and $a_n - b_n \leq (a - b)/2^n$.]

(b) Gauss's identity states that

$$\frac{1}{M(a, b)} = \frac{2}{\pi} \int_0^{\pi/2} \frac{d\theta}{(a^2 \cos^2 \theta + b^2 \sin^2 \theta)^{1/2}}.$$

To prove this relation, show that if $I(a, b)$ denotes the integral on the right-hand side, then it suffices to establish the invariance of I, namely

$$(6) \qquad\qquad I(a, b) = I\left(\frac{a + b}{2}, (ab)^{1/2}\right).$$

Then, observe that the connection with elliptic integrals takes the form

$$I(a, b) = \frac{1}{a} K(k) = \frac{1}{a} \int_0^1 \frac{dx}{\sqrt{(1 - x^2)(1 - k^2 x^2)}} \qquad \text{where } k^2 = 1 - b^2/a^2,$$

and that the relation (6) is a consequence of the identity in (b) of Exercise 24.

9 An Introduction to Elliptic Functions

> The form that Jacobi had given to the theory of elliptic functions was far from perfection; its flaws are obvious. At the base we find three fundamental functions sn, cn and dn. These functions do not have the same periods...
>
> In Weierstrass' system, instead of three fundamental functions, there is only one, $\wp(u)$, and it is the simplest of all having the same periods. It has only one double infinity; and finally its definition is so that it does not change when one replaces one system of periods by another equivalent system.
>
> *H. Poincaré,* 1899

The theory of elliptic functions, which is of interest in several parts of mathematics, initially grew out of the study of elliptic integrals. These can be described generally as integrals of the form $\int R(x, \sqrt{P(x)})\, dx$, where R is a rational function and P a polynomial of degree three or four.[1] These integrals arose in computing the arc-length of an ellipse, or of a lemniscate, and in a variety of other problems. Their early study was centered on their special transformation properties and on the discovery of an inherent double-periodicity. We have seen an example of this latter phenomenon in the mapping function of the half-plane to a rectangle taken up in Section 4.5 of the previous chapter.

It was Jacobi who transformed the subject by initiating the systematic study of doubly-periodic functions (called elliptic functions). In this theory, the theta functions he introduced played a decisive role. Weierstrass after him developed another approach, which in its initial steps is simpler and more elegant. It is based on his $\wp$ function, and in this chapter we shall sketch the beginnings of that theory. We will go as far as to glimpse a possible connection with number theory, by considering the Eisenstein series and their expression involving divisor functions. A number of more direct links with combinatorics and number theory arise from the theta

---

[1] The case when P is a quadratic polynomial is essentially that of "circular functions", and can be reduced to the trigonometric functions $\sin x$, $\cos x$, etc.

functions, which we will take up in the next chapter. The remarkable facts we shall see there attest to the great interest of these functions in mathematics. As such they ought to soften the harsh opinion expressed above about the imperfection of Jacobi's theory.

1 Elliptic functions

We are interested in meromorphic functions f on $\mathbb{C}$ that have two periods; that is, there are two non-zero complex numbers ω_1 and ω_2 such that

$$f(z + \omega_1) = f(z) \quad \text{and} \quad f(z + \omega_2) = f(z),$$

for all $z \in \mathbb{C}$. A function with two periods is said to be **doubly periodic**.

The case when ω_1 and ω_2 are linearly dependent over $\mathbb{R}$, that is $\omega_2/\omega_1 \in \mathbb{R}$, is uninteresting. Indeed, Exercise 1 shows that in this case f is either periodic with a simple period (if the quotient ω_2/ω_1 is rational) or f is constant (if ω_2/ω_1 is irrational). Therefore, we make the following assumption: the periods ω_1 and ω_2 are linearly independent over $\mathbb{R}$.

We now describe a normalization that we shall use extensively in this chapter. Let $\tau = \omega_2/\omega_1$. Since τ and $1/\tau$ have imaginary parts of opposite signs, and since τ is not real, we may assume (after possibly interchanging the roles of ω_1 and ω_2) that $\text{Im}(\tau) > 0$. Observe now that the function f has periods ω_1 and ω_2 if and only if the function $F(z) = f(\omega_1 z)$ has periods 1 and τ, and moreover, the function f is meromorphic if and only if F is meromorphic. Also the properties of f are immediately deducible from those of F. We may therefore assume, without loss of generality, that f is a meromorphic function on $\mathbb{C}$ with periods 1 and τ where $\text{Im}(\tau) > 0$.

Successive applications of the periodicity conditions yield

$$(1) \quad f(z + n + m\tau) = f(z) \quad \text{for all integers } n, m \text{ and all } z \in \mathbb{C},$$

and it is therefore natural to consider the lattice in $\mathbb{C}$ defined by

$$\Lambda = \{n + m\tau : n, m \in \mathbb{Z}\}.$$

We say that 1 and τ **generate** Λ (see Figure 1).

Equation (1) says that f is constant under translations by elements of Λ. Associated to the lattice Λ is the **fundamental parallelogram** defined by

$$P_0 = \{z \in \mathbb{C} : z = a + b\tau \text{ where } 0 \le a < 1 \text{ and } 0 \le b < 1\}.$$

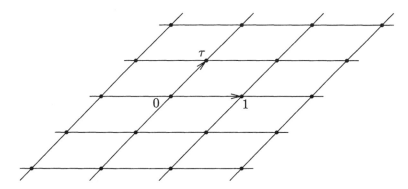

Figure 1. The lattice Λ generated by 1 and τ

The importance of the fundamental parallelogram comes from the fact that f is completely determined by its behavior on P_0. To see this, we need a definition: two complex numbers z and w are **congruent modulo** Λ if

$$z = w + n + m\tau \quad \text{for some } n, m \in \mathbb{Z},$$

and we write $z \sim w$. In other words, z and w differ by a point in the lattice, $z - w \in \Lambda$. By (1) we conclude that $f(z) = f(w)$ whenever $z \sim w$. If we can show that any point in $z \in \mathbb{C}$ is congruent to a unique point in P_0 then we will have proved that f is completely determined by its values in the fundamental parallelogram. Suppose $z = x + iy$ is given, and write $z = a + b\tau$ where $a, b \in \mathbb{R}$. This is possible since 1 and τ form a basis over the reals of the two-dimensional vector space $\mathbb{C}$. Then choose n and m to be the greatest integers $\leq a$ and $\leq b$, respectively. If we let $w = z - n - m\tau$, then by definition $z \sim w$, and moreover $w = (a - n) + (b - m)\tau$. By construction, it is clear that $w \in P_0$. To prove uniqueness, suppose that w and w' are two points in P_0 that are congruent. If we write $w = a + b\tau$ and $w' = a' + b'\tau$, then $w - w' = (a - a') + (b - b')\tau \in \Lambda$, and therefore both $a - a'$ and $b - b'$ are integers. But since $0 \leq a, a' < 1$, we have $-1 < a - a' < 1$, which then implies $a - a' = 0$. Similarly $b - b' = 0$, and we conclude that $w = w'$.

More generally, a **period parallelogram** P is any translate of the fundamental parallelogram, $P = P_0 + h$ with $h \in \mathbb{C}$ (see Figure 2).

Since we can apply the lemma to $z - h$, we conclude that every point in $\mathbb{C}$ is congruent to a unique point in a given period parallelogram. Therefore, f is uniquely determined by its behavior on any period parallelogram.

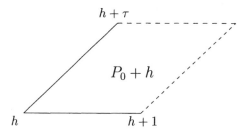

Figure 2. A period parallelogram

Finally, note that Λ and P_0 give rise to a covering (or tiling) of the complex plane

$$(2) \qquad \qquad \mathbb{C} = \bigcup_{n,m \in \mathbb{Z}} (n + m\tau + P_0),$$

and moreover, this union is disjoint. This is immediate from the facts we just collected and the definition of P_0. We summarize what we have seen so far.

Proposition 1.1 *Suppose f is a meromorphic function with two periods 1 and τ which generate the lattice Λ. Then:*

(i) *Every point in $\mathbb{C}$ is congruent to a unique point in the fundamental parallelogram.*

(ii) *Every point in $\mathbb{C}$ is congruent to a unique point in any given period parallelogram.*

(iii) *The lattice Λ provides a disjoint covering of the complex plane, in the sense of (2).*

(iv) *The function f is completely determined by its values in any period parallelogram.*

1.1 Liouville's theorems

We can now see why we assumed from the beginning that f is meromorphic rather than just holomorphic.

Theorem 1.2 *An entire doubly periodic function is constant.*

Proof. The function is completely determined by its values on P_0 and since the closure of P_0 is compact, we conclude that the function is bounded on $\mathbb{C}$, hence constant by Liouville's theorem in Chapter 2.

A non-constant doubly periodic meromorphic function is called an **elliptic function**. Since a meromorphic function can have only finitely many zeros and poles in any large disc, we see that an elliptic function will have only finitely many zeros and poles in any given period parallelogram, and in particular, this is true in the fundamental parallelogram. Of course, nothing excludes f from having a pole or zero on the boundary of P_0.

As usual, we count poles and zeros with multiplicities. Keeping this in mind we can prove the following theorem.

Theorem 1.3 *The total number of poles of an elliptic function in P_0 is always ≥ 2.*

In other words, f cannot have only one simple pole. It must have at least two poles, and this does not exclude the case of a single pole of multiplicity ≥ 2.

Proof. Suppose first that f has no poles on the boundary ∂P_0 of the fundamental parallelogram. By the residue theorem we have

$$\int_{\partial P_0} f(z)\, dz = 2\pi i \sum \operatorname{res} f,$$

and we contend that the integral is 0. To see this, we simply use the periodicity of f. Note that

$$\int_{\partial P_0} f(z)\, dz = \int_0^1 f(z)\, dz + \int_1^{1+\tau} f(z)\, dz + \int_{1+\tau}^{\tau} f(z)\, dz + \int_{\tau}^0 f(z)\, dz,$$

and the integrals over opposite sides cancel out. For instance

$$\int_0^1 f(z)\, dz + \int_{1+\tau}^{\tau} f(z)\, dz = \int_0^1 f(z)\, dz + \int_1^0 f(s+\tau)\, ds$$

$$= \int_0^1 f(z)\, dz + \int_1^0 f(s)\, ds$$

$$= \int_0^1 f(z)\, dz - \int_0^1 f(z)\, dz$$

$$= 0,$$

and similarly for the other pair of sides. Hence $\int_{\partial P_0} f = 0$ and $\sum \operatorname{res} f = 0$. Therefore f must have at least two poles in P_0.

If f has a pole on ∂P_0 choose a small $h \in \mathbb{C}$ so that if $P = h + P_0$, then f has no poles on ∂P. Arguing as before, we find that f must have at least two poles in P, and therefore the same conclusion holds for P_0.

The total number of poles (counted according to their multiplicities) of an elliptic function is called its **order**. The next theorem says that elliptic functions have as many zeros as they have poles, if the zeros are counted with their multiplicities.

Theorem 1.4 *Every elliptic function of order m has m zeros in P_0.*

Proof. Assuming first that f has no zeros or poles on the boundary of P_0, we know by the argument principle in Chapter 3 that

$$\int_{\partial P_0} \frac{f'(z)}{f(z)}\, dz = 2\pi i (\mathcal{N}_{\mathfrak{z}} - \mathcal{N}_{\mathfrak{p}})$$

where $\mathcal{N}_{\mathfrak{z}}$ and $\mathcal{N}_{\mathfrak{p}}$ denote the number of zeros and poles of f in P_0, respectively. By periodicity, we can argue as in the proof of the previous theorem to find that $\int_{\partial P_0} f'/f = 0$, and therefore $\mathcal{N}_{\mathfrak{z}} = \mathcal{N}_{\mathfrak{p}}$.

In the case when a pole or zero of f lies on ∂P_0 it suffices to apply the argument to a translate of P.

As a consequence, if f is elliptic then the equation $f(z) = c$ has as many solutions as the order of f for every $c \in \mathbb{C}$, simply because $f - c$ is elliptic and has as many poles as f.

Despite the rather simple nature of the theorems above, there remains the question of showing that elliptic functions exist. We now turn to a constructive solution of this problem.

1.2 The Weierstrass $\wp$ function

An elliptic function of order two

This section is devoted to the basic example of an elliptic function. As we have seen above, any elliptic function must have at least two poles; we shall in fact construct one whose only singularity will be a double pole at the points of the lattice generated by the periods.

Before looking at the case of doubly-periodic functions, let us first consider briefly functions with only a single period. If one wished to construct a function with period 1 and poles at all the integers, a simple choice would be the sum

$$F(z) = \sum_{n=-\infty}^{\infty} \frac{1}{z+n}.$$

Note that the sum remains unchanged if we replace z by $z + 1$, and the poles are at the integers. However, the series defining F is not absolutely convergent, and to remedy this problem, we sum symmetrically, that is, we define

$$F(z) = \lim_{N \to \infty} \sum_{|n| \leq N} \frac{1}{z+n} = \frac{1}{z} + \sum_{n=1}^{\infty} \left[\frac{1}{z+n} + \frac{1}{z-n} \right].$$

On the far right-hand side, we have paired up the terms corresponding to n and $-n$, a trick which makes the quantity in brackets $O(1/n^2)$, and hence the last sum is absolutely convergent. As a consequence, F is meromorphic with poles precisely at the integers. In fact, we proved earlier in Chapter 5 that $F(z) = \pi \cot \pi z$.

There is a second way to deal with the series $\sum_{-\infty}^{\infty} 1/(z+n)$, which is to write it as

$$\frac{1}{z} + \sum_{n \neq 0} \left[\frac{1}{z+n} - \frac{1}{n} \right],$$

where the sum is taken over all non-zero integers. Notice that $1/(z+n) - 1/n = O(1/n^2)$, which makes this series absolutely convergent. Moreover, since

$$\frac{1}{z+n} + \frac{1}{z-n} = \left(\frac{1}{z+n} - \frac{1}{n} \right) + \left(\frac{1}{z-n} - \frac{1}{-n} \right),$$

we get the same sum as before.

In analogy to this, the idea is to mimic the above to produce our first example of an elliptic function. We would like to write it as

$$\sum_{\omega \in \Lambda} \frac{1}{(z+\omega)^2},$$

but again this series does not converge absolutely. There are several approaches to try to make sense of this series (see Problem 1), but the simplest is to follow the second way we dealt with the cotangent series.

To overcome the non-absolute convergence of the series, let Λ^* denote the lattice minus the origin, that is, $\Lambda^* = \Lambda - \{(0,0)\}$, and consider instead the following series:

$$\frac{1}{z^2} + \sum_{\omega \in \Lambda^*} \left[\frac{1}{(z+\omega)^2} - \frac{1}{\omega^2} \right],$$

where we have subtracted the factor $1/\omega^2$ to make the sum converge. The term in brackets is now

$$\frac{1}{(z+\omega)^2} - \frac{1}{\omega^2} = \frac{-z^2 - 2z\omega}{(z+\omega)^2 \omega^2} = O\left(\frac{1}{\omega^3} \right) \qquad \text{as } |\omega| \to \infty,$$

and the new series will define a meromorphic function with the desired poles once we have proved the following lemma.

Lemma 1.5 *The two series*

$$\sum_{(n,m)\neq(0,0)} \frac{1}{(|n| + |m|)^r} \quad and \quad \sum_{n+m\tau\in\Lambda^*} \frac{1}{|n + m\tau|^r}$$

converge if $r > 2$.

Recall that according to the Note at the end of Chapter 7, the question whether a double series converges absolutely is independent of the order of summation. In the present case, we shall first sum in m and then in n.

For the first series, the usual integral comparison can be applied.[2] For each $n \neq 0$

$$\sum_{m\in\mathbb{Z}} \frac{1}{(|n| + |m|)^r} = \frac{1}{|n|^r} + 2\sum_{m\geq 1} \frac{1}{(|n| + |m|)^r}$$

$$= \frac{1}{|n|^r} + 2\sum_{k\geq |n|+1} \frac{1}{k^r}$$

$$\leq \frac{1}{|n|^r} + 2\int_{|n|}^{\infty} \frac{dx}{x^r}$$

$$\leq \frac{1}{|n|^r} + C\frac{1}{|n|^{r-1}}.$$

Therefore, $r > 2$ implies

$$\sum_{(n,m)\neq(0,0)} \frac{1}{(|n| + |m|)^r} = \sum_{|m|\neq 0} \frac{1}{|m|^r} + \sum_{|n|\neq 0}\sum_{m\in\mathbb{Z}} \frac{1}{(|n| + |m|)^r}$$

$$\leq \sum_{|m|\neq 0} \frac{1}{|m|^r} + \sum_{|n|\neq 0} \left(\frac{1}{|n|^r} + C\frac{1}{|n|^{r-1}}\right)$$

$$< \infty.$$

To prove that the second series also converges, it suffices to show that there is a constant c such that

$$|n| + |m| \leq c|n + \tau m| \quad \text{for all } n, m \in \mathbb{Z}.$$

---

[2] We simply use $1/k^r \leq 1/x^r$ when $k - 1 \leq x \leq k$; see also the first figure in Chapter 8, Book I.

We use the notation $x \lesssim y$ if there exists a positive constant a such that $x \leq ay$. We also write $x \approx y$ if both $x \lesssim y$ and $y \lesssim x$ hold. Note that for any two positive numbers A and B, one has

$$(A^2 + B^2)^{1/2} \approx A + B.$$

On the one hand $A \leq (A^2 + B^2)^{1/2}$ and $B \leq (A^2 + B^2)^{1/2}$, so that $A + B \leq 2(A^2 + B^2)^{1/2}$. On the other hand, it suffices to square both sides to see that $(A^2 + B^2)^{1/2} \leq A + B$.

The proof that the second series in Lemma 1.5 converges is now a consequence of the following observation:

$$|n| + |m| \approx |n + m\tau| \quad \text{whenever } \tau \in \mathbb{H}.$$

Indeed, if $\tau = s + it$ with $s, t \in \mathbb{R}$ and $t > 0$, then

$$|n + m\tau| = [(n + ms)^2 + (mt)^2]^{1/2} \approx |n + ms| + |mt| \approx |n + ms| + |m|,$$

by the previous observation. Then, $|n + ms| + |m| \approx |n| + |m|$, by considering separately the cases when $|n| \leq 2|m||s|$ and $|n| \geq 2|m||s|$.

Remark. The proof above shows that when $r > 2$ the series $\sum |n + m\tau|^{-r}$ converges uniformly in every half-plane $\text{Im}(\tau) \geq \delta > 0$.
In contrast, when $r = 2$ this series fails to converge (Exercise 3).

With this technical point behind us, we may now return to the definition of the **Weierstrass $\wp$ function**, which is given by the series

$$\wp(z) = \frac{1}{z^2} + \sum_{\omega \in \Lambda^*} \left[\frac{1}{(z + \omega)^2} - \frac{1}{\omega^2} \right]$$

$$= \frac{1}{z^2} + \sum_{(n,m) \neq (0,0)} \left[\frac{1}{(z + n + m\tau)^2} - \frac{1}{(n + m\tau)^2} \right].$$

We claim that $\wp$ is a meromorphic function with double poles at the lattice points. To see this, suppose that $|z| < R$, and write

$$\wp(z) = \frac{1}{z^2} + \sum_{|\omega| \leq 2R} \left[\frac{1}{(z + \omega)^2} - \frac{1}{\omega^2} \right] + \sum_{|\omega| > 2R} \left[\frac{1}{(z + \omega)^2} - \frac{1}{\omega^2} \right].$$

The term in the second sum is $O(1/|\omega|^3)$ uniformly for $|z| < R$, so by Lemma 1.5 this second sum defines a holomorphic function in $|z| < R$. Finally, note that the first sum exhibits double poles at the lattice points in the disc $|z| < R$.

Observe that because of the insertion of the terms $-1/\omega^2$, it is no longer obvious whether $\wp$ is doubly periodic. Nevertheless this is true, and $\wp$ has all the properties of an elliptic function of order 2. We gather this result in a theorem.

Theorem 1.6 *The function $\wp$ is an elliptic function that has periods 1 and τ, and double poles at the lattice points.*

Proof. It remains only to prove that $\wp$ is periodic with the correct periods. To do so, note that the derivative is given by differentiating the series for $\wp$ termwise so

$$\wp'(z) = -2 \sum_{n,m \in \mathbb{Z}} \frac{1}{(z+n+m\tau)^3}.$$

This accomplishes two things for us. First, the differentiated series converges absolutely whenever z is not a lattice point, by the case $r = 3$ of Lemma 1.5. Second, the differentiation also eliminates the subtraction term $1/\omega^2$; therefore the series for $\wp'$ is clearly periodic with periods 1 and τ, since it remains unchanged after replacing z by $z+1$ or $z+\tau$.

Hence, there are two constants a and b such that

$$\wp(z+1) = \wp(z) + a \quad \text{and} \quad \wp(z+\tau) = \wp(z) + b.$$

It is clear from the definition, however, that $\wp$ is even, that is, $\wp(z) = \wp(-z)$, since the sum over $\omega \in \Lambda$ can be replaced by the sum over $-\omega \in \Lambda$. Therefore $\wp(-1/2) = \wp(1/2)$ and $\wp(-\tau/2) = \wp(\tau/2)$, and setting $z = -1/2$ and $z = -\tau/2$, respectively, in the two expressions above proves that $a = b = 0$.

A direct proof of the periodicity of $\wp$ can be given without differentiation; see Exercise 4.

Properties of $\wp$

Several remarks are in order. First, we have already observed that $\wp$ is even, and therefore $\wp'$ is odd. Since $\wp'$ is also periodic with periods 1 and τ, we find that

$$\wp'(1/2) = \wp'(\tau/2) = \wp'\left(\frac{1+\tau}{2}\right) = 0.$$

Indeed, one has, for example,

$$\wp'(1/2) = -\wp'(-1/2) = -\wp'(-1/2+1) = -\wp'(1/2).$$

Since $\wp'$ is elliptic and has order 3, the three points $1/2$, $\tau/2$, and $(1+\tau)/2$ (which are called the **half-periods**) are the only roots of $\wp'$ in the fundamental parallelogram, and they have multiplicity 1. Therefore, if we define

$$\wp(1/2) = e_1, \quad \wp(\tau/2) = e_2, \quad \text{and} \quad \wp\left(\frac{1+\tau}{2}\right) = e_3,$$

we conclude that the equation $\wp(z) = e_1$ has a double root at $1/2$. Since $\wp$ has order 2, there are no other solutions to the equation $\wp(z) = e_1$ in the fundamental parallelogram. Similarly the equations $\wp(z) = e_2$ and $\wp(z) = e_3$ have only double roots at $\tau/2$ and $(1+\tau)/2$, respectively. In particular, the three numbers e_1, e_2, and e_3 are distinct, for otherwise $\wp$ would have at least four roots in the fundamental parallelogram, contradicting the fact that $\wp$ has order 2. From these observations we can prove the following theorem.

Theorem 1.7 *The function $(\wp')^2$ is the cubic polynomial in $\wp$*

$$(\wp')^2 = 4(\wp - e_1)(\wp - e_2)(\wp - e_3).$$

Proof. The only roots of $F(z) = (\wp(z) - e_1)(\wp(z) - e_2)(\wp(z) - e_3)$ in the fundamental parallelogram have multiplicity 2 and are at the points $1/2, \tau/2$, and $(1+\tau)/2$. Also, $(\wp')^2$ has double roots at these points. Moreover, F has poles of order 6 at the lattice points, and so does $(\wp')^2$ (because $\wp'$ has poles of order 3 there). Consequently $(\wp')^2/F$ is holomorphic and still doubly-periodic, hence this quotient is constant. To find the value of this constant we note that for z near 0, one has

$$\wp(z) = \frac{1}{z^2} + \cdots \quad \text{and} \quad \wp'(z) = \frac{-2}{z^3} + \cdots,$$

where the dots indicate terms of higher order. Therefore the constant is 4, and the theorem is proved.

We next demonstrate the universality of $\wp$ by showing that every elliptic function is a simple combination of $\wp$ and $\wp'$.

Theorem 1.8 *Every elliptic function f with periods 1 and τ is a rational function of $\wp$ and $\wp'$.*

The theorem will be an easy consequence of the following version of it.

Lemma 1.9 *Every even elliptic function F with periods 1 and τ is a rational funcion of $\wp$.*

Proof. If F has a zero or pole at the origin it must be of even order, since F is an even function. As a consequence, there exists an integer m so that $F\wp^m$ has no zero or pole at the lattice points. We may therefore assume that F itself has no zero or pole on Λ.

Our immediate goal is to use $\wp$ to construct a doubly-periodic function G with precisely the same zeros and poles as F. To achieve this, we recall that $\wp(z) - \wp(a)$ has a single zero of order 2 if a is a half-period, and two distinct zeros at a and $-a$ otherwise. We must therefore carefully count the zeros and poles of F.

If a is a zero of F, then so is $-a$, since F is even. Moreover, a is congruent to $-a$ if and only if it is a half-period, in which case the zero is of even order. Therefore, if the points $a_1, -a_1, \ldots, a_m, -a_m$ counted with multiplicities[3] describe all the zeros of F, then

$$[\wp(z) - \wp(a_1)] \cdots [\wp(z) - \wp(a_m)]$$

has precisely the same roots as F. A similar argument, where $b_1, -b_1, \ldots, b_m, -b_m$ (with multiplicities) describe all the poles of F, then shows that

$$G(z) = \frac{[\wp(z) - \wp(a_1)] \cdots [\wp(z) - \wp(a_m)]}{[\wp(z) - \wp(b_1)] \cdots [\wp(z) - \wp(b_m)]}$$

is periodic and has the same zeros and poles as F. Therefore, F/G is holomorphic and doubly-periodic, hence constant. This concludes the proof of the lemma.

To prove the theorem, we first recall that $\wp$ is even while $\wp'$ odd. We then write f as a sum of an even and an odd function,

$$f(z) = f_{\text{even}}(z) + f_{\text{odd}}(z),$$

where in fact

$$f_{\text{even}}(z) = \frac{f(z) + f(-z)}{2} \quad \text{and} \quad f_{\text{odd}}(z) = \frac{f(z) - f(-z)}{2}.$$

Then, since $f_{\text{odd}}/\wp'$ is even, it is clear from the lemma applied to f_{even} and $f_{\text{odd}}/\wp'$ that f is a rational function of $\wp$ and $\wp'$.

---

[3]If a_j is not a half-period, then a_j and $-a_j$ have the multiplicity of F at these points. If a_j is a half-period, then a_j and $-a_j$ are congruent and each has multiplicity half of the multiplicity of F at this point.

2 The modular character of elliptic functions and Eisenstein series

We shall now study the modular character of elliptic functions, that is, their dependence on τ.

Recall the normalization we made at the beginning of the chapter. We started with two periods ω_1 and ω_2 linearly that are independent over $\mathbb{R}$, and we defined $\tau = \omega_2/\omega_1$. We could then assume that $\mathrm{Im}(\tau) > 0$, and also that the two periods are 1 and τ. Next, we considered the lattice generated by 1 and τ and constructed the function $\wp$, which is elliptic of order 2 with periods 1 and τ. Since the construction of $\wp$ depends on τ, we could write $\wp_\tau$ instead. This leads us to change our point of view and think of $\wp_\tau(z)$ primarily as a function of τ. This approach yields many interesting new insights.

Our considerations are guided by the following observations. First, since 1 and τ generate the periods of $\wp_\tau(z)$, and 1 and $\tau + 1$ generate the same periods, we can expect a close relationship between $\wp_\tau(z)$ and $\wp_{\tau+1}(z)$. In fact, it is easy to see that they are identical. Second, since $\tau = \omega_2/\omega_1$, by the normalization imposed at the beginning of Section 1, we see that $-1/\tau = -\omega_1/\omega_2$ (with $\mathrm{Im}(-1/\tau) > 0$). This corresponds essentially to an interchange of the two periods ω_1 and ω_2, and thus we can also expect an intimate connection between $\wp_\tau$ and $\wp_{-1/\tau}$. In fact, it is easy to verify that $\wp_{-1/\tau}(z) = \tau^2 \wp_\tau(\tau z)$.

So we are led to consider the group of transformations of the upper half-plane $\mathrm{Im}(\tau) > 0$, generated by the two transformations $\tau \mapsto \tau + 1$ and $\tau \mapsto -1/\tau$. This group is called the **modular group**. On the basis of what we said, it can be expected that all quantities intrinsically attached to $\wp_\tau(z)$ reflect the above transformations. We see this clearly when we consider the Eisenstein series.

2.1 Eisenstein series

The **Eisenstein series** of order k is defined by

$$E_k(\tau) = \sum_{(n,m)\neq(0,0)} \frac{1}{(n+m\tau)^k},$$

whenever k is an integer ≥ 3 and τ is a complex number with $\mathrm{Im}(\tau) > 0$. If Λ is the lattice generated by 1 and τ, and if we write $\omega = n + m\tau$, then another expression for the Eisenstein series is $\sum_{\omega \in \Lambda^*} 1/\omega^k$.

Theorem 2.1 *Eisenstein series have the following properties:*

(i) *The series $E_k(\tau)$ converges if $k \geq 3$, and is holomorphic in the upper half-plane.*

(ii) *$E_k(\tau) = 0$ if k is odd.*

(iii) *$E_k(\tau)$ satisfies the following transformation relations:*

$$E_k(\tau + 1) = E_k(\tau) \quad and \quad E_k(\tau) = \tau^{-k} E_k(-1/\tau).$$

The last property is sometimes referred to as the **modular** character of the Eisenstein series. We shall return to these and other modular identities in the next chapter.

Proof. By Lemma 1.5 and the remark after it, the series $E_k(\tau)$ converges absolutely and uniformly in every half-plane $\text{Im}(\tau) \geq \delta > 0$, whenever $k \geq 3$; hence $E_k(\tau)$ is holomorphic in the upper half-plane $\text{Im}(\tau) > 0$.

By symmetry, replacing n and m by $-n$ and $-m$, we see that whenever k is odd the Eisenstein series is identically zero.

Finally, the fact that $E_k(\tau)$ is periodic of period 1 is clear from the fact that $n + m(\tau + 1) = n + m + m\tau$, and that we can rearrange the sum by replacing $n + m$ by n. Also, we have

$$(n + m(-1/\tau))^k = \tau^{-k}(n\tau - m)^k,$$

and again we can rearrange the sum, this time replacing $(-m, n)$ by (n, m). Conclusion (iii) then follows.

Remark. Because of the second property, some authors define the Eisenstein series of order k to be $\sum_{(n,m) \neq (0,0)} 1/(n + m\tau)^{2k}$, possibly also with a constant factor in front.

The connection of the E_k with the Weierstrass $\wp$ function arises when we investigate the series expansion of $\wp$ near 0.

Theorem 2.2 *For z near 0, we have*

$$\wp(z) = \frac{1}{z^2} + 3E_4 z^2 + 5E_6 z^4 + \cdots$$

$$= \frac{1}{z^2} + \sum_{k=1}^{\infty} (2k+1) E_{2k+2} z^{2k}.$$

Proof. From the definition of $\wp$, if we note that we may replace ω by $-\omega$ without changing the sum, we have

$$\wp(z) = \frac{1}{z^2} + \sum_{\omega \in \Lambda^*} \left[\frac{1}{(z+\omega)^2} - \frac{1}{\omega^2} \right] = \frac{1}{z^2} + \sum_{\omega \in \Lambda^*} \left[\frac{1}{(z-\omega)^2} - \frac{1}{\omega^2} \right],$$

where $\omega = n + m\tau$. The identity

$$\frac{1}{(1-w)^2} = \sum_{\ell=0}^{\infty} (\ell+1)w^\ell, \quad \text{for } |w| < 1,$$

which follows from differentiating the geometric series, implies that for all small z

$$\frac{1}{(z-\omega)^2} = \frac{1}{\omega^2} \sum_{\ell=0}^{\infty} (\ell+1) \left(\frac{z}{\omega}\right)^\ell = \frac{1}{\omega^2} + \frac{1}{\omega^2} \sum_{\ell=1}^{\infty} (\ell+1) \left(\frac{z}{\omega}\right)^\ell.$$

Therefore

$$\wp(z) = \frac{1}{z^2} + \sum_{\omega \in \Lambda^*} \sum_{\ell=1}^{\infty} (\ell+1) \frac{z^\ell}{\omega^{\ell+2}}$$

$$= \frac{1}{z^2} + \sum_{\ell=1}^{\infty} (\ell+1) \left(\sum_{\omega \in \Lambda^*} \frac{1}{\omega^{\ell+2}}\right) z^\ell$$

$$= \frac{1}{z^2} + \sum_{\ell=1}^{\infty} (\ell+1) E_{\ell+2} z^\ell$$

$$= \frac{1}{z^2} + \sum_{k=1}^{\infty} (2k+1) E_{2k+2} z^{2k},$$

where we have used the fact that $E_{\ell+2} = 0$ whenever ℓ is odd.

From this theorem, we obtain the following three expansions for z near 0:

$$\wp'(z) = \frac{-2}{z^3} + 6E_4 z + 20E_6 z^3 + \cdots,$$

$$(\wp'(z))^2 = \frac{4}{z^6} - \frac{24E_4}{z^2} - 80E_6 + \cdots,$$

$$(\wp(z))^3 = \frac{1}{z^6} + \frac{9E_4}{z^2} + 15E_6 + \cdots.$$

From these, one sees that the difference $(\wp'(z))^2 - 4(\wp(z))^3 + 60E_4\wp(z) + 140E_6$ is holomorphic near 0, and in fact equal to 0 at the origin. Since this difference is also doubly periodic, we conclude by Theorem 1.2 that it is constant, and hence identically 0. This proves the following corollary.

Corollary 2.3 *If $g_2 = 60E_4$ and $g_3 = 140E_6$, then*

$$(\wp')^2 = 4\wp^3 - g_2\wp - g_3.$$

Note that this identity is another version of Theorem 1.7, and it allows one to express the symmetric functions of the e_j's in terms of the Eisenstein series.

2.2 Eisenstein series and divisor functions

We will describe now the link between Eisenstein series and some number-theoretic quantities. This relation comes about if we consider the Fourier coefficients in the Fourier expansion of the periodic function $E_k(\tau)$. Equivalently, we can write $\mathcal{E}(z) = E_k(\tau)$ with $z = e^{2\pi i \tau}$, and investigate the Laurent expansion of $\mathcal{E}$ as a function of z.

We begin with a lemma.

Lemma 2.4 *If $k \geq 2$ and $\mathrm{Im}(\tau) > 0$, then*

$$\sum_{n=-\infty}^{\infty} \frac{1}{(n+\tau)^k} = \frac{(-2\pi i)^k}{(k-1)!} \sum_{\ell=1}^{\infty} \ell^{k-1} e^{2\pi i \tau \ell}.$$

Proof. This identity follows from applying the Poisson summation formula to $f(z) = 1/(z+\tau)^k$; see Exercise 7 in Chapter 4.

An alternate proof consists of noting that it first suffices to establish the formula for $k = 2$, since the other cases are then obtained by differentiating term by term. To prove this special case, we differentiate the formula for the cotangent derived in Chapter 5

$$\sum_{n=-\infty}^{\infty} \frac{1}{n+\tau} = \pi \cot \pi \tau.$$

This yields

$$\sum_{n=-\infty}^{\infty} \frac{1}{(n+\tau)^2} = \frac{\pi^2}{\sin^2(\pi\tau)}.$$

Now use Euler's formula for the sine and the fact that

$$\sum_{r=1}^{\infty} r w^r = \frac{w}{(1-w)^2} \qquad \text{with } w = e^{2\pi i \tau}$$

to obtain the desired result.

As a consequence of this lemma, we can draw a connection between the Eisenstein series, the zeta function, and the divisor functions. The

divisor function $\sigma_\ell(r)$ that arises here is defined as the sum of the ℓth powers of the divisors of r, that is,

$$\sigma_\ell(r) = \sum_{d|r} d^\ell.$$

Theorem 2.5 *If $k \geq 4$ is even, and $\mathrm{Im}(\tau) > 0$, then*

$$E_k(\tau) = 2\zeta(k) + \frac{2(-1)^{k/2}(2\pi)^k}{(k-1)!} \sum_{r=1}^{\infty} \sigma_{k-1}(r) e^{2\pi i r \tau}.$$

Proof. First observe that $\sigma_{k-1}(r) \leq r r^{k-1} = r^k$. If $\mathrm{Im}(\tau) = t$, then whenever $t \geq t_0$ we have $|e^{2\pi i r \tau}| \leq e^{-2\pi r t_0}$, and we see that the series in the theorem is absolutely convergent in any half-plane $t \geq t_0$, by comparison with $\sum_{r=1}^{\infty} r^k e^{-2\pi r t_0}$. To establish the formula, we use the definition of E_k, that of ζ, the fact that k is even, and the previous lemma (with τ replaced by $m\tau$) to get successively

$$E_k(\tau) = \sum_{(n,m)\neq(0,0)} \frac{1}{(n+m\tau)^k}$$

$$= \sum_{n\neq0} \frac{1}{n^k} + \sum_{m\neq0} \sum_{n=-\infty}^{\infty} \frac{1}{(n+m\tau)^k}$$

$$= 2\zeta(k) + \sum_{m\neq0} \sum_{n=-\infty}^{\infty} \frac{1}{(n+m\tau)^k}$$

$$= 2\zeta(k) + 2 \sum_{m>0} \sum_{n=-\infty}^{\infty} \frac{1}{(n+m\tau)^k}$$

$$= 2\zeta(k) + 2 \sum_{m>0} \frac{(-2\pi i)^k}{(k-1)!} \sum_{\ell=1}^{\infty} \ell^{k-1} e^{2\pi i m \tau \ell}$$

$$= 2\zeta(k) + \frac{2(-1)^{k/2}(2\pi)^k}{(k-1)!} \sum_{m>0} \sum_{\ell=1}^{\infty} \ell^{k-1} e^{2\pi i \tau m \ell}$$

$$= 2\zeta(k) + \frac{2(-1)^{k/2}(2\pi)^k}{(k-1)!} \sum_{r=1}^{\infty} \sigma_{k-1}(r) e^{2\pi i r \tau}.$$

This proves the desired formula.

Finally, we turn to the *forbidden* case $k = 2$. The series we have in mind $\sum_{(n,m)\neq(0,0)} 1/(n+m\tau)^2$ no longer converges absolutely, but we seek to give it a meaning anyway. We define

$$F(\tau) = \sum_m \left(\sum_n \frac{1}{(n + m\tau)^2} \right)$$

summed in the indicated order with $(n, m) \neq (0, 0)$. The argument given in the above theorem proves that the double sum converges, and in fact has the expected expression.

Corollary 2.6 *The double sum defining F converges in the indicated order. We have*

$$F(\tau) = 2\zeta(2) - 8\pi^2 \sum_{r=1}^{\infty} \sigma(r) e^{2\pi i r \tau},$$

where $\sigma(r) = \sum_{d|r} d$ is the sum of the divisors of r.

It can be seen that $F(-1/\tau)\tau^{-2}$ does *not* equal $F(\tau)$, and this is the same as saying that the double series for F gives a different value ($\tilde{F}$, the reverse of F) when we sum first in m and then in n. It turns out that nevertheless the **forbidden Eisenstein series** $F(\tau)$ can be used in a crucial way in the proof of the celebrated theorem about representing an integer as the sum of four squares. We turn to these matters in the next chapter.

3 Exercises

1. Suppose that a meromorphic function f has two periods ω_1 and ω_2, with $\omega_2/\omega_1 \in \mathbb{R}$.

(a) Suppose ω_2/ω_1 is rational, say equal to p/q, where p and q are relatively prime integers. Prove that as a result the periodicity assumption is equivalent to the assumption that f is periodic with the simple period $\omega_0 = \frac{1}{q}\omega_1$. [Hint: Since p and q are relatively prime, there exist integers m and n such that $mq + np = 1$ (Corollary 1.3, Chapter 8, Book I).]

(b) If ω_2/ω_1 is irrational, then f is constant. To prove this, use the fact that $\{m - n\tau\}$ is dense in $\mathbb{R}$ whenever τ is irrational and m, n range over the integers.

2. Suppose that $a_1, \ldots, a_r$ and $b_1, \ldots, b_r$ are the zeros and poles, respectively, in the fundamental parallelogram of an elliptic function f. Show that

$$a_1 + \cdots + a_r - b_1 - \cdots - b_r = n\omega_1 + m\omega_2$$

for some integers n and m.

[Hint: If the boundary of the parallelogram contains no zeros or poles, simply integrate $zf'(z)/f(z)$ over that boundary, and observe that the integral of $f'(z)/f(z)$ over a side is an integer multiple of $2\pi i$. If there are zeros or poles on the side of the parallelogram, translate it by a small amount to reduce the problem to the first case.]

3. In contrast with the result in Lemma 1.5, prove that the series

$$\sum_{n+m\tau\in\Lambda^*} \frac{1}{|n+m\tau|^2} \qquad \text{where } \tau \in \mathbb{H}$$

does not converge. In fact, show that

$$\sum_{1\le n^2+m^2\le R^2} 1/(n^2+m^2) = 2\pi \log R + O(1) \qquad \text{as } R \to \infty.$$

4. By rearranging the series

$$\frac{1}{z^2} + \sum_{w\in\Lambda^*} \left[\frac{1}{(z+w)^2} - \frac{1}{w^2} \right],$$

show directly, without differentiation, that $\wp(z+w) = \wp(z)$ whenever $w \in \Lambda$.

[Hint: For R sufficiently large, note that $\wp(z) = \wp^R(z) + O(1/R)$, where $\wp^R(z) = z^{-2} + \sum_{0<|w|<R}((z+w)^{-2} - w^{-2})$. Next, observe that both $\wp^R(z+1) - \wp^R(z)$ and $\wp^R(z+\tau) - \wp^R(z)$ are $O(\sum_{R-c<|w|<R+c} |w|^{-2}) = O(1/R)$.]

5. Let $\sigma(z)$ be the canonical product

$$\sigma(z) = z \prod_{j=1}^{\infty} E_2(z/\tau_j),$$

where τ_j is an enumeration of the periods $\{n+m\tau\}$ with $(n,m) \ne (0,0)$, and $E_2(z) = (1-z)e^{z+z^2/2}$.

(a) Show that $\sigma(z)$ is an entire function of order 2 that has simple zeros at all the periods $n + m\tau$, and vanishes nowhere else.

(b) Show that

$$\frac{\sigma'(z)}{\sigma(z)} = \frac{1}{z} + \sum_{(n,m)\ne(0,0)} \left[\frac{1}{z-n-m\tau} + \frac{1}{n+m\tau} + \frac{z}{(n+m\tau)^2} \right],$$

and that this series converges whenever z is not a lattice point.

(c) Let $L(z) = -\sigma'(z)/\sigma(z)$. Then

$$L'(z) = \frac{(\sigma'(z))^2 - \sigma(z)\sigma''(z)}{(\sigma(z))^2} = \wp(z).$$

6. Prove that $\wp''$ is a quadratic polynomial in $\wp$.

7. Setting $\tau = 1/2$ in the expression

$$\sum_{m=-\infty}^{\infty} \frac{1}{(m+\tau)^2} = \frac{\pi^2}{\sin^2(\pi\tau)},$$

deduce that

$$\sum_{m\geq 1,\ m\ \text{odd}} \frac{1}{m^2} = \frac{\pi^2}{8} \qquad \text{and} \qquad \sum_{m\geq 1} \frac{1}{m^2} = \frac{\pi^2}{6} = \zeta(2).$$

Similarly, using $\sum 1/(m+\tau)^4$ deduce that

$$\sum_{m\geq 1,\ m\ \text{odd}} \frac{1}{m^4} = \frac{\pi^4}{96} \qquad \text{and} \qquad \sum_{m\geq 1} \frac{1}{m^4} = \frac{\pi^4}{90} = \zeta(4).$$

These results were already obtained using Fourier series in the exercises at the end of Chapters 2 and 3 in Book I.

8. Let

$$E_4(\tau) = \sum_{(n,m)\neq(0,0)} \frac{1}{(n+m\tau)^4}$$

be the Eisenstein series of order 4.

(a) Show that $E_4(\tau) \to \pi^4/45$ as $\text{Im}(\tau) \to \infty$.

(b) More precisely,

$$\left| E_4(\tau) - \frac{\pi^4}{45} \right| \leq ce^{-2\pi t} \qquad \text{if } \tau = x + it \text{ and } t \geq 1.$$

(c) Deduce that

$$\left| E_4(\tau) - \tau^{-4}\frac{\pi^4}{45} \right| \leq ct^{-4}e^{-2\pi/t} \qquad \text{if } \tau = it \text{ and } 0 < t \leq 1.$$

4 Problems

1. Besides the approach in Section 1.2, there are several alternate ways of dealing with the sum $\sum 1/(z+w)^2$, where $w = n + m\tau$. For example, one may sum either (a) circularly, (b) first in n then in m, (c) or first in m then in n.

(a) Prove that if $z \notin \Lambda$, then

$$\lim_{R \to \infty} \sum_{n^2+m^2 \leq R^2} \frac{1}{(z+n+m\tau)^2} = S_1(z)$$

exists and $S_1(z) = \wp(z) + c_1$.

(b) Similarly,

$$\sum_m \left(\sum_n \frac{1}{(z+n+m\tau)^2} \right) = S_2(z)$$

exists and $S_2(z) = \wp(z) + c_2$, where $c_2 = F(\tau)$, and F is the forbidden Eisenstein series.

(c) Also

$$\sum_n \left(\sum_m \frac{1}{(z+n+m\tau)^2} \right) = S_3(z)$$

exists with $S_3(z) = \wp(z) + c_3$, and $c_3 = \tilde{F}(\tau)$, the reverse of F.

[Hint: To prove (a), it suffices to show that $\lim_{R \to \infty}$, $\displaystyle\sum_{1 \leq n^2+m^2 \leq R^2} 1/(n+m\tau)^2 = c_1$ exists. This is proved by a comparision with $\int_{1 \leq x^2+y^2 \leq R^2} \frac{dx}{(x+y\tau)^2} = I(R)$. It can be shown that $I(R) = 0$, which follows because $(x+y\tau)^{-2} = -(\partial/\partial x)(x+y\tau)^{-1}$.]

2. Show that

$$\wp(z) = c + \pi^2 \sum_{m=-\infty}^{\infty} \frac{1}{\sin^2((z+m\tau)\pi)}$$

where c is an appropriate constant. In fact, by part (b) of the previous problem $c = -F(\tau)$.

3. * Suppose Ω is a simply connected domain that excludes the three roots of the polynomial $4z^3 - g_2z - g_3$. For $w_0 \in \Omega$ and w_0 fixed, define the function I on Ω by

$$I(w) = \int_{w_0}^{w} \frac{dz}{\sqrt{4z^3 - g_2z - g_3}} \qquad w \in \Omega.$$

Then the function I has an inverse given by $\wp(z+\alpha)$ for some constant α; that is,

$$I(\wp(z+\alpha)) = z$$

for appropriate α.

[Hint: Prove that $(I(\wp(z+\alpha)))' = \pm 1$, and use the fact that $\wp$ is even.]

4.* Suppose τ is purely imaginary, say $\tau = it$ with $t > 0$. Consider the division of the complex plane into congruent rectangles obtained by considering the lines $x = n/2$, $y = tm/2$ as n and m range over the integers. (An example is the rectangle whose vertices are $0, 1/2, 1/2 + \tau/2$, and $\tau/2$.)

   (a) Show that $\wp$ is real-valued on all these lines, and hence on the boundaries of all these rectangles.

   (b) Prove that $\wp$ maps the interior of each rectangle conformally to the upper (or lower) half-plane.

10 Applications of Theta Functions

The problem of the representation of an integer n as the sum of a given number k of integral squares is one of the most celebrated in the theory of numbers. Its history may be traced back to Diophantus, but begins effectively with Girard's (or Fermat's) theorem that a prime $4m + 1$ is the sum of two squares. Almost every arithmetician of note since Fermat has contributed to the solution of the problem, and it has its puzzles for us still.

G. H. Hardy, 1940

This chapter is devoted to a closer look at the theory of theta functions and some of its applications to combinatorics and number theory.

The theta function is given by the series

$$\Theta(z|\tau) = \sum_{n=-\infty}^{\infty} e^{\pi i n^2 \tau} e^{2\pi i n z},$$

which converges for all $z \in \mathbb{C}$, and τ in the upper half-plane.

A remarkable feature of the theta function is its dual nature. When viewed as a function of z, we see it in the arena of elliptic functions, since Θ is periodic with period 1 and "quasi-period" τ. When considered as a function of τ, Θ reveals its modular nature and close connection with the partition function and the problem of representation of integers as sums of squares.

The two main tools allowing us to exploit these links are the triple-product for Θ and its transformation law. Once we have proved these theorems, we give a brief introduction to the connection with partitions, and then pass to proofs of the celebrated theorems about representation of integers as sums of two or four squares.

1 Product formula for the Jacobi theta function

In its most elaborate form, Jacobi's **theta function** is defined for $z \in \mathbb{C}$ and $\tau \in \mathbb{H}$ by

$$(1) \qquad \Theta(z|\tau) = \sum_{n=-\infty}^{\infty} e^{\pi i n^2 \tau} e^{2\pi i n z}.$$

Two significant special cases (or variants) are $\theta(\tau)$ and $\vartheta(t)$, which are defined by

$$\theta(\tau) = \sum_{n=-\infty}^{\infty} e^{\pi i n^2 \tau}, \qquad \tau \in \mathbb{H},$$

$$\vartheta(t) = \sum_{n=-\infty}^{\infty} e^{-\pi n^2 t}, \qquad t > 0.$$

In fact, the relation between these various functions is given by $\theta(\tau) = \Theta(0|\tau)$ and $\vartheta(t) = \theta(it)$, with of course, $t > 0$.

We have already encountered these functions several times. For example, in the study of the heat diffusion equation for the circle, in Chapter 4 of Book I, we found that the heat kernel was given by

$$H_t(x) = \sum_{n=-\infty}^{\infty} e^{-4\pi^2 n^2 t} e^{2\pi i n x},$$

and therefore $H_t(x) = \Theta(x|4\pi i t)$.

Another instance was the occurence of ϑ in the study of the zeta function. In fact, we proved in Chapter 6 that the functional equation of ϑ implied that of ζ, which then led to the analytic continuation of the zeta function.

We begin our closer look at Θ as a function of z, with τ fixed, by recording its basic structural properties, which to a large extent characterize it.

Proposition 1.1 *The function Θ satisfies the following properties:*

(i) Θ *is entire in $z \in \mathbb{C}$ and holomorphic in $\tau \in \mathbb{H}$.*

(ii) $\Theta(z + 1|\tau) = \Theta(z|\tau)$.

(iii) $\Theta(z + \tau|\tau) = \Theta(z|\tau) e^{-\pi i \tau} e^{-2\pi i z}$.

(iv) $\Theta(z|\tau) = 0$ *whenever $z = 1/2 + \tau/2 + n + m\tau$ and $n, m \in \mathbb{Z}$.*

Proof. Suppose that $\text{Im}(\tau) = t \geq t_0 > 0$ and $z = x + iy$ belongs to a bounded set in $\mathbb{C}$, say $|z| \leq M$. Then, the series defining Θ is absolutely and uniformly convergent, since

$$\sum_{n=-\infty}^{\infty} |e^{\pi i n^2 \tau} e^{2\pi i n z}| \leq C \sum_{n \geq 0} e^{-\pi n^2 t_0} e^{2\pi n M} < \infty.$$

Therefore, for each fixed $\tau \in \mathbb{H}$ the function $\Theta(\cdot|\tau)$ is entire, and for each fixed $z \in \mathbb{C}$ the function $\Theta(z|\cdot)$ is holomorphic in the upper half-plane.

Since the exponential $e^{2\pi i n z}$ is periodic of period 1, property (ii) is immediate from the definition of Θ.

To show the third property we may complete the squares in the expression for $\Theta(z + \tau|\tau)$. In detail, we have

$$\Theta(z + \tau|\tau) = \sum_{n=-\infty}^{\infty} e^{\pi i n^2 \tau} e^{2\pi i n(z+\tau)}$$

$$= \sum_{n=-\infty}^{\infty} e^{\pi i (n^2 + 2n)\tau} e^{2\pi i n z}$$

$$= \sum_{n=-\infty}^{\infty} e^{\pi i (n+1)^2 \tau} e^{-\pi i \tau} e^{2\pi i n z}$$

$$= \sum_{n=-\infty}^{\infty} e^{\pi i (n+1)^2 \tau} e^{-\pi i \tau} e^{2\pi i (n+1) z} e^{-2\pi i z}$$

$$= \Theta(z|\tau) e^{-\pi i \tau} e^{-2\pi i z}.$$

Thus we see that $\Theta(z|\tau)$, as a function of z, is periodic with period 1 and "quasi-periodic" with period τ.

To establish the last property it suffices, by what was just shown, to prove that $\Theta(1/2 + \tau/2|\tau) = 0$. Again, we use the interplay between n and n^2 to get

$$\Theta(1/2 + \tau/2|\tau) = \sum_{n=-\infty}^{\infty} e^{\pi i n^2 \tau} e^{2\pi i n(1/2 + \tau/2)}$$

$$= \sum_{n=-\infty}^{\infty} (-1)^n e^{\pi i (n^2 + n)\tau}.$$

To see that this last sum is identically zero, it suffices to match $n \geq 0$ with $-n - 1$, and to observe that they have opposite parity, and that $(-n - 1)^2 + (-n - 1) = n^2 + n$. This completes the proof of the proposition.

We consider next a product $\Pi(z|\tau)$ that enjoys the same structural properties as $\Theta(z|\tau)$ as a function of z. This product is defined for $z \in \mathbb{C}$ and $\tau \in \mathbb{H}$ by

$$\Pi(z|\tau) = \prod_{n=1}^{\infty}(1 - q^{2n})(1 + q^{2n-1}e^{2\pi i z})(1 + q^{2n-1}e^{-2\pi i z}),$$

where we have used the notation that is standard in the subject, namely $q = e^{\pi i \tau}$. The function $\Pi(z|\tau)$ is sometimes referred to as the **triple-product**.

Proposition 1.2 *The function $\Pi(z|\tau)$ satisfies the following properties:*

(i) $\Pi(z, \tau)$ *is entire in $z \in \mathbb{C}$ and holomorphic for $\tau \in \mathbb{H}$.*

(ii) $\Pi(z + 1|\tau) = \Pi(z|\tau)$.

(iii) $\Pi(z + \tau|\tau) = \Pi(z|\tau)e^{-\pi i \tau}e^{-2\pi i z}$.

(iv) $\Pi(z|\tau) = 0$ *whenever $z = 1/2 + \tau/2 + n + m\tau$ and $n, m \in \mathbb{Z}$. Moreover, these points are simple zeros of $\Pi(\cdot|\tau)$, and $\Pi(\cdot|\tau)$ has no other zeros.*

Proof. If $\text{Im}(\tau) = t \geq t_0 > 0$ and $z = x + iy$, then $|q| \leq e^{-\pi t_0} < 1$ and

$$(1 - q^{2n})(1 + q^{2n-1}e^{2\pi i z})(1 + q^{2n-1}e^{-2\pi i z}) = 1 + O\left(|q|^{2n-1}e^{2\pi|z|}\right).$$

Since the series $\sum |q|^{2n-1}$ converges, the results for infinite products in Chapter 5 guarantee that $\Pi(z|\tau)$ defines an entire function of z with $\tau \in \mathbb{H}$ fixed, and a holomorphic function for $\tau \in \mathbb{H}$ with $z \in \mathbb{C}$ fixed.

Also, it is clear from the definition that $\Pi(z|\tau)$ is periodic of period 1 in the z variable.

To prove the third property, we first observe that since $q^2 = e^{2\pi i \tau}$ we have

$$\Pi(z + \tau|\tau) = \prod_{n=1}^{\infty}(1 - q^{2n})(1 + q^{2n-1}e^{2\pi i(z+\tau)})(1 + q^{2n-1}e^{-2\pi i(z+\tau)})$$

$$= \prod_{n=1}^{\infty}(1 - q^{2n})(1 + q^{2n+1}e^{2\pi i z})(1 + q^{2n-3}e^{-2\pi i z}).$$

Comparing this last product with $\Pi(z|\tau)$, and isolating the factors that are either missing or extra leads to

$$\Pi(z + \tau|\tau) = \Pi(z|\tau)\left(\frac{1 + q^{-1}e^{-2\pi i z}}{1 + qe^{2\pi i z}}\right).$$

Hence (iii) follows because $(1 + x)/(1 + x^{-1}) = x$, whenever $x \neq -1$.

Finally, to find the zeros of $\Pi(z|\tau)$ we recall that a product that converges vanishes only if at least one of its factors is zero. Clearly, the factor $(1 - q^n)$ never vanishes since $|q| < 1$. The second factor $(1 + q^{2n-1}e^{2\pi i z})$ vanishes when $q^{2n-1}e^{2\pi i z} = -1 = e^{\pi i}$. Since $q = e^{\pi i \tau}$, we then have[1]

$$(2n - 1)\tau + 2z = 1 \pmod{2}.$$

Hence,

$$z = 1/2 + \tau/2 - n\tau \pmod{1},$$

and this takes care of the zeros of the type $1/2 + \tau/2 - n\tau + m$ with $n \geq 1$ and $m \in \mathbb{Z}$. Similarly, the third factor vanishes if

$$(2n - 1)\tau - 2z = 1 \pmod{2}$$

which implies that

$$z = -1/2 - \tau/2 + n\tau \pmod{1}$$
$$= 1/2 + \tau/2 + n'\tau \pmod{1},$$

where $n' \geq 0$. This exhausts the zeros of $\Pi(\cdot|\tau)$. Finally, these zeros are simple, since the function $e^w - 1$ vanishes at the origin to order 1 (a fact obvious from a power series expansion or a simple differentiation).

The importance of the product Π comes from the following theorem, called the product formula for the theta function. The fact that $\Theta(z|\tau)$ and $\Pi(z|\tau)$ satisfy similar properties hints at a close connection between the two. This is indeed the case.

Theorem 1.3 (Product formula) *For all $z \in \mathbb{C}$ and $\tau \in \mathbb{H}$ we have the identity $\Theta(z|\tau) = \Pi(z|\tau)$.*

Proof. Fix $\tau \in \mathbb{H}$. We claim first that there exists a constant $c(\tau)$ such that

(2) $$\Theta(z|\tau) = c(\tau)\Pi(z|\tau).$$

In fact, consider the quotient $F(z) = \Theta(z|\tau)/\Pi(z|\tau)$, and note that by the previous two propositions, the function F is entire and doubly periodic with periods 1 and τ. This implies that F is constant as claimed.

---

[1] We use the standard short-hand, $a = b \pmod{c}$, to mean that $a - b$ is an integral multiple of c.

We must now prove that $c(\tau) = 1$ for all τ, and the main point is to establish that $c(\tau) = c(4\tau)$. If we put $z = 1/2$ in (2), so that $e^{2i\pi z} = e^{-2i\pi z} = -1$, we obtain

$$\sum_{n=-\infty}^{\infty} (-1)^n q^{n^2} = c(\tau) \prod_{n=1}^{\infty} (1 - q^{2n})(1 - q^{2n-1})(1 - q^{2n-1})$$

$$= c(\tau) \prod_{n=1}^{\infty} \left[(1 - q^{2n-1})(1 - q^{2n}) \right] (1 - q^{2n-1})$$

$$= c(\tau) \prod_{n=1}^{\infty} (1 - q^n)(1 - q^{2n-1}).$$

Hence

(3) $$c(\tau) = \frac{\sum_{n=-\infty}^{\infty}(-1)^n q^{n^2}}{\prod_{n=1}^{\infty}(1 - q^n)(1 - q^{2n-1})}.$$

Next, we put $z = 1/4$ in (2), so that $e^{2i\pi z} = i$. On the one hand, we have

$$\Theta(1/4|\tau) = \sum_{n=-\infty}^{\infty} q^{n^2} i^n,$$

and due to the fact that $1/i = -i$, only the terms corresponding to $n = $ even $= 2m$ are not cancelled; thus

$$\Theta(1/4|\tau) = \sum_{m=-\infty}^{\infty} q^{4m^2}(-1)^m.$$

On the other hand,

$$\Pi(1/4|\tau) = \prod_{m=1}^{\infty} (1 - q^{2m})(1 + iq^{2m-1})(1 - iq^{2m-1})$$

$$= \prod_{m=1}^{\infty} (1 - q^{2m})(1 + q^{4m-2})$$

$$= \prod_{n=1}^{\infty} (1 - q^{4n})(1 - q^{8n-4}),$$

where the last line is obtained by considering separately the two cases $2m = 4n - 4$ and $2m = 4n - 2$ in the first factor. Hence

(4) $$c(\tau) = \frac{\sum_{n=-\infty}^{\infty}(-1)^n q^{4n^2}}{\prod_{n=1}^{\infty}(1 - q^{4n})(1 - q^{8n-4})},$$

and combining (3) and (4) establishes our claim that $c(\tau) = c(4\tau)$. Successive applications of this identity give $c(\tau) = c(4^k\tau)$, and since $q^{4^k} = e^{i\pi 4^k \tau} \to 0$ as $k \to \infty$, we conclude from (2) that $c(\tau) = 1$. This proves the theorem.

The product formula for the function Θ specializes to its variant $\theta(\tau) = \Theta(0|\tau)$, and this provides a proof that θ is non-vanishing in the upper half-plane.

Corollary 1.4 *If* $\operatorname{Im}(\tau) > 0$ *and* $q = e^{\pi i \tau}$, *then*

$$\theta(\tau) = \prod_{n=1}^{\infty}(1 - q^{2n})(1 + q^{2n-1})^2.$$

Thus $\theta(\tau) \neq 0$ *for* $\tau \in \mathbb{H}$.

The next corollary shows that the properties of the function Θ now yield the construction of an elliptic function (which is in fact closely related to the Weierstrass $\wp$ function).

Corollary 1.5 *For each fixed* $\tau \in \mathbb{H}$, *the quotient*

$$(\log \Theta(z|\tau))'' = \frac{\Theta(z|\tau)\Theta''(z|\tau) - (\Theta'(z|\tau))^2}{\Theta(z|\tau)^2}$$

is an elliptic function of order 2 with periods 1 and τ, and with a double pole at $z = 1/2 + \tau/2$.

In the above, the primes $'$ denote differentiation with respect to the z variable.

Proof. Let $F(z) = (\log \Theta(z|\tau))' = \Theta(z|\tau)'/\Theta(z|\tau)$. Differentiating the identities (ii) and (iii) of Proposition 1.1 gives $F(z + 1) = F(z)$, $F(z + \tau) = F(z) - 2\pi i$, and differentiating again shows that $F'(z)$ is doubly periodic. Since $\Theta(z|\tau)$ vanishes only at $z = 1/2 + \tau/2$ in the fundamental parallelogram, the function $F(z)$ has only a single pole, and thus $F'(z)$ has only a double pole there.

The precise connection between $(\log \Theta(z|\tau))''$ and $\wp_\tau(z)$ is stated in Exercise 1.

For an analogy between Θ and the Weierstrass σ function, see Exercise 5 of the previous chapter.

1.1 Further transformation laws

We now come to the study of the transformation relations in the τ-variable, that is, to the modular character of Θ.

Recall that in the previous chapter, the modular character of the Weierstrass $\wp$ function and Eisenstein series E_k was reflected by the two transformations

$$\tau \mapsto \tau + 1 \quad \text{and} \quad \tau \mapsto -1/\tau,$$

which preserve the upper half-plane. In what follows, we shall denote these two transformations by T_1 and S, respectively.

When looking at the Θ function, however, it will be natural to consider instead the transformations

$$T_2 : \tau \mapsto \tau + 2 \quad \text{and} \quad S : \tau \mapsto -1/\tau,$$

since $\Theta(z|\tau + 2) = \Theta(z|\tau)$, but $\Theta(z|\tau + 1) \neq \Theta(z|\tau)$.

Our first task is to study the transformation of $\Theta(z|\tau)$ under the mapping $\tau \mapsto -1/\tau$.

Theorem 1.6 *If $\tau \in \mathbb{H}$, then*

$$(5) \qquad \Theta(z|-1/\tau) = \sqrt{\frac{\tau}{i}} e^{\pi i \tau z^2} \Theta(z\tau|\tau) \quad \text{for all } z \in \mathbb{C}.$$

Here $\sqrt{\tau/i}$ denotes the branch of the square root defined on the upper half-plane, that is positive when $\tau = it$, $t > 0$.

Proof. It suffices to prove this formula for $z = x$ real and $\tau = it$ with $t > 0$, since for each fixed $x \in \mathbb{R}$, the two sides of equation (5) are holomorphic functions in the upper half-plane which then agree on the positive imaginary axis, and hence must be equal everywhere. Also, for a fixed $\tau \in \mathbb{H}$ the two sides define holomorphic functions in z that agree on the real axis, and hence must be equal everywhere.

With x real and $\tau = it$ the formula becomes

$$\sum_{n=-\infty}^{\infty} e^{-\pi n^2/t} e^{2\pi i n x} = t^{1/2} e^{-\pi t x^2} \sum_{n=-\infty}^{\infty} e^{-\pi n^2 t} e^{-2\pi n x t}.$$

Replacing x by a, we find that we must prove

$$\sum_{n=-\infty}^{\infty} e^{-\pi t(n+a)^2} = \sum_{n=-\infty}^{\infty} t^{-1/2} e^{-\pi n^2/t} e^{2\pi i n a}.$$

However, this is precisely equation (3) in Chapter 4, which was derived from the Poisson summation formula.

In particular, by setting $z = 0$ in the theorem, we find the following.

Corollary 1.7 *If* $\text{Im}(\tau) > 0$, *then* $\theta(-1/\tau) = \sqrt{\tau/i}\,\theta(\tau)$.

Note that if $\tau = it$, then $\theta(\tau) = \vartheta(t)$, and the above relation is precisely the functional equation for ϑ which appeared in Chapter 4.

The transformation law $\theta(-1/\tau) = (\tau/i)^{1/2}\theta(\tau)$ gives us very precise information about the behavior when $\tau \to 0$. The next corollary will be used later, when we need to analyze the behavior of $\theta(\tau)$ as $\tau \to 1$.

Corollary 1.8 *If* $\tau \in \mathbb{H}$, *then*

$$\theta(1 - 1/\tau) = \sqrt{\frac{\tau}{i}} \sum_{n=-\infty}^{\infty} e^{\pi i (n+1/2)^2 \tau}$$

$$= \sqrt{\frac{\tau}{i}} \left(2e^{\pi i \tau/4} + \cdots \right).$$

The second identity means that $\theta(1 - 1/\tau) \sim \sqrt{\tau/i}\,2e^{i\pi\tau/4}$ *as* $\text{Im}(\tau) \to \infty$.

Proof. First, we note that n and n^2 have the same parity, so

$$\theta(1 + \tau) = \sum_{n=-\infty}^{\infty} (-1)^n e^{i\pi n^2 \tau} = \Theta(1/2|\tau),$$

hence $\theta(1 - 1/\tau) = \Theta(1/2| - 1/\tau)$. Next, we use Theorem 1.6 with $z = 1/2$, and the result is

$$\theta(1 - 1/\tau) = \sqrt{\frac{\tau}{i}} e^{\pi i \tau/4} \Theta(\tau/2|\tau)$$

$$= \sqrt{\frac{\tau}{i}} e^{\pi i \tau/4} \sum_{n=-\infty}^{\infty} e^{\pi i n^2 \tau} e^{\pi i n \tau}$$

$$= \sqrt{\frac{\tau}{i}} \sum_{n=-\infty}^{\infty} e^{\pi i (n+1/2)^2 \tau}.$$

The terms corresponding to $n = 0$ and $n = -1$ contribute $2e^{\pi i \tau/4}$, which has absolute value $2e^{-\pi t/4}$ where $\tau = \sigma + it$. Finally, the sum of the other terms $n \neq 0, -1$ is of order

$$O\left(\sum_{k=1}^{\infty} e^{-(k+1/2)^2 \pi t} \right) = O\left(e^{-9\pi t/4} \right).$$

Our final corollary of the transformation law pertains to the **Dedekind eta function**, which is defined for $\text{Im}(\tau) > 0$ by

$$\eta(\tau) = e^{\frac{\pi i \tau}{12}} \prod_{n=1}^{\infty}(1 - e^{2\pi i n \tau}).$$

The functional equation for η given below will be relevant to our discussion of the four-square theorem, and in the theory of partitions.

Proposition 1.9 *If* $\text{Im}(\tau) > 0$, *then* $\eta(-1/\tau) = \sqrt{\tau/i}\,\eta(\tau)$.

This identity is deduced by differentiating the relation in Theorem 1.6 and evaluating it at $z_0 = 1/2 + \tau/2$. The details are as follows.

Proof. From the product formula for the theta function, we may write with $q = e^{\pi i \tau}$,

$$\Theta(z|\tau) = (1 + qe^{-2\pi i z}) \prod_{n=1}^{\infty}(1 - q^{2n})(1 + q^{2n-1}e^{2\pi i z})(1 + q^{2n+1}e^{-2\pi i z}),$$

and since the first factor vanishes at $z_0 = 1/2 + \tau/2$, we see that

$$\Theta'(z_0|\tau) = 2\pi i H(\tau), \quad \text{where } H(\tau) = \prod_{n=1}^{\infty}(1 - e^{2\pi i n \tau})^3.$$

Next, we observe that with $-1/\tau$ replaced by τ in (5), we obtain

$$\Theta(z|\tau) = \sqrt{i/\tau}e^{-\pi i z^2/\tau}\Theta(-z/\tau| - 1/\tau).$$

If we differentiate this expression and then evaluate it at the point $z_0 = 1/2 + \tau/2$, we find

$$2\pi i H(\tau) = \sqrt{i/\tau}e^{-\frac{\pi i}{4\tau}}e^{-\frac{\pi i}{2}}e^{-\frac{\pi i \tau}{4}}\left(\frac{-2\pi i}{\tau}\right)H(-1/\tau).$$

Hence

$$e^{\frac{\pi i \tau}{4}}H(\tau) = \left(\frac{i}{\tau}\right)^{3/2}e^{-\frac{\pi i}{4\tau}}H(-1/\tau).$$

We note that when $\tau = it$, with $t > 0$, the function $\eta(\tau)$ is positive, and thus taking the cube root of the above gives $\eta(\tau) = \sqrt{i/\tau}\,\eta(-1/\tau)$; therefore this identity holds for all $\tau \in \mathbb{H}$ by analytic continuation.

A connection between the function η and the theory of elliptic functions is given in Problem 5.

2 Generating functions

Given a sequence $\{F_n\}_{n=0}^{\infty}$, which may arise either combinatorially, recursively, or in terms of some number-theoretic law, an important tool in its study is the passage to its **generating function**, defined by

$$F(x) = \sum_{n=0}^{\infty} F_n x^n.$$

Often times, the defining properties of the sequence $\{F_n\}$ imply interesting algebraic or analytic properties of the function $F(x)$, and exploiting these can eventually lead us back to new insights about the sequence $\{F_n\}$. A very simple-minded example is given by the Fibonacci sequence. (See Exercise 2). Here we want to study less elementary examples of this idea, related to the Θ function.

We shall first discuss very briefly the theory of partitions.

The **partition function** is defined as follows: if n is a positive integer, we let $p(n)$ denote the numbers of ways n can be written as a sum of positive integers. For instance, $p(1) = 1$, and $p(2) = 2$ since $2 = 2 + 0 = 1 + 1$. Also, $p(3) = 3$ since $3 = 3 + 0 = 2 + 1 = 1 + 1 + 1$. We set $p(0) = 1$ and collect some further values of $p(n)$ in the following table.

n	0	1	2	3	4	5	6	7	8	$\cdots$	12
$p(n)$	1	1	2	3	5	7	11	15	22	$\cdots$	77

The first theorem is Euler's identity for the generating function of the partition sequence $\{p(n)\}$, which is reminiscent of the product formula for the zeta function.

Theorem 2.1 *If $|x| < 1$, then* $\displaystyle\sum_{n=0}^{\infty} p(n)x^n = \prod_{k=1}^{\infty} \frac{1}{1 - x^k}.$

Formally, we can write each fraction as

$$\frac{1}{1 - x^k} = \sum_{m=0}^{\infty} x^{km},$$

and multiply these out together to obtain $p(n)$ as the coefficient of x^n. Indeed, when we group together equal integers in a partition of n, this partition can be written as

$$n = m_1 k_1 + \cdots + m_r k_r,$$

where $k_1, \ldots, k_r$ are distinct positive integers. This partition corresponds to the term

$$(x^{k_1})^{m_1} \cdots (x^{k_r})^{m_r}$$

that arises in the product.

The justification of this formal argument proceeds as in the proof of the product formula for the zeta function (Section 1, Chapter 7); this is based on the convergence of the product $\prod 1/(1 - x^k)$. This convergence in turn follows from the fact that for each fixed $|x| < 1$ one has

$$\frac{1}{1 - x^k} = 1 + O(x^k).$$

A similar argument shows that the product $\prod 1/(1 - x^{2n-1})$ is equal to the generating function for $p_o(n)$, the number of partitions of n into odd parts. Also, $\prod (1 + x^n)$ is the generating function for $p_u(n)$, the number of partitions of n into unequal parts. Remarkably, $p_o(n) = p_u(n)$ for all n, and this translates into the identity

$$\prod_{n=1}^{\infty} \left(\frac{1}{1 - x^{2n-1}} \right) = \prod_{n=1}^{\infty} (1 + x^n).$$

To prove this note that $(1 + x^n)(1 - x^n) = 1 - x^{2n}$, and therefore

$$\prod_{n=1}^{\infty} (1 + x^n) \prod_{n=1}^{\infty} (1 - x^n) = \prod_{n=1}^{\infty} (1 - x^{2n}).$$

Moreover, taking into account the parity of integers, we have

$$\prod_{n=1}^{\infty} (1 - x^{2n}) \prod_{n=1}^{\infty} (1 - x^{2n-1}) = \prod_{n=1}^{\infty} (1 - x^n),$$

which combined with the above proves the desired identity.

The proposition that follows is deeper, and in fact involves the Θ function directly. Let $p_{e,u}(n)$ denote the number of partitions of n into an even number of unequal parts, and $p_{o,u}(n)$ the number of partitions of n into an odd number of unequal parts. Then, Euler proved that, unless n is a pentagonal number, one has $p_{e,u}(n) = p_{o,u}(n)$. By definition, **pentagonal numbers**[2] are integers n of the form $k(3k + 1)/2$, with $k \in \mathbb{Z}$. For

---

[2]The traditional definition is as follows. Integers of the form $n = k(k - 1)/2$, $k \in \mathbb{Z}$, are "triangular numbers"; those of the form $n = k^2$ are "squares"; and those of the form $k(3k + 1)/2$ are "pentagonal numbers." In general, numbers of the form $(k/2)((\ell - 2)k + \ell - 4)$ are associated with an ℓ-sided polygon.

example, the first few pentagonal numbers are $1, 2, 5, 7, 12, 15, 22, 26, \ldots$.
In fact, if n is pentagonal, then

$$p_{e,u}(n) - p_{o,u}(n) = (-1)^k, \quad \text{if } n = k(3k+1)/2.$$

To prove this result, we first observe that

$$\prod_{n=1}^{\infty}(1 - x^n) = \sum_{n=1}^{\infty}[p_{e,u}(n) - p_{o,u}(n)]x^n.$$

This follows since multiplying the terms in the product, we obtain terms
of the form $(-1)^r x^{n_1 + \cdots + n_r}$ where the integers $n_1, \ldots, n_r$ are distinct.
Hence in the coefficient of x^n, each partition $n_1 + \cdots + n_r$ of n into an
even number of unequal parts contributes for $+1$ (r is even), and each
partition into an odd number of unequal parts contributes -1 (r is odd).
This gives precisely the coefficient $p_{e,u}(n) - p_{o,u}(n)$.

With the above identity, we see that Euler's theorem is a consequence
of the following proposition.

Proposition 2.2 $\displaystyle\prod_{n=1}^{\infty}(1 - x^n) = \sum_{k=-\infty}^{\infty}(-1)^k x^{\frac{k(3k+1)}{2}}.$

Proof. If we set $x = e^{2\pi i u}$, then we can write

$$\prod_{n=1}^{\infty}(1 - x^n) = \prod_{n=1}^{\infty}(1 - e^{2\pi i n u})$$

in terms of the triple product

$$\prod_{n=1}^{\infty}(1 - q^{2n})(1 + q^{2n-1}e^{2\pi i z})(1 + q^{2n-1}e^{-2\pi i z})$$

by letting $q = e^{3\pi i u}$ and $z = 1/2 + u/2$. This is because

$$\prod_{n=1}^{\infty}(1 - e^{2\pi i 3nu})(1 - e^{2\pi i (3n-1)u})(1 - e^{2\pi i (3n-2)u}) = \prod_{n=1}^{\infty}(1 - e^{2\pi i n u}).$$

By Theorem 1.3 the product equals

$$\sum_{n=-\infty}^{\infty} e^{3\pi i n^2 u}(-1)^n e^{2\pi i n u/2} = \sum_{n=-\infty}^{\infty}(-1)^n e^{\pi i n(3n+1)u}$$

$$= \sum_{n=-\infty}^{\infty} (-1)^n x^{n(3n+1)/2},$$

which was to be proved.

We make a final comment about the partition function $p(n)$. The nature of its growth as $n \to \infty$ can be analyzed in terms of the behavior of $1/\prod_{n=1}^{\infty}(1-x)^n$ as $|x| \to 1$. In fact, by elementary considerations, we can get a rough order of growth of $p(n)$ from the growth of the generating function as $x \to 1$; see Exercises 5 and 6. A more refined analysis requires the transformation properties of the generating function which goes back to the corresponding Proposition 1.9 about η. This leads to a very good asymptotic formula for $p(n)$. It may be found in Appendix A.

3 The theorems about sums of squares

The ancient Greeks were fascinated by triples of integers (a, b, c) that occurred as sides of right triangles. These are the "Pythagorean triples," which satisfy $a^2 + b^2 = c^2$. According to Diophantus of Alexandria (ca. 250 AD), if c is an integer of the above kind, and a and b have no common factors (a case to which one may easily reduce), then c is the sum of two squares, that is, $c = m^2 + n^2$ with $m, n \in \mathbb{Z}$; and conversely, any such c arises as the hypotenuse of a triangle whose sides are given by a Pythagorean triple (a, b, c). (See Exercise 8.) Therefore, it is natural to ask the following question: which integers can be written as the sum of two squares? It is easy to see that no number of the form $4k + 3$ can be so written, but to determine which integers can be expressed in this way is not obvious.

Let us pose the question in a more quantitative form. We define $r_2(n)$ to be the number of ways n can be written as the sum of two squares, counting obvious repetitions; that is, $r_2(n)$ is the number of pairs (x, y), $x, y \in \mathbb{Z}$, so that

$$n = x^2 + y^2.$$

For example, $r_2(3) = 0$, but $r_2(5) = 8$ because $5 = (\pm 2)^2 + (\pm 1)^2$, and also $5 = (\pm 1)^2 + (\pm 2)^2$. Hence, our first problem can be posed as follows:

> **Sum of two squares:** Which integers can be written as a sum of two squares? More precisely, can one determine an expression for $r_2(n)$?

Next, since not every positive integer can be expressed as the sum of two squares, we may ask if three squares, or possibly four squares suffice.

However, the fact is that there are infinitely many integers that cannot be written as the sum of three squares, since it is easy to check that no integer of the form $8k + 7$ can be so written. So we turn to the question of four squares and define, in analogy with $r_2(n)$, the function $r_4(n)$ to be the number of ways of expressing n as a sum of four squares. Therefore, a second problem that arises is:

> **Sum of four squares:** Can every positive integer be written as a sum of four squares? More precisely, determine a formula for $r_4(n)$.

It turns out that the problems of two squares and four squares, which go back to the third century, were not resolved until about 1500 years later, and their full solution was first given by the use of Jacobi's theory of theta functions!

3.1 The two-squares theorem

The problem of representing an integer as the sum of two squares, while obviously additive in nature, has a nice multiplicative aspect: if n and m are two integers that can be written as the sum of two squares, then so can their product nm. Indeed, suppose $n = a^2 + b^2$, $m = c^2 + d^2$, and consider the complex number

$$x + iy = (a + ib)(c + id).$$

Clearly, x and y are integers since $a, b, c, d \in \mathbb{Z}$, and by taking absolute values on both sides we see that

$$x^2 + y^2 = (a^2 + b^2)(c^2 + d^2),$$

and it follows that $nm = x^2 + y^2$.

For these reasons the divisibility properties of n play a crucial role in determining $r_2(n)$. To state the basic result we define two new **divisor functions**: we let $d_1(n)$ denote the number of divisors of n of the form $4k + 1$, and $d_3(n)$ the number of divisors of n of the form $4k + 3$. The main result of this section provides a complete answer to the two-squares problem:

Theorem 3.1 *If $n \geq 1$, then $r_2(n) = 4(d_1(n) - d_3(n))$.*

A direct consequence of the above formula for $r_2(n)$ may be stated as follows. If $n = p_1^{a_1} \cdots p_r^{a_r}$ is the prime factorization of n where $p_1, \ldots, p_r$ are distinct, then:

The positive integer n can be represented as the sum of two squares if and only if every prime p_j of the form $4k + 3$ that occurs in the factorization of n has an even exponent a_j.

The proof of this deduction is outlined in Exercise 9.

To prove the theorem, we first establish a crucial relationship that identifies the generating function of the sequence $\{r_2(n)\}_{n=1}^{\infty}$ with the square of the θ function, namely

$$
(6) \qquad \theta(\tau)^2 = \sum_{n=0}^{\infty} r_2(n)q^n,
$$

whenever $q = e^{\pi i \tau}$ with $\tau \in \mathbb{H}$. The proof of this identity relies simply on the definition of r_2 and θ. Indeed, if we first recall that $\theta(\tau) = \sum_{-\infty}^{\infty} q^{n^2}$, then we obtain

$$
\theta(\tau)^2 = \left(\sum_{n_1=-\infty}^{\infty} q^{n_1^2} \right) \left(\sum_{n_2=-\infty}^{\infty} q^{n_2^2} \right)
$$

$$
= \sum_{(n_1,n_2)\in\mathbb{Z}\times\mathbb{Z}} q^{n_1^2+n_2^2}
$$

$$
= \sum_{n=0}^{\infty} r_2(n)q^n,
$$

since $r_2(n)$ counts the number of pairs (n_1, n_2) with $n_1^2 + n_2^2 = n$.

Proposition 3.2 *The identity $r_2(n) = 4(d_1(n) - d_3(n))$, $n \geq 1$, is equivalent to the identities*

$$
(7) \qquad \theta(\tau)^2 = 2 \sum_{n=-\infty}^{\infty} \frac{1}{q^n + q^{-n}} = 1 + 4 \sum_{n=1}^{\infty} \frac{q^n}{1 + q^{2n}},
$$

whenever $q = e^{\pi i \tau}$ and $\tau \in \mathbb{H}$.

Proof. We note first that both series converge absolutely since $|q| < 1$, and the first equals the second, because $1/(q^n + q^{-n}) = q^{|n|}/(1 + q^{2|n|})$.

Since $(1 + q^{2n})^{-1} = (1 - q^{2n})/(1 - q^{4n})$, the right-hand side of (7) equals

$$
1 + 4 \sum_{n=1}^{\infty} \left(\frac{q^n}{1 - q^{4n}} - \frac{q^{3n}}{1 - q^{4n}} \right).
$$

However, since $1/(1 - q^{4n}) = \sum_{m=0}^{\infty} q^{4nm}$, we have

$$\sum_{n=1}^{\infty} \frac{q^n}{1 - q^{4n}} = \sum_{n=1}^{\infty} \sum_{m=0}^{\infty} q^{n(4m+1)} = \sum_{k=1}^{\infty} d_1(k)q^k,$$

because $d_1(k)$ counts the number of divisors of k that are of the form $4m + 1$. Observe that the series $\sum d_1(k)q^k$ converges since $d_1(k) \leq k$.

A similar argument shows that

$$\sum_{n=1}^{\infty} \frac{q^{3n}}{1 - q^{4n}} = \sum_{k=1}^{\infty} d_3(k)q^k,$$

and the proof of the proposition is complete.

In effect, we see that the identity (6) links the original problem in arithmetic with the problem in complex analysis of establishing the relation (7).

We shall now find it convenient to use $\mathcal{C}(\tau)$ to denote[3]

$$(8) \qquad \mathcal{C}(\tau) = 2 \sum_{n=-\infty}^{\infty} \frac{1}{q^n + q^{-n}} = \sum_{n=-\infty}^{\infty} \frac{1}{\cos(n\pi\tau)},$$

where $q = e^{\pi i \tau}$ and $\tau \in \mathbb{H}$. Our work then becomes to prove the identity $\theta(\tau)^2 = \mathcal{C}(\tau)$.

What is truly remarkable are the different yet parallel ways that the functions θ and $\mathcal{C}$ arise. The genesis of the function θ may be thought to be the heat diffusion equation on the real line; the corresponding heat kernel is given in terms of the Gaussian $e^{-\pi x^2}$ which is its own Fourier transform; and finally the transformation rule for θ results from the Poisson summation formula.

The parallel with $\mathcal{C}$ is that it arises from another differential equation: the steady-state heat equation in a strip; there, the corresponding kernel is $1/\cosh \pi x$ (Section 1.3, Chapter 8), which again is its own Fourier transform (Example 3, Chapter 3). The transformation rule for $\mathcal{C}$ results, once again, from the Poisson summation formula.

To prove the identity $\theta^2 = \mathcal{C}$ we will first show that these two functions satisfy the same structural properties. For θ^2 we had the transformation law $\theta(\tau)^2 = (i/\tau)\theta(-1/\tau)^2$ (Corollary 1.7).

---

[3]We denote the function by $\mathcal{C}$ because we are summing a series of cosines.

An identical transformation law holds for $C(\tau)$! Indeed, if we set $a = 0$ in the relation (5) of Chapter 4 we obtain

$$\sum_{n=-\infty}^{\infty} \frac{1}{\cosh(\pi n t)} = \frac{1}{t} \sum_{n=-\infty}^{\infty} \frac{1}{\cosh(\pi n/t)}.$$

This is precisely the identity

$$C(\tau) = (i/\tau)\, C(-1/\tau)$$

for $\tau = it$, $t > 0$, which therefore also holds for all $\tau \in \mathbb{H}$ by analytic continuation.

It is also obvious from their definitions that both $\theta(\tau)^2$ and $C(\tau)$ tend to 1 as $\text{Im}(\tau) \to \infty$. The last property we want to examine is the behavior of both functions at the "cusp" $\tau = 1$.[4]

For θ^2 we shall invoke Corollary 1.8 to see that $\theta(1 - 1/\tau)^2 \sim 4(\tau/i)e^{\pi i\tau/2}$ as $\text{Im}(\tau) \to \infty$.

For C we can do the same, again using the Poisson summation formula. In fact, if we set $a = 1/2$ in equation (5), Chapter 4, we find

$$\sum_{n=-\infty}^{\infty} \frac{(-1)^n}{\cosh(\pi n/t)} = t \sum_{n=-\infty}^{\infty} \frac{1}{\cosh(\pi(n + 1/2)t)}.$$

Therefore, by analytic continuation we deduce that

$$C(1 - 1/\tau) = \left(\frac{\tau}{i}\right) \sum_{n=-\infty}^{\infty} \frac{1}{\cos(\pi(n + 1/2)\tau)}.$$

The main terms of this sum are those for $n = -1$ and $n = 0$. This easily gives

$$C(1 - 1/\tau) = 4\left(\frac{\tau}{i}\right) e^{\pi i\tau/2} + O\left(|\tau|e^{-3\pi t/2}\right), \quad \text{as } t \to \infty,$$

and where $\tau = \sigma + it$. We summarize our conclusions in a proposition.

Proposition 3.3 *The function* $C(\tau) = \sum 1/\cos(\pi n\tau)$, *defined in the upper half-plane, satisfies*

(i) $C(\tau + 2) = C(\tau)$.

(ii) $C(\tau) = (i/\tau)C(-1/\tau)$.

---

[4]Why we refer to the point $\tau = 1$ as a cusp, and the reason for its importance, will become clear later on.

(iii) $\mathcal{C}(\tau) \to 1$ *as* $\operatorname{Im}(\tau) \to \infty$.

(iv) $\mathcal{C}(1 - 1/\tau) \sim 4(\tau/i)e^{\pi i \tau/2}$ *as* $\operatorname{Im}(\tau) \to \infty$.

Moreover, $\theta(\tau)^2$ *satisfies the same properties.*

With this proposition, we prove the identity of $\theta(\tau)^2 = \mathcal{C}(\tau)$ with the aid of the following theorem, in which we shall ultimately set $f = \mathcal{C}/\theta^2$.

Theorem 3.4 *Suppose f is a holomorphic function in the upper half-plane that satisfies:*

(i) $f(\tau + 2) = f(\tau)$,

(ii) $f(-1/\tau) = f(\tau)$,

(iii) $f(\tau)$ *is bounded,*

then f is constant.

For the proof of this theorem, we introduce the following subset of the closed upper half-plane, which is defined by

$$\mathcal{F} = \{\tau \in \overline{\mathbb{H}} : |\operatorname{Re}(\tau)| \le 1 \text{ and } |\tau| \ge 1\},$$

and illustrated in Figure 1.

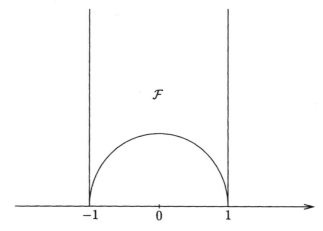

Figure 1. The domain $\mathcal{F}$

The points corresponding to $\tau = \pm 1$ are called **cusps**. They are equivalent under the mapping $\tau \mapsto \tau + 2$.

Lemma 3.5 *Every point in the upper half-plane can be mapped into $\mathcal{F}$ using repeatedly one or another of the following fractional linear transformations or their inverses:*

$$T_2 : \tau \mapsto \tau + 2, \qquad S : \tau \mapsto -1/\tau.$$

For this reason, $\mathcal{F}$ is called the **fundamental domain**[5] for the group of transformations generated by T_2 and S.

In fact, we let G denote the group generated by T_2 and S. Since T_2 and S are fractional linear transformations, we may represent an element $g \in G$ by a matrix

$$g = \begin{pmatrix} a & b \\ c & d \end{pmatrix},$$

with the understanding that

$$g(\tau) = \frac{a\tau + b}{c\tau + d}.$$

Since the matrices representing T_2 and S have integer coefficients and determinant 1, the same is true for all matrices of elements in G. In particular, if $\tau \in \mathbb{H}$, then

$$(9) \qquad\qquad \mathrm{Im}(g(\tau)) = \frac{\mathrm{Im}(\tau)}{|c\tau + d|^2}.$$

Proof of Lemma 3.5. Let $\tau \in \mathbb{H}$. If $g \in G$ with $g(\tau) = (a\tau + b)/(c\tau + d)$, then c and d are integers, and by (9) we may choose a $g_0 \in G$ such that $\mathrm{Im}(g_0(\tau))$ is maximal. Since the translations T_2 and their inverses do not change imaginary parts, we may apply finitely many of them to see that there exists $g_1 \in G$ with $|\mathrm{Re}(g_1(\tau))| \leq 1$ and $\mathrm{Im}(g_1(\tau))$ is maximal. It now suffices to prove that $|g_1(\tau)| \geq 1$ to conclude that $g_1(\tau) \in \mathcal{F}$. If this were not true, that is, $|g_1(\tau)| < 1$, then $\mathrm{Im}(Sg_1(\tau))$ would be greater than $\mathrm{Im}(g_1(\tau))$ since

$$\mathrm{Im}(Sg_1(\tau)) = \mathrm{Im}(-1/g_1(\tau)) = -\frac{\mathrm{Im}(\overline{g_1(\tau)})}{|g_1(\tau)|^2} > \mathrm{Im}(g_1(\tau)),$$

and this contradicts the maximality of $\mathrm{Im}(g_1(\tau))$.

We can now prove the theorem. Suppose f is not constant, and let $g(z) = f(\tau)$ where $z = e^{\pi i \tau}$. The function g is well defined for z in the

---

[5]Strictly speaking, the notion of a fundamental domain requires that every point have a unique representative in the domain. In the present case, ambiguity arises only for points that are on the boundary of $\mathcal{F}$.

punctured unit disc, since f is periodic of period 2, and moreover, g is bounded near the origin by assumption (iii) of the theorem. Hence 0 is a removable singularity for g, and $\lim_{z\to 0} g(z) = \lim_{\mathrm{Im}(\tau)\to\infty} f(\tau)$ exists. So by the maximum modulus principle,

$$\lim_{\mathrm{Im}(\tau)\to\infty} |f(\tau)| < \sup_{\tau\in\mathcal{F}} |f(\tau)|.$$

Now we must investigate the behavior of f at the points $\tau = \pm 1$. Since $f(\tau+2) = f(\tau)$, it suffices to consider the point $\tau = 1$. We claim that

$$\lim_{\mathrm{Im}(\tau)\to\infty} f(1-1/\tau)$$

exists and moreover

$$\lim_{\mathrm{Im}(\tau)\to\infty} |f(1-1/\tau)| < \sup_{\tau\in\mathcal{F}} |f(\tau)|.$$

The argument is essentially the same as the one above, except that we first need to interchange $\tau = 1$ with the point at infinity. In other words, we wish to investigate the behavior of $F(\tau) = f(1-1/\tau)$ for τ near ∞. The important step is to prove that F is periodic. To this end, we consider the fractional linear transformation associated to the matrix

$$U_n = \begin{pmatrix} 1-n & n \\ -n & 1+n \end{pmatrix},$$

that is,

$$\tau \mapsto \frac{(1-n)\tau + n}{-n\tau + (1+n)},$$

which maps 1 to 1. Now let $\mu(\tau) = 1/(1-\tau)$ which maps 1 to ∞, and whose inverse $\mu^{-1}(\tau) = 1 - 1/\tau$ takes ∞ to 1. Then

$$U_n = \mu^{-1}T_n\mu,$$

where T_n is the translation $T_n(\tau) = \tau + n$. As a consequence,

$$U_n U_m = U_{n+m},$$

and

$$U_{-1} = \begin{pmatrix} 2 & -1 \\ 1 & 0 \end{pmatrix} = T_2 S.$$

Thus any U_n can be obtained by finitely many applications of T_2, S, or their inverses. Since f is invariant under T_2 and S, it is also invariant under U_m. So we find that

$$f(\mu^{-1}T_n\mu(\tau)) = f(\tau).$$

Therefore, if we let $F(\tau) = f(\mu^{-1}(\tau)) = f(1 - 1/\tau)$, we find that F is periodic of period 1, that is,

$$F(T_n\tau) = F(\tau) \quad \text{for every integer } n.$$

Now, by the previous argument, if we set $h(z) = F(\tau)$ with $z = e^{2\pi i \tau}$, we see that h has a removable singularity at $z = 0$, and the desired inequality follows by the maximum principle.

We conclude from this analysis that f attains its maximum in the interior of the upper half-plane, and this contradicts the maximum principle.

The proof of the two-squares theorem is now only one step away.

We consider the function $f(\tau) = \mathcal{C}(\tau)/\theta(\tau)^2$. Since we know by the product formula that $\theta(\tau)$ does not vanish in the upper half-plane (Corollary 1.4), we find that f is holomorphic in $\mathbb{H}$. Moreover, by Proposition 3.3, f is invariant under the transformations T_2 and S, that is, $f(\tau + 2) = f(\tau)$ and $f(-1/\tau) = f(\tau)$. Finally, in the fundamental domain $\mathcal{F}$, the function $f(\tau)$ is bounded, and in fact tends to 1 as $\text{Im}(\tau)$ tends to infinity, or as τ tends to the cusps ± 1. This is because of properties (iii) and (iv) in Proposition 3.3, which are verified by both $\mathcal{C}$ and θ^2. Thus f is bounded in $\mathbb{H}$. The result is that f is a constant, which must be 1, proving that $\theta(\tau)^2 = \mathcal{C}(\tau)$, and with it the two-squares theorem.

3.2 The four-squares theorem

Statement of the theorem

In the rest of this chapter, we shall consider the case of four squares. More precisely, we will prove that every positive integer is the sum of four squares, and moreover we will determine a formula for $r_4(n)$ that describes the number of ways this can be done.

We need to introduce another divisor function, which we denote by $\sigma_1^*(n)$, and which equals the sum of divisors of n that are not divisible by 4. The main theorem we shall prove is the following.

Theorem 3.6 *Every positive integer is the sum of four squares, and moreover*

$$r_4(n) = 8\sigma_1^*(n) \quad \text{for all } n \geq 1.$$

As before, we relate the sequence $\{r_4(n)\}$ via its generating function to an appropriate power of the function θ, which in this case is its fourth power. The result is that

$$\theta(\tau)^4 = \sum_{n=0}^{\infty} r_4(n)q^n$$

whenever $q = e^{\pi i \tau}$ with $\tau \in \mathbb{H}$.

The next step is to find the modular function whose equality with $\theta(\tau)^4$ expresses the identity $r_4(n) = 8\sigma_1^*(n)$. Unfortunately, here there is nothing as simple as the function $C(\tau)$ that arose in the two-squares theorem; instead we shall need to construct a rather subtle variant of the Eisenstein series considered in the previous chapter. In fact, we define

$$E_2^*(\tau) = \sum_m \sum_n \frac{1}{\left(\frac{m\tau}{2} + n\right)^2} - \sum_m \sum_n \frac{1}{\left(m\tau + \frac{n}{2}\right)^2}$$

for $\tau \in \mathbb{H}$. The indicated order of summation is critical, since the above series do not converge absolutely. The following reduces the four-squares theorem to the modular properties of E_2^*.

Proposition 3.7 *The assertion $r_4(n) = 8\sigma_1^*(n)$ is equivalent to the identity*

$$\theta(\tau)^4 = \frac{-1}{\pi^2} E_2^*(\tau), \quad \text{where } \tau \in \mathbb{H}.$$

Proof. It suffices to prove that if $q = e^{\pi i \tau}$, then

$$\frac{-1}{\pi^2} E_2^*(\tau) = 1 + \sum_{k=1}^{\infty} 8\sigma_1^*(k)q^k.$$

First, recall the forbidden Eisenstein series that we considered in the last section of the previous chapter, and which is defined by

$$F(\tau) = \sum_m \left[\sum_n \frac{1}{(m\tau + n)^2} \right],$$

where the term $n = m = 0$ is omitted. Since the sum above is not absolutely convergent, the order of summation, first in n and then in m, is crucial. With this in mind, the definitions of E_2^* and F give immediately

(10) $$E_2^*(\tau) = F\left(\frac{\tau}{2}\right) - 4F(2\tau).$$

In Corollary 2.6 (and Exercise 7) of the last chapter, we proved that

$$F(\tau) = \frac{\pi^2}{3} - 8\pi^2 \sum_{k=1}^{\infty} \sigma_1(k)e^{2\pi i k \tau},$$

where $\sigma_1(k)$ is the sum of the divisors of k.

Now observe that

$$\sigma_1^*(n) = \begin{cases} \sigma_1(n) & \text{if } n \text{ is not divisible by 4,} \\ \sigma_1(n) - 4\sigma_1(n/4) & \text{if } n \text{ is divisible by 4.} \end{cases}$$

Indeed, if n is not divisible by 4, then no divisors of n are divisible by 4. If $n = 4\tilde{n}$, and d is a divisor of n that is divisible by 4, say $d = 4\tilde{d}$, then $\tilde{d}$ divides $\tilde{n}$. This gives the second formula. Therefore, from this observation and (10) we find that

$$E_2^*(\tau) = -\pi^2 - 8\pi^2 \sum_{k=1}^{\infty} \sigma_1^*(k)e^{\pi i k \tau},$$

and the proof of the proposition is complete.

We have therefore reduced Theorem 3.6 to the identity $\theta^4 = -\pi^{-2}E_2^*$, and the key to establish this relation is that E_2^* satisfies the same modular properties as $\theta(\tau)^4$.

Proposition 3.8 *The function $E_2^*(\tau)$ defined in the upper half-plane has the following properties:*

(i) $E_2^*(\tau + 2) = E_2^*(\tau)$.

(ii) $E_2^*(\tau) = -\tau^{-2}E_2^*(-1/\tau)$.

(iii) $E_2^*(\tau) \to -\pi^2$ *as* $\text{Im}(\tau) \to \infty$.

(iv) $|E_2^*(1 - 1/\tau)| = O(|\tau^2 e^{\pi i \tau}|)$ *as* $\text{Im}(\tau) \to \infty$.

Moreover $-\pi^2\theta^4$ *has the same properties.*

The periodicity (i) of E_2^* is immediate from the definition. The proofs of the other properties of E_2^* are a little more involved.

Consider the forbidden Eisenstein series F and its reverse $\tilde{F}$, which is obtained from reversing the order of summation:

$$F(\tau) = \sum_m \sum_n \frac{1}{(m\tau + n)^2} \quad \text{and} \quad \tilde{F}(\tau) = \sum_n \sum_m \frac{1}{(m\tau + n)^2}.$$

In both cases, the term $n = m = 0$ is omitted.

Lemma 3.9 *The functions F and $\tilde{F}$ satisfy:*

(a) $F(-1/\tau) = \tau^2 \tilde{F}(\tau)$,

(b) $F(\tau) - \tilde{F}(\tau) = 2\pi i/\tau$,

(c) $F(-1/\tau) = \tau^2 F(\tau) - 2\pi i \tau$.

Proof. Property (a) follows directly from the definitions of F and $\tilde{F}$, and the identity

$$(n + m(-1/\tau))^2 = \tau^{-2}(-m + n\tau)^2.$$

To prove property (b), we invoke the functional equation for the Dedekind eta function which was established earlier:

$$\eta(-1/\tau) = \sqrt{\tau/i}\,\eta(\tau),$$

where $\eta(\tau) = q^{1/12} \prod_{n=1}^{\infty}(1 - q^{2n})$, and $q = e^{\pi i \tau}$.

First, we take the logarithmic derivative of η with respect to the variable τ to find (by Proposition 3.2 in Chapter 5)

$$(\eta'/\eta)(\tau) = \frac{\pi i}{12} - 2\pi i \sum_{n=1}^{\infty} \frac{nq^{2n}}{1 - q^{2n}}.$$

However, if $\sigma_1(k)$ denotes the sum of the divisors of k, then one sees that

$$\sum_{n=1}^{\infty} \frac{nq^{2n}}{1 - q^{2n}} = \sum_{n=1}^{\infty}\sum_{\ell=0}^{\infty} nq^{2n}q^{2\ell n}$$

$$= \sum_{n=1}^{\infty}\sum_{m=1}^{\infty} nq^{2nm}$$

$$= \sum_{k=1}^{\infty} \sigma_1(k)q^{2k}.$$

If we recall that $F(\tau) = \pi^2/3 - 8\pi^2 \sum_{k=1}^{\infty} \sigma_1(k)q^{2k}$, we find

$$(\eta'/\eta)(\tau) = \frac{i}{4\pi} F(\tau).$$

By the chain rule, the logarithmic derivative of $\eta(-1/\tau)$ is $\tau^{-2}(\eta'/\eta)(-1/\tau)$, and using property (a), we see that the logarithmic derivative of $\eta(-1/\tau)$

equals $(i/4\pi)\tilde{F}(\tau)$. Therefore, taking the logarithmic derivative of the functional equation for η we find

$$\frac{i}{4\pi}\tilde{F}(\tau) = \frac{1}{2\tau} + \frac{i}{4\pi}F(\tau),$$

and this gives $\tilde{F}(\tau) = -2\pi i/\tau + F(\tau)$, as desired.

Finally, (c) is a consequence of (a) and (b).

To prove the transformation formula (ii) for E_2^* under $\tau \mapsto -1/\tau$, we begin with

$$E_2^*(\tau) = F(\tau/2) - 4F(2\tau).$$

Then

$$
\begin{aligned}
E_2^*(-1/\tau) &= F(-1/(2\tau)) - 4F(-2/\tau) \\
&= [4\tau^2 F(2\tau) - 4\pi i\tau] - 4[(\tau/2)^2 F(\tau/2) - \pi i\tau] \\
&= 4\tau^2 F(2\tau) - 4(\tau^2/4)F(\tau/2) \\
&= -\tau^2(F(\tau/2) - 4F(2\tau)) \\
&= -\tau^2 E_2^*(\tau),
\end{aligned}
$$

as desired. To prove the third property recall that

$$F(\tau) = \frac{\pi^2}{3} - 8\pi^2 \sum_{k=1}^{\infty} \sigma_1(k)e^{2\pi i k\tau},$$

where the sum goes to 0 as $\text{Im}(\tau) \to \infty$. Then, if we use the fact that

$$E_2^*(\tau) = F(\tau/2) - 4F(2\tau),$$

we conclude that $E_2^*(\tau) \to -\pi^2$ as $\text{Im}(\tau) \to \infty$.

To prove the final property, we begin by showing that

$$(11) \qquad E_2^*(1 - 1/\tau) = \tau^2 \left[F\left(\frac{\tau - 1}{2}\right) - F(\tau/2) \right].$$

From the transformation formulas for F we have

$$
\begin{aligned}
F(1/2 - 1/2\tau) &= F\left(\frac{\tau - 1}{2\tau}\right) \\
&= \left(\frac{2\tau}{\tau - 1}\right)^2 F\left(\frac{2\tau}{1 - \tau}\right) - 2\pi i\frac{2\tau}{1 - \tau},
\end{aligned}
$$

and

$$F\left(\frac{2\tau}{1-\tau}\right) = F(-2 + 2/(1-\tau))$$

$$= F(2/(1-\tau))$$

$$= \left(\frac{1-\tau}{2}\right)^2 F\left(\frac{\tau-1}{2}\right) - 2\pi i\left(\frac{\tau-1}{2}\right).$$

Hence,

$$F(1/2 - 1/2\tau) = \tau^2 F\left(\frac{\tau-1}{2}\right) - \frac{2\pi i 2\tau}{1-\tau} - 2\pi i\frac{(2\tau)^2}{(\tau-1)^2}\left(\frac{\tau-1}{2}\right).$$

But $F(2 - 2/\tau) = F(-2/\tau) = (\tau^2/4)F(\tau/2) - 2\pi i\tau/2$, and hence

$$E_2^*(1 - 1/\tau) = F(1/2 - 1/2\tau) - 4F(2 - 2/\tau)$$

$$= \tau^2\left[F\left(\frac{\tau-1}{2}\right) - F(\tau/2)\right] - 2\pi i\left(\frac{2\tau}{1-\tau} + \frac{2\tau^2}{\tau-1}\right) + 4\pi i\tau$$

$$= \tau^2\left[F\left(\frac{\tau-1}{2}\right) - F(\tau/2)\right].$$

This proves (11). Then, the last property follows from it and the fact that

$$F(\tau) = \frac{\pi^2}{3} - 8\pi^2\sum_{k=1}^{\infty}\sigma_1(k)e^{2\pi i k\tau}.$$

Thus Proposition 3.8 is proved.

We can now conclude the proof of the four-squares theorem by considering the quotient $f(\tau) = E_2^*(\tau)/\theta(\tau)^4$, and applying Theorem 3.4, as in the two-squares theorem. Recall $\theta(\tau)^4 \to 1$ and $\theta(1 - 1/\tau)^4 \sim 16\tau^2 e^{\pi i\tau}$, as $\text{Im}(\tau) \to \infty$. The result is that $f(\tau)$ is a constant, which equals $-\pi^2$ by Proposition 3.8. This completes the proof of the four-squares theorem.

4 Exercises

1. Prove that

$$\frac{(\Theta'(z|\tau))^2 - \Theta(z|\tau)\Theta''(z|\tau)}{\Theta(z|\tau)^2} = \wp_\tau(z - 1/2 - \tau/2) + c_\tau,$$

where c_τ can be expressed in terms of the first three derivatives of $\Theta(z|\tau)$, with respect to z, at $z = 1/2 + \tau/2$. Compare this formula with the result in Exercise 5 in the previous chapter.

2. Consider the **Fibonacci numbers** $\{F_n\}_{n=0}^\infty$, defined by the two initial values $F_0 = 0$, $F_1 = 1$ and the recursion relation

$$F_n = F_{n-1} + F_{n-2} \quad \text{for } n \geq 2.$$

(a) Consider the generating function $F(x) = \sum_{n=0}^\infty F_n x^n$ associated to $\{F_n\}$, and prove that

$$F(x) = x^2 F(x) + x F(x) + x$$

for all x in a neighborhood of 0.

(b) Show that the polynomial $q(x) = 1 - x - x^2$ can be factored as

$$q(x) = (1 - \alpha x)(1 - \beta x),$$

where α and β are the roots of the polynomial $p(x) = x^2 - x - 1$.

(c) Expand the expression for F in partial fractions and obtain

$$F(x) = \frac{x}{1 - x - x^2} = \frac{x}{(1 - \alpha x)(1 - \beta x)} = \frac{A}{1 - \alpha x} + \frac{B}{1 - \beta x},$$

where $A = 1/(\alpha - \beta)$ and $B = 1/(\beta - \alpha)$.

(d) Conclude that $F_n = A\alpha^n + B\beta^n$ for $n \geq 0$. The two roots of p are actually

$$\alpha = \frac{1 + \sqrt{5}}{2} \quad \text{and} \quad \beta = \frac{1 - \sqrt{5}}{2},$$

so that $A = 1/\sqrt{5}$ and $B = -1/\sqrt{5}$.

The number $1/\alpha = (\sqrt{5} - 1)/2$, which is known as the **golden mean**, satisfies the following property: given a line segment $[AC]$ of unit length (Figure 2), there exists a unique point B on this segment so that the following proportion holds

$$\frac{AC}{AB} = \frac{AB}{BC}.$$

If $\ell = AB$, this reduces to the equation $\ell^2 + \ell - 1 = 0$, whose only positive solution is the golden mean. This ratio arises also in the construction of the regular pentagon. It has played a role in architecture and art, going back to the time of ancient Greece.

3. More generally, consider the difference equation given by the initial values u_0 and u_1, and the recurrence relation $u_n = a u_{n-1} + b u_{n-2}$ for $n \geq 2$. Define the generating function associated to $\{u_n\}_{n=0}^\infty$ by $U(x) = \sum_{n=0}^\infty u_n x^n$. The recurrence relation implies that $U(x)(1 - ax - bx^2) = u_0 + (u_1 - a u_0)x$ in a neighborhood of

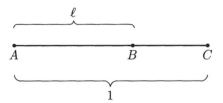

Figure 2. Appearance of the golden mean

the origin. If α and β denote the roots of the polynomial $p(x) = x^2 - ax - b$, then we may write

$$U(x) = \frac{u_0 + (u_1 - au_0)x}{(1 - \alpha x)(1 - \beta x)} = \frac{A}{1 - \alpha x} + \frac{B}{(1 - \beta x)} = A\sum_{n=0}^{\infty} \alpha^n x^n + B\sum_{n=0}^{\infty} \beta^n x^n,$$

where it is an easy matter to solve for A and B. Finally, this gives $u_n = A\alpha^n + B\beta^n$. Note that this approach yields a solution to our problem if the roots of p are distinct, namely $\alpha \neq \beta$. A variant of the formula holds if $\alpha = \beta$.

4. Using the generating formula for $p(n)$, prove the recurrence formula

$$\begin{aligned}
p(n) &= p(n-1) + p(n-2) - p(n-5) - p(n-7) - \cdots \\
&= \sum_{k\neq 0} (-1)^{k+1} p\left(n - \frac{k(3k+1)}{2}\right),
\end{aligned}$$

where the right-hand side is the finite sum taken over those $k \in \mathbb{Z}$, $k \neq 0$, with $k(3k+1)/2 \leq n$. Use this formula to calculate $p(5)$, $p(6)$, $p(7)$, $p(8)$, $p(9)$, and $p(10)$; check that $p(10) = 42$.

The next two exercises give elementary results related to the asymptotics of the partition function. More refined statements can be found in Appendix A.

5. Let

$$F(x) = \sum_{n=0}^{\infty} p(n)x^n = \prod_{n=1}^{\infty} \frac{1}{1 - x^n}$$

be the generating function for the partitions. Show that

$$\log F(x) \sim \frac{\pi^2}{6(1-x)} \qquad \text{as } x \to 1, \text{ with } 0 < x < 1.$$

[Hint: Use $\log F(x) = \sum \log(1/(1 - x^n))$ and $\log(1/(1 - x^n)) = \sum (1/m)x^{nm}$, so

$$\log F(x) = \sum \frac{1}{m} \frac{x^m}{1 - x^m}.$$

Use also $mx^{m-1}(1 - x) < 1 - x^m < m(1 - x)$.]

6. Show as a consequence of Exercise 5 that

$$e^{c_1 n^{1/2}} \leq p(n) \leq e^{c_2 n^{1/2}}$$

for two positive constants c_1 and c_2.

[Hint: $F(e^{-y}) = \sum p(n)e^{-ny} \leq Ce^{c/y}$ as $y \to 0$. So $p(n)e^{-ny} \leq ce^{c/y}$. Take $y = 1/n^{1/2}$ to get $p(n) \leq c'e^{c'n^{1/2}}$. In the opposite direction

$$\sum_{n=0}^{m} p(n)e^{-ny} \geq C(e^{c/y} - \sum_{n=m+1}^{\infty} e^{cn^{1/2}}e^{-ny}),$$

and it suffices to take $y = Am^{-1/2}$ where A is a large constant, and use the fact that the sequence $p(n)$ is increasing.]

7. Use the product formula for Θ to prove:

   (a) The "triangular number" identity

$$\prod_{n=0}^{\infty} (1 + x^n)(1 - x^{2n+2}) = \sum_{n=-\infty}^{\infty} x^{n(n+1)/2},$$

   which holds for $|x| < 1$.

   (b) The "septagonal number" identity

$$\prod_{n=0}^{\infty} (1 - x^{5n+1})(1 - x^{5n+4})(1 - x^{5n+5}) = \sum_{n=-\infty}^{\infty} (-1)^n x^{n(5n+3)/2},$$

   which holds for $|x| < 1$.

8. Consider Pythagorean triples (a, b, c) with $a^2 + b^2 = c^2$, and with $a, b, c \in \mathbb{Z}$. Suppose moreover that a and b have no common factors.

   (a) Show that either a or b must be odd, and the other even.

   (b) Show in this case (assuming a is odd and b even) that there are integers m, n so that $a = m^2 - n^2$, $b = 2mn$, and $c = m^2 + n^2$. [Hint: Note that $b^2 = (c - a)(c + a)$, and prove that $(c - a)/2$ and $(c + a)/2$ are relatively prime integers.]

   (c) Conversely, show that whenever c is a sum of two-squares, then there exist integers a and b such that $a^2 + b^2 = c^2$.

9. Use the formula for $r_2(n)$ to prove the following:

   (a) If $n = p$, where p is a prime of the form $4k + 1$, then $r_2(n) = 8$. This implies that n can be written in a unique way as $n = n_1^2 + n_2^2$, except for the signs and reordering of n_1 and n_2.

   (b) If $n = q^a$, where q is prime of the form $4k + 3$ and a is a positive integer, then $r_2(n) > 0$ if and only if a is even.

(c) In general, n can be represented as the sum of two squares if and only if all the primes of the form $4k + 3$ that arise in the prime decomposition of n occur with even exponents.

10. Observe the following irregularities of the functions $r_2(n)$ and $r_4(n)$ as n becomes large:

(a) $r_2(n) = 0$ for infinitely many n, while $\lim\sup_{n\to\infty} r_2(n) = \infty$.

(b) $r_4(n) = 24$ for infinitely many n while $\lim\sup_{n\to\infty} r_4(n)/n = \infty$.

[Hint: For (a) consider $n = 5^k$; for (b) consider alternatively $n = 2^k$, and $n = q^k$ where q is odd and large.]

11. Recall from Problem 2 in Chapter 2, that

$$\sum_{n=1}^{\infty} d(n)z^n = \sum_{n=1}^{\infty} \frac{z^n}{1 - z^n}, \qquad |z| < 1$$

where $d(n)$ denotes the number of divisors of n.
  More generally, show that

$$\sum_{n=1}^{\infty} \sigma_\ell(n)z^n = \sum_{n=1}^{\infty} \frac{n^\ell z^n}{1 - z^n}, \qquad |z| < 1$$

where $\sigma_\ell(n)$ is the sum of the ℓth powers of divisors of n.

12. Here we give another identity involving θ^4, which is equivalent to the four-squares theorem.

(a) Show that for $|q| < 1$

$$\sum_{n=1}^{\infty} \frac{nq^n}{1 - q^n} = \sum_{n=1}^{\infty} \frac{q^n}{(1 - q^n)^2}.$$

[Hint: The left-hand side is $\sum \sigma_1(n)q^n$. Use $x/(1 - x)^2 = \sum_{n=1}^{\infty} nx^n$.]

(b) Show as a result that

$$\sum_{n=1}^{\infty} \frac{nq^n}{1 - q^n} - \sum_{n=1}^{\infty} \frac{4nq^{4n}}{1 - q^{4n}} = \sum_{n=1}^{\infty} \frac{q^n}{(1 - q^n)^2} - 4\sum_{n=1}^{\infty} \frac{q^{4n}}{(1 - q^{4n})^2} = \sum \sigma_1^*(n)q^n$$

where $\sigma_1^*(n)$ is the sum of the divisors of d that are not divisible by 4.

(c) Show that the four-squares theorem is equivalent to the identity

$$\theta(\tau)^4 = 1 + 8\sum_{n=1}^{\infty} \frac{q^n}{(1 + (-1)^n q^n)^2}, \qquad q = e^{\pi i \tau}.$$

5 Problems

1.* Suppose n is of the form $n = 4^a(8k + 7)$, where a and k are positive integers. Show that n cannot be written as the sum of three-squares. The converse, that every n that is *not* of that form *can* be written as the sum of three-squares, is a difficult theorem of Legendre and Gauss.

2. Let $\mathrm{SL}_2(\mathbb{Z})$ denote the set of 2×2 matrices with integer entries and determinant 1, that is,

$$\mathrm{SL}_2(\mathbb{Z}) = \left\{ g = \begin{pmatrix} a & b \\ c & d \end{pmatrix} : a, b, c, d \in \mathbb{Z} \text{ and } ad - bc = 1 \right\}.$$

This group acts on the upper half-plane by the fractional linear transformation $g(\tau) = (a\tau + b)/(c\tau + d)$. Together with this action comes the so-called fundamental domain $\mathcal{F}_1$ in the complex plane defined by

$$\mathcal{F}_1 = \{ \tau \in \mathbb{C} : |\tau| \geq 1, \ |\mathrm{Re}(\tau)| \leq 1/2 \text{ and } |\mathrm{Im}(\tau)| \geq 0 \}.$$

It is illustrated in Figure 3.

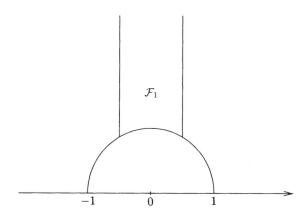

Figure 3. The fundamental domain $\mathcal{F}_1$

Consider the two elements in $\mathrm{SL}_2(\mathbb{Z})$ defined by $S(\tau) = -1/\tau$ and $T_1(\tau) = \tau + 1$. These correspond (for example) to the matrices

$$\begin{pmatrix} 0 & -1 \\ 1 & 0 \end{pmatrix} \quad \text{and} \quad \begin{pmatrix} 1 & 1 \\ 0 & 1 \end{pmatrix},$$

respectively. Let $\mathfrak{g}$ be the subgroup of $\mathrm{SL}_2(\mathbb{Z})$ generated by S and T_1.

  (a) Show that for every $\tau \in \mathbb{H}$ there exists $g \in \mathfrak{g}$ such that $g(\tau) \in \mathcal{F}_1$.

  (b) We say that two points τ and τ' are congruent if there exists $g \in \mathrm{SL}_2(\mathbb{Z})$ such that $g(\tau) = w$. Prove that if $\tau, w \in \mathcal{F}_1$ are congruent, then either $\mathrm{Re}(\tau) =$

$\pm 1/2$ and $\tau' = \tau \mp 1$ or $|\tau| = 1$ and $\tau' = -1/\tau$. [Hint: Say $\tau' = g(\tau)$. Why can one assume that $\text{Im}(\tau') \geq \text{Im}(\tau)$, and therefore $|c\tau + d| \leq 1$? Now consider separately the possibilities $c = -1$, $c = 0$, or $c = 1$.]

(c) Prove that S and T_1 generate the modular group in the sense that every fractional linear transformation corresponding to $g \in SL_2(\mathbb{Z})$ is a composition of finitely many S's and T_1's, and their inverses. Strictly speaking, the matrices associated to S and T_1 generate the projective special linear group $PSL_2(\mathbb{Z})$, which equals $SL_2(\mathbb{Z})$ modulo $\pm I$. [Hint: Observe that $2i$ is in the interior of $\mathcal{F}_1$. Now map $g(2i)$ back into $\mathcal{F}_1$ by using part (a). Use part (b) to conclude.]

3. In this problem, consider the group G of matrices $\begin{pmatrix} a & b \\ c & d \end{pmatrix}$ with integer entries, determinant 1, and such that a and d have the same parity, b and c have the same parity, and c and d have opposite parity. This group also acts on the upper half-plane by fractional linear transformations. To the group G corresponds the fundamental domain $\mathcal{F}$ defined by $|\tau| \geq 1$, $|\text{Re}(\tau)| \leq 1$, and $\text{Im}(\tau) \geq 0$ (see Figure 1). Also, let

$$ S(\tau) = -1/\tau \leftrightarrow \begin{pmatrix} 0 & -1 \\ 1 & 0 \end{pmatrix} \quad \text{and} \quad T_2(\tau) = \tau + 2 \leftrightarrow \begin{pmatrix} 1 & 2 \\ 0 & 1 \end{pmatrix}. $$

Prove that every fractional linear transformation corresponding to $g \in G$ is a composition of finitely many S, T_2 and their inverses, in analogy with the previous problem.

4. Let G denote the group of matrices given in the previous problem. Here we give an alternate proof of Theorem 3.4, that states that a function in $\mathbb{H}$ which is holomorphic, bounded, and invariant under G must be constant.

(a) Suppose that $f : \mathbb{H} \to \mathbb{C}$ is holomorphic, bounded, and that there exists a sequence of complex numbers $\tau_k = x_k + iy_k$ such that

$$ f(\tau_k) = 0, \quad \sum_{k=1}^{\infty} y_k = \infty, \quad 0 < y_k \leq 1, \quad \text{and} \quad |x_k| \leq 1. $$

Then $f = 0$. [Hint: When $x_k = 0$ see Problem 5 in Chapter 8.]

(b) Given two relatively prime integers c and d with different parity, show that there exist integers a and b such that $\begin{pmatrix} a & b \\ c & d \end{pmatrix} \in G$. [Hint: All the solutions of $xc + dy = 1$ take the form $x_0 + dt$ and $y_0 - ct$ where x_0, y_0 is a particular solution and $t \in \mathbb{Z}$.]

(c) Prove that $\sum 1/(c^2 + d^2) = \infty$ where the sum is taken over all c and d that are relatively prime and of opposite parity. [Hint: Suppose not, and prove that $\sum_{(a,b)=1} 1/(a^2 + b^2) < \infty$ where the sum is over all relatively prime integers a and b. To do so, note that if a and b are both odd and relatively prime, then the two numbers c and d defined by $c = (a + b)/2$

and $d = (a - b)/2$ are relatively prime and of opposite parity. Moreover, $c^2 + d^2 \leq A(a^2 + b^2)$ for some universal constant A. Therefore

$$\sum_{n \neq 0} \frac{1}{n^2} \sum_{(a,b)=1} \frac{1}{a^2 + b^2} < \infty,$$

hence $\sum 1/(k^2 + \ell^2) < \infty$, where the sum is over all integers k and ℓ such that $k, \ell \neq 0$. Why is this a contradiction?]

(d) Prove that if $F : \mathbb{H} \to \mathbb{C}$ is holomorphic, bounded, and invariant under G, then F is constant. [Hint: Replace $F(\tau)$ by $F(\tau) - F(\alpha i)$, where $\alpha > 0$ and α^2 is irrational. For each appropriate c, d, choose $g \in G$ so that $g(\alpha i) = x_{c,d} + \alpha i/(c^2 \alpha^2 + d^2)$. Then the $\{c^2 \alpha^2 + d^2\}$ are distinct, while $c^2 \alpha^2 + d^2 \approx c^2 + d^2$.]

5.* In Chapter 9 we proved that the Weierstrass $\wp$ function satisfies the cubic equation

$$(\wp')^2 = 4\wp^3 - g_2\wp - g_3,$$

where $g_2 = 60E_4$, $g_3 = 140E_6$, with E_k is the Eisenstein series of order k. The discriminant of the cubic $y^2 = 4x^3 - g_2 x - g_3$ is defined by $\triangle = g_2^3 - 27g_3^2$. Prove that

$$\triangle(\tau) = (2\pi)^{12} \eta^{24}(\tau) \quad \text{for all } \tau \in \mathbb{H}.$$

[Hint: $\triangle$ and η^{24} satisfy the same transformation laws under $\tau \mapsto \tau + 1$ and $\tau \mapsto -1/\tau$. Because of the fundamental domain described in Problem 2, it suffices then to investigate the behavior at the only cusp, which is at infinity.]

6.* Here we will deduce the formula for $r_8(n)$, which counts the number of representations of n as a sum of eight squares. The method is parallel to that of $r_4(n)$, but the details are less delicate.

Theorem: $r_8(n) = 16\sigma_3^*(n)$.

Here $\sigma_3^*(n) = \sigma_3(n) = \sum_{d|n} d^3$, when n is odd. Also, when n is even

$$\sigma_3^*(n) = \sum_{d|n} (-1)^d d^3 = \sigma_3^e(n) - \sigma_3^o(n),$$

where $\sigma_3^e(n) = \sum_{d|n, \, d \text{ even}} d^3$ and $\sigma_3^o(n) = \sum_{d|n, \, d \text{ odd}} d^3$.

Consider the appropriate Eisenstein series

$$E_4^*(\tau) = \sum \frac{1}{(n + m\tau)^4},$$

where the sum is over integers n and m with opposite parity. Recall the standard Eisenstein series

$$E_4(\tau) = \sum_{(n,m) \neq (0,0)} \frac{1}{(n + m\tau)^4}.$$

Notice that the series defining E_4^* is absolutely convergent, in distinction to $E_2^*(\tau)$, which arose when considering $r_4(n)$. This makes some of the considerations below quite a bit simpler.

(a) Prove that $r_8(n) = 16\sigma_3^*(n)$ is equivalent to the identity $\theta(\tau)^8 = 48\pi^{-4}E_4^*(\tau)$. [Hint: Use the fact that $E_4(\tau) = 2\zeta(4) + \frac{(2\pi)^4}{3}\sum_{k=1}^{\infty}\sigma_3(k)e^{2\pi ik\tau}$ and $\zeta(4) = \pi^4/90$.]

(b) Note that $E_4^*(\tau) = E_4(\tau) - 2^{-4}E_4((\tau-1)/2)$.

(c) $E_4^*(\tau+2) = E_4^*(\tau)$.

(d) $E_4^*(\tau) = \tau^{-4}E_4^*(-1/\tau)$.

(e) $(48/\pi^4)E_4^*(\tau) \to 1$ as $\tau \to \infty$.

(f) $|E_4^*(1-1/\tau)| \approx |\tau|^4|e^{2\pi i\tau}|$, as $\text{Im}(\tau) \to \infty$. [Hint: Verify that $E_4^*(1-1/\tau) = \tau^4(E_4(\tau) - E_4(2\tau))$.]

Since $\theta(\tau)^8$ satisfies properties similar to (c), (d), (e) and (f) above, it follows that the invariant function $48\pi^{-4}E_4^*(\tau)/\theta(\tau)^8$ is bounded and hence a constant, which must be 1. This gives the desired result.

Appendix A: Asymptotics

On the numerical computation of the definite integral $\int_w \cos \frac{\pi}{2} (w^3 - m.w)$, between the limits 0 and $\frac{1}{0}$.

The simplicity of the form of this differential coefficient induces me to suppose that the integral may possibly be expressible by some of the integrals whose values have been tabulated. After many attempts however, I have not succeeded in reducing it to any known integral: and I have therefore computed its value by actual summation to a considerable extent and by series for the remainder.

G. B. Airy, 1838

In a number of problems in analysis the solution is given by a function whose explicit calculation is not tractable. Often a useful substitute (and the only recourse) is to study the asymptotic behavior of this function near the point of interest. Here we shall investigate several related types of asymptotics, where the ideas of complex analysis are of crucial help. These typically center about the behavior for large values of the variable s of an integral of the form

$$(1) \qquad I(s) = \int_a^b e^{-s\Phi(x)} \, dx.$$

We organize our presentation by formulating three guiding principles.

(i) **Deformation of contour.** The function Φ is in general complex-valued, therefore, for large s the integrand in (1) may oscillate rapidly, so that the resulting cancellations mask the true behavior of $I(s)$. When Φ is holomorphic (which is often the case) one can hope to change the contour of integration so that as far as possible, on the new contour Φ is essentially real-valued. If this is possible, one can then hope to read off the behavior of $I(s)$ in a rather direct manner. This idea will be illustrated first in the context of Bessel functions.

(ii) **Laplace's method.** In the case when Φ is real-valued on the contour and s is positive, the maximum contribution to $I(s)$ comes from the integration near a minimum of Φ, and this leads to a satisfactory expansion in terms of the quadratic behavior of Φ near its minimum. We apply these ideas to present the asymptotics of the gamma function (Stirling's formula), and also those of the Airy function.

(iii) **Generating functions.** If $\{F_n\}$ is a number-theoretic or combinatorial sequence, we have already seen in several examples that one can exploit analytic properties of the generating function, $F(u) = \sum F_n u^n$, to obtain interesting conclusions regarding $\{F_n\}$. In fact the asymptotic behavior of F_n, as $n \to \infty$, can also be analyzed this way, via the formula

$$F_n = \int_\gamma F(e^{2\pi i z}) e^{-2\pi i n z} \, dz.$$

Here γ is an appropriate segment of unit length in the upper half-plane. This formula can then be studied as a variant of the integral (1). We shall show how these ideas apply in an important particular case to obtain an asymptotic formula for $p(n)$, the number of partitions of n.

1 Bessel functions

Bessel functions appear naturally in many problems that exhibit rotational symmetries. For instance, the Fourier transform of a spherical function in $\mathbb{R}^d$ is neatly expressed in terms of a Bessel function of order $(d/2) - 1$. See Chapter 6 in Book I.

The Bessel functions can be defined by a number of alternative formulas. We take the one that is valid for all order $\nu > -1/2$, given by

$$(2) \qquad J_\nu(s) = \frac{(s/2)^\nu}{\Gamma(\nu + 1/2)\Gamma(1/2)} \int_{-1}^{1} e^{isx}(1 - x^2)^{\nu - 1/2} \, dx.$$

If we also write $J_{-1/2}(s)$ for $\lim_{\nu \to -1/2} J_\nu(s)$, we see that it equals $\sqrt{\frac{2}{\pi s}} \cos s$; observe in addition that $J_{1/2}(s) = \sqrt{\frac{2}{\pi s}} \sin s$. However, $J_\nu(s)$ has an expression in terms of elementary functions only when ν is half-integral, and understanding this function in general requires further analysis. Its behavior for large s is suggested by the two examples above.

Theorem 1.1 $J_\nu(s) = \sqrt{\dfrac{2}{\pi s}} \cos\left(s - \dfrac{\pi\nu}{2} - \dfrac{\pi}{4}\right) + O\left(s^{-3/2}\right)$ *as* $s \to \infty$.

In view of the formula for $J_\nu(s)$ it suffices to investigate

(3) $$I(s) = \int_{-1}^{1} e^{isx}(1-x^2)^{\nu-1/2}\,dx,$$

and to this end we consider the analytic function $f(z) = e^{isz}(1-z^2)^{\nu-1/2}$ in the complex plane slit along the rays $(-\infty,-1)\cup(1,\infty)$; for $(1-z^2)^{\nu-1/2}$ we choose that branch that is positive when $z = x \in (-1,1)$. With $s > 0$ fixed, we apply Cauchy's theorem to see that

$$I(s) = -I_-(s) - I_+(s),$$

where the integrals $I(s)$, $I_-(s)$, and $I_+(s)$ are taken over the lines shown in Figure 1. This is established by using the fact that $\int_{\gamma_{\epsilon,R}} f(z)\,dz = 0$ where $\gamma_{\epsilon,R}$ is the second contour of Figure 1, and letting $\epsilon \to 0$ and $R \to \infty$.

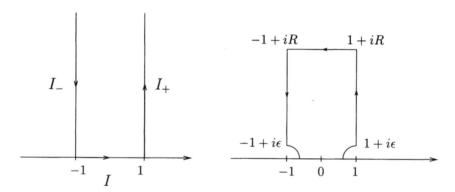

Figure 1. Contours of integration of $I(s)$, $I_-(s)$, $I_+(s)$, and the contour $\gamma_{\epsilon,R}$

On the contour for $I_+(s)$ we have $z = 1 + iy$, so

(4) $$I_+(s) = ie^{is}\int_0^\infty e^{-sy}(1-(1+iy)^2)^{\nu-1/2}\,dy.$$

There is a similar expression for $I_-(s)$.

What has the passage from $I(s)$ to $-(I_-(s)+I_+(s))$ gained us? Observe that for large positive s, the exponential e^{isx} in (3) oscillates rapidly, so the estimation of that integral is not obvious at first glance. However, in (4) the corresponding exponential is e^{-sy}, and it decreases rapidly as $s \to \infty$, except when $y = 0$. Thus in this case one sees immediately that the main contribution to the integral comes from the integration near $y = 0$, and this allows one readily to approximate this integral. This idea is made precise in the following observation.

Proposition 1.2 *Suppose a and m are fixed, with $a > 0$ and $m > -1$. Then as $s \to \infty$*

$$\text{(5)} \qquad \int_0^a e^{-sx} x^m \, dx = s^{-m-1} \Gamma(m+1) + O(e^{-cs}),$$

for some positive c.

Proof. The fact that $m > -1$ guarantees that $\int_0^a e^{-sx} x^m \, dx = \lim_{\epsilon \to 0} \int_\epsilon^a e^{-sx} x^m \, dx$ exists. Then, we write

$$\int_0^a e^{-sx} x^m \, dx = \int_0^\infty e^{-sx} x^m \, dx - \int_a^\infty e^{-sx} x^m \, dx.$$

The first integral on the right-hand side can be seen to equal $s^{-m-1} \Gamma(m+1)$, if we make the change of variables $x \mapsto x/s$. For the second integral we note that

$$\text{(6)} \qquad \int_a^\infty e^{-sx} x^m \, dx = e^{-cs} \int_a^\infty e^{-s(x-c)} x^m \, dx = O(e^{-cs}),$$

as long as $c < a$, and so the proposition is proved.

We return to the integral (4) and observe that

$$(1 - (1+iy)^2)^{\nu - 1/2} = (-2iy)^{\nu - 1/2} + O(y^{\nu + 1/2}), \qquad \text{for } 0 \le y \le 1,$$

while

$$(1 - (1+iy)^2)^{\nu - 1/2} = O(y^{\nu - 1/2} + y^{2\nu - 1}), \qquad \text{for } 1 \le y.$$

So, applying the proposition with $a = 1$ and $m = \nu \mp 1/2$, as well as (6), gives

$$I_+(s) = i(-2i)^{\nu - 1/2} e^{is} s^{-\nu - 1/2} \Gamma(\nu + 1/2) + O(s^{-\nu - 3/2}).$$

Similarly,

$$I_-(s) = i(2i)^{\nu - 1/2} e^{is} s^{-\nu - 1/2} \Gamma(\nu + 1/2) + O(s^{-\nu - 3/2}).$$

If we recall that

$$J_\nu(s) = \frac{(s/2)^\nu}{\Gamma(\nu + 1/2)\Gamma(1/2)}[-I_-(s) - I_+(s)],$$

and the fact that $\Gamma(1/2) = \sqrt{\pi}$, we see that we have obtained the proof of the theorem.

For later purposes it is interesting to point out that under certain restricted circumstances, the gist of the conclusion in Proposition 1.2 extends to the complex half-plane $\text{Re}(s) \geq 0$.

Proposition 1.3 *Suppose* a *and* m *are fixed, with* $a > 0$ *and* $-1 < m < 0$. *Then as* $|s| \to \infty$ *with* $\text{Re}(s) \geq 0$,

$$\int_0^a e^{-sx} x^m \, dx = s^{-m-1}\Gamma(m + 1) + O(1/|s|).$$

(Here s^{-m-1} is the branch of that function that is positive for $s > 0$).

Proof. We begin by showing that when $\text{Re}(s) \geq 0$, $s \neq 0$,

$$\int_0^\infty e^{-sx} x^m \, dx = \lim_{N \to \infty} \int_0^N e^{-sx} x^m \, dx$$

exists and equals $s^{-m-1}\Gamma(m + 1)$. If N is large, we first write

$$\int_0^N e^{-sx} x^m \, dx = \int_0^a e^{-sx} x^m \, dx + \int_a^N e^{-sx} x^m \, dx.$$

Since $m > -1$, the first integral on the right-hand side defines an analytic function everywhere. For the second integral, we note that $-\frac{1}{s}\frac{d}{dx}(e^{-sx}) = e^{-sx}$, so an integration by parts gives

$$(7) \qquad \int_a^N e^{-sx} x^m \, dx = \frac{m}{s} \int_a^N e^{-sx} x^{m-1} \, dx - \left[\frac{e^{-sx}}{s} x^m\right]_a^N.$$

This identity, together with the convergence of the integral $\int_a^\infty x^{m-1} dx$, shows that $\int_a^\infty e^{-sx} x^m \, dx$ defines an analytic function on $\text{Re}(s) > 0$ that is continuous on $\text{Re}(s) \geq 0$, $s \neq 0$. Thus $\int_0^\infty e^{-sx} x^m \, dx$ is analytic on the half-plane $\text{Re}(s) > 0$ and continuous on $\text{Re}(s) \geq 0$, $s \neq 0$. Since it equals $s^{-m-1}\Gamma(m + 1)$ when s is positive, we deduce that $\int_0^\infty e^{-sx} x^m \, dx = s^{-m-1}\Gamma(m + 1)$ when $\text{Re}(s) \geq 0$, $s \neq 0$.

However, we now have

$$\int_0^a e^{-sx} x^m \, dx = \int_0^\infty e^{-sx} x^m \, dx - \int_a^\infty e^{-sx} x^m \, dx.$$

It is clear from (7), and from the fact that $m < 0$, that if we let $N \to \infty$, then $\int_a^\infty e^{-sx} x^{m-1} \, dx = O(1/|s|)$. The proposition if therefore proved.

Note. If one wants to obtain a better error term in Proposition 1.3, or for that matter extend the range of m, then one needs to mitigate the effect of the contribution of the end-point $x = a$. This can be done by introducing suitable smooth cut-offs. See Problem 1.

2 Laplace's method; Stirling's formula

We have already mentioned that when Φ is real-valued, the main contribution to $\int_a^b e^{-s\Phi(x)} \, dx$ as $s \to \infty$ comes from the point where Φ takes its minimum value. A situation where this minimum is attained at one of the end-points, a or b, was considered in Proposition 1.2. We now turn to the important case when the minimum is achieved in the interior of $[a, b]$.

Consider

$$\int_a^b e^{-s\Phi(x)} \psi(x) \, dx$$

where the **phase** Φ is real-valued, and both it and the **amplitude** ψ are assumed for simplicity to be indefinitely differentiable. Our hypothesis regarding the minimum of Φ is that there is an $x_0 \in (a, b)$ so that $\Phi'(x_0) = 0$, but $\Phi''(x) > 0$ throughout $[a, b]$ (Figure 2 illustrates the situation.)

Proposition 2.1 *Under the above assumptions, with $s > 0$ and $s \to \infty$,*

$$(8) \qquad \int_a^b e^{-s\Phi(x)} \psi(x) \, dx = e^{-s\Phi(x_0)} \left[\frac{A}{s^{1/2}} + O\left(\frac{1}{s}\right) \right],$$

where

$$A = \sqrt{2\pi} \frac{\psi(x_0)}{(\Phi''(x_0))^{1/2}}.$$

Proof. By replacing $\Phi(x)$ by $\Phi(x) - \Phi(x_0)$ we may assume that $\Phi(x_0) = 0$. Since $\Phi'(x_0) = 0$, we note that

$$\frac{\Phi(x)}{(x - x_0)^2} = \frac{\Phi''(x_0)}{2} \varphi(x),$$

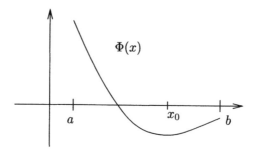

Figure 2. The function Φ, with its minimum at x_0

where φ is smooth, and $\varphi(x) = 1 + O(x - x_0)$ as $x \to x_0$. We can therefore make the smooth change of variables $x \mapsto y = (x - x_0)(\varphi(x))^{1/2}$ in a small neighborhood of $x = x_0$, and observe that $dy/dx|_{x_0} = 1$, and thus $dx/dy = 1 + O(y)$ as $y \to 0$. Moreover, we have $\psi(x) = \tilde{\psi}(y)$ with $\tilde{\psi}(y) = \psi(x_0) + O(y)$ as $y \to 0$. Hence if $[a', b']$ is a sufficiently small interval containing x_0 in its interior, by making the indicated change of variables we obtain

(9)
$$\int_{a'}^{b'} e^{-s\Phi(x)} \psi(x)\, dx = \psi(x_0) \int_{\alpha}^{\beta} e^{-s\frac{\Phi''(x_0)}{2}y^2}\, dy + O\left(\int_{\alpha}^{\beta} e^{-s\frac{\Phi''(x_0)}{2}y^2} |y|\, dy \right),$$

where $\alpha < 0 < \beta$. We now make the further change of variables $y^2 = X$, $dy = \frac{1}{2}X^{-1/2}\, dX$, and we see by (5) that the first integral on the right-hand side in (9) is

$$\int_0^{a_0} e^{-s\frac{\Phi''(x_0)}{2}X} X^{-1/2}\, dX + O(e^{-\delta s}) = s^{-1/2} \left(\frac{2\pi}{\Phi''(x_0)} \right)^{1/2} + O(e^{-\delta s}),$$

for some $\delta > 0$. By the same argument, the second integral is $O(1/s)$. What remains are the integrals of $e^{-s\Phi(x)}\psi(x)$ over $[a, a']$ and $[b', b]$; but these integrals decay exponentially as $s \to \infty$, since $\Phi(x) \geq c > 0$ in these two sub-intervals. Altogether, this establishes (8) and the proposition.

It is important to realize that the asymptotic relation (8) extends to all complex s with $\operatorname{Re}(s) \geq 0$. The proof, however, requires a somewhat different argument: here we must take into account the oscillations of $e^{-s\Phi(x)}$ when $|s|$ is large but $\operatorname{Re}(s)$ is small, and this is achieved by a simple integration by parts.

Proposition 2.2 *With the same assumptions on* Φ *and* ψ, *the relation* (8) *continues to hold if* $|s| \to \infty$ *with* $\mathrm{Re}(s) \geq 0$.

Proof.   We proceed as before to the equation (9), and obtain the appropriate asymptotic for the first term, by virtue of Proposition 1.3, with $m = -1/2$. To deal with the rest we start with an observation. If Ψ and ψ are given on an interval $[\bar{a}, \bar{b}]$, are indefinitely differentiable, and $\Psi(x) \geq 0$, while $|\Psi'(x)| \geq c > 0$, then if $\mathrm{Re}(s) \geq 0$,

$$(10) \qquad \int_{\bar{a}}^{\bar{b}} e^{-s\Psi(x)} \psi(x)\, dx = O\left(\frac{1}{|s|}\right) \qquad \text{as } |s| \to \infty.$$

Indeed, the integral equals

$$-\frac{1}{s} \int_{\bar{a}}^{\bar{b}} \frac{d}{dx}\left(e^{-s\Psi(x)}\right) \frac{\psi(x)}{\Psi'(x)}\, dx,$$

which by integration by parts gives

$$\frac{1}{s} \int_{\bar{a}}^{\bar{b}} e^{-s\Psi(x)} \frac{d}{dx}\left(\frac{\psi(x)}{\Psi'(x)}\right) dx - \frac{1}{s}\left[e^{-s\Psi(x)} \frac{\psi(x)}{\Psi'(x)}\right]_{\bar{a}}^{\bar{b}}.$$

The assertion (10) follows immediately since $|e^{-s\Psi(x)}| \leq 1$, when $\mathrm{Re}(s) \geq 0$. This allows us to deal with the integrals of $e^{-s\Phi(x)}\psi(x)$ in the complementary intervals $[a, a']$ and $[b', b]$, because in each, $|\Phi'(x)| \geq c > 0$, since $\Phi'(x_0) = 0$ and $\Phi''(x) \geq c_1 > 0$.

Finally, for the second term on the right-hand side of (9) we observe that it is actually of the form

$$\int_{\alpha}^{\beta} e^{-s\frac{\Phi''(x_0)}{2}y^2} y\eta(y)\, dy,$$

where $\eta(y)$ is differentiable. Then we can again estimate this term by integration by parts, once we write it as

$$-\frac{1}{s\Phi''(x_0)} \int_{\alpha}^{\beta} \frac{d}{dy}\left(e^{-s\frac{\Phi''(x_0)}{2}y^2}\right) \eta(y)\, dy,$$

obtaining the bound $O(1/|s|)$.

The special case of Proposition 2.2 when s is purely imaginary, $s = it$, $t \to \pm\infty$, is often treated separately; the argument in this situation is usually referred to as the method of **stationary phase**. The points x_0 for which $\Phi'(x_0) = 0$ are called the **critical points**.

Our first application will be to the asymptotic behavior of the gamma function Γ, given by Stirling's formula. This formula will be valid in any sector of the complex plane that omits the negative real axis. For any $\delta > 0$ we set $S_\delta = \{s : |\arg s| \leq \pi - \delta\}$, and denote by $\log s$ the principal branch of the logarithm that is given in the plane slit along the negative real axis.

Theorem 2.3 *If* $|s| \to \infty$ *with* $s \in S_\delta$, *then*

$$(11) \qquad \Gamma(s) = e^{s\log s} e^{-s} \frac{\sqrt{2\pi}}{s^{1/2}} \left(1 + O\left(\frac{1}{|s|^{1/2}}\right)\right).$$

Remark. With a little extra effort one can improve the error term to $O(1/|s|)$, and in fact obtain a complete asymptotic expansion in powers of $1/s$; see Problem 2. Also, we note that (11) implies $\Gamma(s) \sim \sqrt{2\pi} s^{s-1/2} e^{-s}$, which is how Stirling's formula is often stated.

To prove the theorem we first establish (11) in the right half-plane. We shall show that the formula holds whenever $\text{Re}(s) > 0$, and in addition that the error term is uniform on the closed half-plane, once we omit a neighborhood of the origin (say $|s| < 1$). To see this, start with $s > 0$, and write

$$\Gamma(s) = \int_0^\infty e^{-x} x^s \frac{dx}{x} = \int_0^\infty e^{-x+s\log x} \frac{dx}{x}.$$

Upon making the change of variables $x \mapsto sx$, the above equals

$$\int_0^\infty e^{-sx+s\log sx} \frac{dx}{x} = e^{s\log s} e^{-s} \int_0^\infty e^{-s\Phi(x)} \frac{dx}{x},$$

where $\Phi(x) = x - 1 - \log x$. By analytic continuation this identity continues to hold, and we have when $\text{Re}(s) > 0$,

$$\Gamma(s) = e^{s\log s} e^{-s} I(s)$$

with

$$I(s) = \int_0^\infty e^{-s\Phi(x)} \frac{dx}{x}.$$

It now suffices to see that

$$(12) \qquad I(s) = \frac{\sqrt{2\pi}}{s^{1/2}} + O\left(\frac{1}{|s|}\right) \qquad \text{for} \qquad \text{Re}(s) > 0.$$

Observe first that $\Phi(1) = \Phi'(1) = 0$, $\Phi''(x) = 1/x^2 > 0$ whenever $0 < x < \infty$, and $\Phi''(1) = 1$. Thus Φ is convex, attains its minimum at $x = 1$, and is positive.

We apply the complex version of the Laplace method, Proposition 2.2, in this situation. Here the critical point is $x_0 = 1$ and $\psi(x) = 1/x$. We choose for convenience the interval $[a, b]$ to be $[1/2, 2]$. Then for $\int_a^b e^{-s\Phi(x)}\psi(x)\,dx$ we get the asymptotic (12). It remains to bound the error terms, those corresponding to integration over $[0, 1/2]$, and $[2, \infty)$. Here the device of integration by parts, which has served us so well, can be applied again. Indeed, since $\Phi'(x) = 1 - 1/x$, we have

$$\int_\epsilon^{1/2} e^{-s\Phi(x)}\,\frac{dx}{x} = -\frac{1}{s}\int_\epsilon^{1/2}\frac{d}{dx}\left(e^{-s\Phi(x)}\right)\frac{dx}{\Phi'(x)x}$$

$$= -\frac{1}{s}\left[\frac{e^{-s\Phi(x)}}{x-1}\right]_\epsilon^{1/2} - \frac{1}{s}\int_\epsilon^{1/2} e^{-s\Phi(x)}\frac{dx}{(x-1)^2}.$$

Noting that $\Phi(\epsilon) \to +\infty$ as $\epsilon \to 0$, and $|e^{-s\Phi(x)}| \le 1$, we find in the limit that

$$\int_0^{1/2} e^{-s\Phi(x)}\,\frac{dx}{x} = \frac{2}{s}e^{-s\Phi(1/2)} - \frac{1}{s}\int_0^{1/2} e^{-s\Phi(x)}\frac{dx}{(x-1)^2}.$$

Thus the left-hand side is $O(1/|s|)$ in the half-plane $\mathrm{Re}(s) \ge 0$.

The integral $\int_2^\infty e^{-s\Phi(x)}\,\frac{dx}{x}$ is treated analogously, once we note that $\int_2^\infty (x-1)^{-2}\,dx$ converges.

Since these estimates are uniform, (12) and thus (11) are proved for $\mathrm{Re}(s) \ge 0$, $|s| \to \infty$.

To pass from $\mathrm{Re}(s) \ge 0$ to $\mathrm{Re}(s) \le 0$, $s \in S_\delta$, we record the following fact about the principal branch of $\log s$: whenever $\mathrm{Re}(s) \ge 0$, $s = \sigma + it$, $t \ne 0$, then

$$\log(-s) = \begin{cases} \log s - i\pi & \text{if } t > 0, \\ \log s + i\pi & \text{if } t < 0. \end{cases}$$

Hence if $G(s) = e^{s\log s}e^{-s}$, $\mathrm{Re}(s) \ge 0$, $t \ne 0$, then

(13)
$$G(-s)^{-1} = \begin{cases} e^{s\log s}e^{-s}e^{-si\pi} & \text{if } t > 0, \\ e^{s\log s}e^{-s}e^{si\pi} & \text{if } t < 0. \end{cases}$$

Next,

(14)
$$\Gamma(s)\Gamma(-s) = \frac{\pi}{-s\sin\pi s},$$

which follows from the fact that $\Gamma(s)\Gamma(1-s) = \pi/\sin\pi s$, and $\Gamma(1-s) = -s\Gamma(-s)$ (see Theorem 1.4 and Lemma 1.2 in Chapter 6). The combination of (13) and (14), together with the fact that for large s, $\left(1 + O(1/|s|^{1/2})\right)^{-1} = 1 + O(1/|s|^{1/2})$, allows us then to extend (11) to the whole sector S_δ, thereby completing the proof of the theorem.

3 The Airy function

The Airy function appeared first in optics, and more precisely, in the analysis of the intensity of light near a caustic; it was an important early instance in the study of asymptotics of integrals, and it continues to arise in a number of other problems. The **Airy function** Ai is defined by

$$(15) \qquad \text{Ai}(s) = \frac{1}{2\pi} \int_{-\infty}^{\infty} e^{i(x^3/3+sx)}\, dx, \qquad \text{with } s \in \mathbb{R}.$$

Let us first see that because of the rapid oscillations of the integrand as $|x| \to \infty$, the integral converges and represents a continuous function of s. In fact, note that

$$\frac{1}{i(x^2+s)}\frac{d}{dx}\left(e^{i(x^3/3+sx)}\right) = e^{i(x^3/3+sx)},$$

so if $a \geq 2|s|^{1/2}$, we can write the integral $\int_a^R e^{i(x^3/3+sx)}\, dx$ as

$$(16) \qquad \int_a^R \frac{1}{i(x^2+s)}\frac{d}{dx}\left(e^{i(x^3/3+sx)}\right)\, dx.$$

We may now integrate by parts and let $R \to \infty$, to see that the integral converges uniformly, and that as a result $\int_a^\infty e^{i(x^3/3+sx)}\, dx$ is also continuous for $|s| \leq a^2/4$. The same argument works for the integral from $-\infty$ to $-a$ and our assertion regarding Ai(s) is established.

A better insight into Ai(s) is given by deforming the contour of integration in (15). A choice of an optimal contour will appear below, but for now let us notice that as soon as we replace the x-axis of integration in (15) by the parallel line $L_\delta = \{x + i\delta,\ x \in \mathbb{R}\}$, $\delta > 0$, matters improve dramatically.

In fact, we may apply the Cauchy theorem to $f(z) = e^{i(z^3/3+sz)}$ over the rectangle γ_R shown in Figure 3.

One observes that $f(z) = O(e^{-\delta x^2})$ on L_δ, while $f(z) = O(e^{-yR^2})$ on the vertical sides of the rectangle. Thus since $\int_0^\delta e^{-yR^2}\, dy \to 0$ as

$z = x + i\delta$

Figure 3. The line L_δ and the contour γ_R

$R \to \infty$, we see that

$$\mathrm{Ai}(s) = \frac{1}{2\pi} \int_{L_\delta} e^{i(z^3/3+sz)}\, dz.$$

Now the majorization $f(z) = O(e^{-\delta x^2})$ continues to hold for each complex s, and hence because of the (rapid) convergence of the integral, $\mathrm{Ai}(s)$ extends to an entire function of s.

We note next that $\mathrm{Ai}(s)$ satisfies the differential equation

$$(17) \qquad\qquad\qquad \mathrm{Ai}''(s) = s\mathrm{Ai}(s).$$

This simple and natural equation helps to explain the ubiquity of the Airy function. To prove (17) observe that

$$\mathrm{Ai}''(s) - s\mathrm{Ai}(s) = \frac{1}{2\pi} \int_{L_\delta} (-z^2 - s)e^{i(z^3/3+sz)}\, dz.$$

But $-(z^2 + s)e^{i(z^3/3+sz)} = i\frac{d}{dz}(e^{i(z^3/3+sz)})$, so

$$\mathrm{Ai}''(s) - s\mathrm{Ai}(s) = \frac{i}{2\pi} \int_{L_\delta} \frac{d}{dz}(f(z))\, dz = 0,$$

since $f(z) = e^{i(z^3/3+sz)}$ vanishes as $|z| \to \infty$ along L_δ.

We now turn to our main problem, the asymptotics of $\mathrm{Ai}(s)$ for large (real) values of s. The differential equation (17) shows us that we may expect different behaviors of the Airy function when $|s|$ is large, depending on whether s is positive or negative. To see this, we compare the equation with a simple analogue

$$(18) \qquad\qquad\qquad y''(s) = Ay(s),$$

where A is a large constant, with A positive when considering s positive and A negative in the other case. The solutions of (18) are of course $e^{\sqrt{A}s}$ and $e^{-\sqrt{A}s}$, the first growing rapidly, and the second decreasing rapidly as $s \to \infty$, if $A > 0$. A glance at the integration by parts following (16) shows that $\mathrm{Ai}(s)$ remains bounded when $s \to \infty$. So the comparison with $e^{\sqrt{A}s}$ must be dismissed, and we might reasonably guess that $\mathrm{Ai}(s)$ is rapidly decreasing in this case. When $s < 0$ we take $A < 0$ in (18). The exponentials $e^{\sqrt{A}s}$ and $e^{-\sqrt{A}s}$ are now oscillating, and we can therefore presume that $\mathrm{Ai}(s)$ should have an oscillatory character as $s \to -\infty$.

Theorem 3.1 *Suppose $u > 0$. Then as $u \to \infty$,*

(i) $\mathrm{Ai}(-u) = \pi^{-1/2} u^{-1/4} \cos(\tfrac{2}{3} u^{3/2} - \tfrac{\pi}{4})(1 + O(1/u^{3/4}))$.

(ii) $\mathrm{Ai}(u) = \dfrac{1}{2\pi^{1/2}} u^{-1/4} e^{-\frac{2}{3} u^{3/2}}(1 + O(1/u^{3/4}))$.

To consider the first case, we make the change of variables $x \mapsto u^{1/2}x$ in the defining integral with $s = -u$. This gives

$$\mathrm{Ai}(-u) = u^{1/2} I_-(u^{3/2}),$$

where

(19) $$I_-(t) = \frac{1}{2\pi} \int_{-\infty}^{\infty} e^{it(x^3/3 - x)} \, dx.$$

Now write

$$I_-(s) = \frac{1}{2\pi} \int_{-\infty}^{\infty} e^{-s\Phi(x)} \, dx,$$

where $\Phi(x) = \Phi_-(x) = x^3/3 - x$, and we shall apply Proposition 2.2, which in this case, since s is purely imaginary, is the method of stationary phase. Note that $\Phi'(x) = x^2 - 1$, so there are two critical points, $x_0 = \pm 1$; observe that $\Phi''(x) = 2x$; also $\Phi(\pm 1) = \mp 2/3$.

We break up the range of integration in (19) into two intervals $[-2, 0]$ and $[0, 2]$ each containing one critical point, and two complementary integrals, $(-\infty, -2]$ and $[2, \infty)$.

Now we apply Proposition 2.2 to the interval $[0, 2]$ with $s = -it$, $x_0 = 1$ $\psi = 1/2\pi$, $\Phi(1) = -2/3$, $\Phi''(1) = 2$, and get a contribution of

$$\frac{1}{2\sqrt{\pi}} e^{-i\frac{2}{3}t} \left(\frac{1}{(-it)^{1/2}} + O\left(\frac{1}{|t|}\right) \right),$$

in view of (8). Similarly the integral over $[-2, 0]$ contributes

$$\frac{1}{2\sqrt{\pi}} e^{i\frac{2}{3}t} \left(\frac{1}{(it)^{1/2}} + O\left(\frac{1}{|t|}\right) \right).$$

Finally, consider the complementary integrals. The first is

$$\int_{-\infty}^{-2} e^{it\Phi(x)}\, dx = \lim_{N\to\infty} \int_{-N}^{-2} e^{it\Phi(x)}\, dx = \lim_{N\to\infty} \frac{1}{it} \int_{-N}^{-2} \frac{d}{dx}\left(e^{it\Phi(x)}\right) \frac{dx}{\Phi'(x)},$$

where $\Phi'(x) = x^2 - 1$. So an integration by parts shows that this is $O(1/|t|)$. The integral over $[2, \infty)$ is treated similarly. Adding these four contributions, and inserting them in the identity $\mathrm{Ai}(-u) = u^{1/2}I_-(u^{3/2})$, proves conclusion (i) of the theorem.[6]

To deal with the conclusion (ii) of the theorem, we make the change of variables $x \mapsto u^{1/2}x$ in the integral (15), with $s = u$. This gives us, for $u > 0$,

$$\mathrm{Ai}(u) = u^{1/2}I_+(u^{3/2}),$$

where

(20) $$I_+(s) = \frac{1}{2\pi} \int_{-\infty}^{\infty} e^{-sF(x)}\, dx$$

and $F(x) = -i(x^3/3 + x)$. Now when $s \to \infty$, the integrand in (20) again oscillates rapidly, but here in distinction to the previous case, there is no critical point on the real axis, since the derivative of $x^3/3 + x$ does not vanish. A repeated integration by parts argument (such as we have used before) shows that actually the integral $I_+(s)$ has fast decay as $s \to \infty$. But what is the exact nature and order of this decrease? To answer this question, we would have to take into account the precise cancellations inherent in (20), and doing this by the above method does not seem feasible.

A better way is to follow the guiding principle used in the asymptotics of the Bessel function, and to deform the line of integration in (20) to a contour on which the imaginary part of $F(z)$ vanishes; having done this, one might then hope to apply Laplace's method, Proposition 2.1, to find the true asymptotic behavior of $I_+(s)$, as $s \to \infty$.

We describe the idea in the more general situation in which we assume only that $F(z)$ is holomorphic. To follow the approach suggested, we seek a contour Γ so that:

---

[6]An alternative derivation of this conclusion can be given as a consequence of the relation of the Airy function with the Bessel functions. See Problem 3 below.

(a) $\mathrm{Im}(F) = 0$ on Γ.

(b) $\mathrm{Re}(F)$ has a minimum on Γ at some point z_0, and this function is non-degenerate in the sense that the second derivative of $\mathrm{Re}(F)$ along Γ is strictly positive at z_0.

Conditions (a) and (b) imply of course that $F'(z_0) = 0$. If as above, $F''(z_0) \neq 0$, then there are two curves Γ_1 and Γ_2 passing through z_0 which are orthogonal, so that $F|_{\Gamma_i}$ is real for $i = 1, 2$, with $\mathrm{Re}(F)$ restricted to Γ_1 having a minimum at z_0; and $\mathrm{Re}(F)$ restricted to Γ_2 having a maximum at z_0 (see Exercise 2 in Chapter 8). We therefore try to deform our original contour of integration to $\Gamma = \Gamma_1$. This approach is usually referred to as the method of **steepest descent**, because at z_0 the function $-\mathrm{Re}(F(z))$ has a saddle point, and starting at this point and following the path of Γ_1, one has the greatest decrease of this function.

Let us return to our special case, $F(z) = -i(z^3/3 + z)$. We note that

$$\begin{cases} \mathrm{Re}(F) & = & x^2 y - y^3/3 + y, \\ \mathrm{Im}(F) & = & -x^3/3 + xy^2 - x. \end{cases}$$

We observe also that $F'(z) = -i(z^2 + 1)$, so we have two non-real critical points $z_0 = \pm i$ at which $F'(z_0) = 0$. If we choose $z_0 = i$, then the two curves passing through this point where $\mathrm{Im}(F) = 0$ are

$$\Gamma_1 = \{(x, y) : y^2 = x^2/3 + 1\} \quad \text{and} \quad \Gamma_2 = \{(x, y) : x = 0\}.$$

On Γ_2, the function $\mathrm{Re}(F)$ clearly has a maximum at the point $z_0 = i$, and so we reject this curve. We choose $\Gamma = \Gamma_1$, which is a branch of a hyperbola, and which can be written as $y = (x^2/3 + 1)^{1/2}$; it is asymptotic to the rays $z = re^{i\pi/6}$, and $z = re^{i5\pi/6}$ at infinity. See Figure 4.

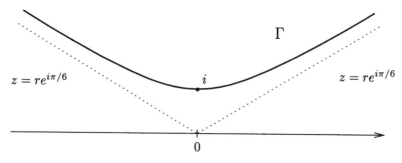

Figure 4. The curve of steepest descent

Next, we see that

$$\text{(21)} \qquad \frac{1}{2\pi} \int_{-\infty}^{\infty} e^{-sF(x)} \, dx = \frac{1}{2\pi} \int_{\Gamma} e^{-sF(z)} \, dz.$$

This identity is justified by applying the Cauchy theorem to $e^{-sF(z)}$ on the contour Γ_R that consists of four arc segments: the parts of the real axis and Γ that lie inside the circle of radius R, and the two arcs of this circle joining the axis with Γ. Since in this region $e^{-sF(z)} = O(e^{-cyx^2})$ as $x \to \pm\infty$, the contributions of the two arcs of the circle are $O(\int_0^{\pi} e^{-cR^2 \sin\theta} \, d\theta) = O(1/R)$, and letting $R \to \infty$ establishes (21).

We now observe that on Γ

$$\Phi(x) = \operatorname{Re}(F) = y(x^2 - y^2/3 + 1) = (\frac{8}{9}x^2 + \frac{2}{3})(x^2/3 + 1)^{1/2},$$

since $y^2 = x^2/3 + 1$ there. Also, on Γ we have that $dz = dx + idy = dx + i(x/3)(x^2/3 + 1)^{-1/2}dx$. Thus,

$$\text{(22)} \qquad \frac{1}{2\pi} \int_{\Gamma} e^{-sF(z)} \, dz = \frac{1}{2\pi} \int_{-\infty}^{\infty} e^{-s\Phi(x)} \, dx,$$

in view of the fact that $\Phi(x)$ is even, while $x(x^2/3 + 1)^{-1/2}$ is odd.

We note next that since $(1 + u)^{1/2} = 1 + u/2 + O(u^2)$ as $u \to 0$,

$$\Phi(x) = (\frac{8}{9}x^2 + \frac{2}{3}) + \frac{2}{3}\frac{1}{2}\frac{x^2}{3} + O(x^4) = x^2 + \frac{2}{3} + O(x^4),$$

and so $\Phi''(0) = 2$. We now apply Proposition 2.1 to estimate the main part of the right-hand side of (22), by

$$\frac{1}{2\pi} \int_{-c}^{c} e^{-s\Phi(x)} \, dx,$$

where c is a small positive constant. Since $\Phi(0) = 2/3$, $\Phi''(0) = 2$, and $\psi(0) = 1/2\pi$, we obtain that this term contributes

$$e^{-\frac{2}{3}s} \left[\frac{1}{2\pi^{1/2}} \frac{1}{s^{1/2}} + O\left(\frac{1}{s}\right) \right].$$

The term $\int_c^{\infty} e^{-s\Phi(x)} \, dx$ is dominated by $e^{-2s/3} \int_c^{\infty} e^{-c_1 sx^2} \, dx$, which is $O(e^{-2s/3}e^{-\delta s})$ for some $\delta > 0$, as soon as $c > 0$. A similar estimate holds for $\int_{-\infty}^{-c} e^{-s\Phi(x)} \, dx$. Altogether, then,

$$I_+(s) = e^{-\frac{2}{3}s} \left[\frac{1}{2\pi^{1/2}} \frac{1}{s^{1/2}} + O\left(\frac{1}{s}\right) \right] \qquad \text{as } s \to \infty,$$

and this gives the desired asymptotic (ii) for the Airy function.

4 The partition function

Our last illustration of the techniques developed in this appendix is in their application to the partition function $p(n)$, which was discussed in Chapter 10. We derive for it the main term of the remarkable asymptotic formula of Hardy-Ramanujan.

Theorem 4.1 *If p denotes the partition function, then*

(i) $p(n) \sim \dfrac{1}{4\sqrt{3}n} e^{Kn^{1/2}}$ *as $n \to \infty$, where $K = \pi\sqrt{\frac{2}{3}}$.*

(ii) *A much more precise assertion is that*

$$p(n) = \frac{1}{2\pi\sqrt{2}} \frac{d}{dn} \left(\frac{e^{K(n-\frac{1}{24})^{1/2}}}{(n-\frac{1}{24})^{1/2}} \right) + O(e^{\frac{K}{2}n^{1/2}}).$$

Note. Observe that $(n - \frac{1}{24})^{1/2} - n^{1/2} = O(n^{-1/2})$, by the mean-value theorem; hence $e^{K(n-\frac{1}{24})^{1/2}} = e^{Kn^{1/2}}(1 + O(n^{-1/2}))$, thus $e^{K(n-\frac{1}{24})^{1/2}} \sim e^{Kn^{1/2}}$, as $n \to \infty$. Of course, clearly $(n - \frac{1}{24})^{1/2} \sim n^{1/2}$, and in particular (ii) implies (i).

We shall discuss first, in a more general setting, how we might derive the asymptotic behavior of a sequence $\{F_n\}$ from the analytic properties of its generating function $F(w) = \sum_{n=0}^{\infty} F_n w^n$. Assuming for the sake of simplicity that $\sum F_n w^n$ has the unit disc as its disc of convergence, we can set forth the following heuristic principle: the asymptotic behavior of F_n is determined by the location and nature of the "singularities" of F on the unit circle, and the contribution to the asymptotic formula due to each singularity corresponds in magnitude to the "order" of that singularity.

A very simple example in which this principle is unambiguous and can be verified occurs when F is meromorphic in a larger disc, but has only one singularity on the circle, a pole of order r at the point $w = 1$. Then there is a polynomial P of degree $r - 1$ so that $F_n = P(n) + O(e^{-\epsilon n})$ as $n \to \infty$, for some $\epsilon > 0$. In fact, $\sum_{n=0}^{\infty} P(n)w^n$ is a good approximation to $F(w)$ near $w = 1$; it is the principal part of the pole of F. (See also Problem 4.)

For the partition function the analysis is not as simple as this example, but the principle stated above is still applicable when properly interpreted. To this task we now turn.

We recall the formula

$$\sum_{n=0}^{\infty} p(n)\, w^n = \prod_{n=1}^{\infty} \frac{1}{1 - w^n},$$

established in Theorem 2.1, Chapter 10. This identity implies that the generating function is holomorphic in the unit disc. In what follows, it will be convenient to pass from the unit disc to the upper half-plane by writing $w = e^{2\pi i z}$, $z = x + iy$, and taking $y > 0$. We therefore have

$$\sum_{n=0}^{\infty} p(n) e^{2\pi i n z} = f(z),$$

with

$$f(z) = \prod_{n=1}^{\infty} \frac{1}{1 - e^{2\pi i n z}},$$

and

(23) $$p(n) = \int_{\gamma} f(z) e^{-2\pi i n z} \, dz.$$

Here γ is the segment in the upper half-plane joining $-1/2 + i\delta$ to $1/2 + i\delta$, with $\delta > 0$; the height δ will be fixed later in terms of n.

To proceed further, we look first at where the main contribution to the integral (23) might be, in terms of the relative size of $f(x + iy)$, as $y \to 0$. Notice that f is largest near $z = 0$. This is because $|f(x + iy)| \leq f(iy)$, and moreover $f(iy)$ increases as y decreases, in view of the fact that the coefficients $p(n)$ are positive. Alternatively, we observe that each factor $1 - e^{2\pi i n z}$, appearing in the product for f, vanishes as $z \to 0$, but the same is true for any other point (mod 1) on the real axis. Thus in analogy with the simple example considered above, we seek an elementary function f_1, which has much the same behavior as f at $z = 0$, and try to replace f by f_1 in (23).

It is here that we are very fortunate, because the generating function is just a variant of the Dedekind eta function,

$$\eta(z) = e^{i\pi z/12} \prod_{n=1}^{\infty} (1 - e^{2\pi i n z}).$$

From this, it is obvious that

$$f(z) = e^{\frac{i\pi z}{12}} (\eta(z))^{-1}.$$

(Incidentally, the fraction $1/12$ arising above will explain the occurrence of the fraction $1/24$ in the asymptotic formula for $p(n)$.)

Since η satisfies the functional equation $\eta(-1/z) = \sqrt{z/i}\,\eta(z)$ (see Proposition 1.9 in Chapter 10), it follows that

$$(24) \qquad\qquad f(z) = \sqrt{z/i}\, e^{\frac{i\pi}{12z}} e^{\frac{i\pi z}{12}} f(-1/z).$$

Notice also that if z is appropriately restricted and $z \to 0$, then $\text{Im}(-1/z) \to \infty$, from which it follows that $f(-1/z) \to 1$ rapidly, because

$$(25) \qquad\qquad f(z) = 1 + O(e^{-2\pi y}), \qquad z = x + iy,\ y \geq 1.$$

Thus it is natural to choose $f_1(z) = \sqrt{z/i}\, e^{\frac{i\pi}{12z}} e^{\frac{i\pi z}{12}}$ as the function which approximates well the generating function $f(z)$ (at $z = 0$), and write (because of (24))

$$p(n) = p_1(n) + E(n),$$

with

$$
\begin{cases}
p_1(n) &= \displaystyle\int_\gamma \sqrt{z/i}\, e^{\frac{i\pi}{12z}} e^{\frac{i\pi z}{12}} e^{-2\pi i n z}\, dz, \\[2em]
E(n) &= \displaystyle\int_\gamma \sqrt{z/i}\, e^{\frac{i\pi}{12z}} e^{\frac{i\pi z}{12}} e^{-2\pi i n z} (f(-1/z) - 1)\, dz.
\end{cases}
$$

We first take care of the error term $E(n)$, and in doing so we specify γ by choosing its height in terms of n. In estimating $E(n)$ we replace its integrand by its absolute value and note that if $z \in \gamma$, then

$$(26) \qquad \left| \sqrt{z/i}\, e^{\frac{i\pi}{12z}} e^{\frac{i\pi z}{12}} e^{-2\pi i n z} \right| \leq c e^{2\pi n \delta} e^{\frac{\pi}{12} \frac{\delta}{\delta^2 + x^2}},$$

since $z = x + iy$, and $\text{Re}(i/z) = \delta/(\delta^2 + x^2)$.

On the other hand, we can make two estimates for $f(-1/z) - 1$. The first arises from (25) by replacing z by $-1/z$, and gives

$$(27) \qquad |f(-1/z) - 1| \leq c e^{-2\pi \frac{\delta}{\delta^2 + x^2}} \qquad \text{if } \frac{\delta}{\delta^2 + x^2} \geq 1.$$

For the second, we observe that $|f(z)| \leq f(iy) \leq C e^{\frac{\pi}{12y}}$, when $y \leq 1$, because of the functional equation (24), and hence

$$(28) \qquad |f(-1/z) - 1| \leq O\left(e^{\frac{\pi}{12} \frac{\delta^2 + x^2}{\delta}} \right) = O\left(e^{\frac{\pi}{48\delta}} \right)$$

if $\frac{\delta}{\delta^2 + x^2} \leq 1$, since $|x| \leq 1/2$.

Therefore in the integral defining $E(n)$ we use (26) and (27) when $\frac{\delta}{\delta^2+x^2} \geq 1$, and (26) and (28) when $\frac{\delta}{\delta^2+x^2} \leq 1$. The first leads to a contribution of $O(e^{2\pi n \delta})$, since $2\pi > \pi/12$. The second gives a contribution of $O(e^{2\pi n \delta} e^{\frac{\pi}{48\delta}})$. Hence $E(n) = O(e^{2\pi n \delta} e^{\frac{\pi}{48\delta}})$, and we choose δ so as to minimize the right-hand side, that is, $2\pi n \delta = \frac{\pi}{48\delta}$; this means we take $\delta = \frac{1}{4\sqrt{6}\, n^{1/2}}$, and we get

$$E(n) = O\left(e^{\frac{4\pi}{4\sqrt{6}} n^{1/2}}\right) = O\left(e^{\frac{K}{2} n^{1/2}}\right),$$

which is the desired size of the error term.

We turn to the main term $p_1(n)$. To simplify later calculations we "improve" the contour γ by adding to it two small end-segments; these are the segment joining $-1/2$ to $-1/2 + i\delta$ and that joining $1/2 + i\delta$ to $1/2$. We call this new contour γ' (see Figure 5).

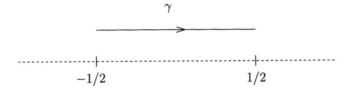

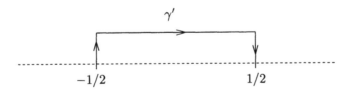

Figure 5. γ and the improved contour γ'

Notice that since $\sqrt{z/i}\, e^{\frac{i\pi}{12z}}$ is $O(1)$ on the two added segments (for the integral defining $p_1(n)$), the modification contributes $O(e^{2\pi n \delta}) = O(e^{\frac{2\pi}{4\sqrt{6}} n^{1/2}}) = O(e^{\frac{K}{4} n^{1/2}})$, which is even smaller than the allowed error, and therefore can be incorporated in $E(n)$. So without introducing further notation we will rewrite $p_1(n)$ replacing the contour γ by γ' in the integration defining p_1, namely

(29) $$p_1(n) = \int_{\gamma'} \sqrt{z/i}\; e^{\frac{i\pi}{12z}} e^{\frac{i\pi z}{12}} e^{-2\pi i n z}\, dz.$$

Next we simplify the triad of exponentials appearing in (29) by making a change of variables $z \mapsto \mu z$ so that their combination takes the form

$$e^{Ai\left(\frac{1}{z} - z\right)}.$$

This can be achieved under the two conditions $A = 2\pi\mu(n - \frac{1}{24})$ and $A = \frac{\pi}{12\mu}$, which means that

$$A = \frac{\pi}{\sqrt{6}}\left(n - \frac{1}{24}\right)^{1/2} \quad \text{and} \quad \mu = \frac{1}{2\sqrt{6}}\left(n - \frac{1}{24}\right)^{-1/2}.$$

Making the indicated change of variables we now have

(30)
$$p_1(n) = \mu^{3/2} \int_\Gamma e^{-sF(z)} \sqrt{z/i} \; dz,$$

with $F(z) = i(z - 1/z)$, $s = \frac{\pi}{\sqrt{6}}\left(n - \frac{1}{24}\right)^{1/2}$. The curve Γ (see Figure 6). is now the union of three segments $[-a_n, -a_n + i\delta']$, $[-a_n + i\delta', a_n + i\delta']$, and $[a_n + i\delta', a_n]$; we can write $\Gamma = \mu^{-1}\gamma'$.

$$\Gamma$$

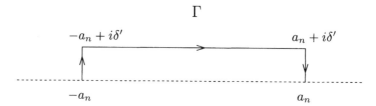

Figure 6. The curve Γ

Here $\quad a_n = \frac{1}{2}\mu^{-1} = \sqrt{6}\left(n - \frac{1}{24}\right)^{1/2} \approx n^{1/2}$, $\quad$ while $\quad \delta' = \delta\mu^{-1} = \frac{2\sqrt{6}}{4\sqrt{6}n^{1/2}}\left(n - \frac{1}{24}\right)^{1/2} \sim 1/2$, as $n \to \infty$.

We apply the method of steepest descent to the integral (30). In doing this, we note that $F(z) = i(z - 1/z)$ has one (complex) critical point $z = i$, in the upper half-plane. Moreover, the two curves passing through i on which F is real are: the imaginary axis, on which F has a maximum at $z = i$, which we reject, and the unit circle, on which F has a minimum at $z = i$. Thus using Cauchy's theorem we replace the integration on Γ by the integration over our final curve Γ^*, which consists of the segment $[-a_n, -1]$, $[1, a_n]$, together with the upper semicircle joining -1 to 1.

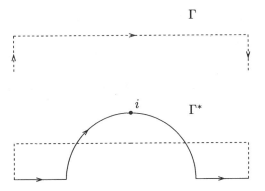

Figure 7. The final curve Γ^*

We therefore have

$$p_1(n) = \mu^{3/2} \int_{\Gamma^*} e^{-sF(z)} \sqrt{z/i}\, dz.$$

The contributions on the segments $[-a_n, -1]$ and $[1, a_n]$ are relatively very small, because on the real axis the exponential has absolute value 1, and hence the integrand is bounded by $\sup_{|z| \le a_n} |z|^{1/2}$, and this leads to two terms which are $O(a_n^{3/2} \mu^{3/2}) = O(1)$.

Finally, we come to the principal part, which is the integration over the semicircle, taken with the orientation on the figure. Here we write $z = e^{i\theta}$, $dz = ie^{i\theta}\, d\theta$. Since $i(z - 1/z) = -2\sin\theta$, this gives a contribution

$$-\mu^{3/2} \int_0^{\pi} e^{2s \sin\theta} e^{i3\theta/2} \sqrt{i}\, d\theta = \mu^{3/2} \int_{-\pi/2}^{\pi/2} e^{2s \cos\theta} (\cos(3\theta/2) + i\sin(3\theta/2)) d\theta.$$

In applying Proposition 2.1, Laplace's method, we take $\Phi(\theta) = -\cos\theta$, $\theta_0 = 0$, so $\Phi(\theta_0) = -1$, $\Phi''(\theta_0) = 1$ and we choose $\psi(\theta) = \cos(3\theta/2) + i\sin(3\theta/2)$, so that $\psi(\theta_0) = 1$. Therefore, the above contributes

$$\mu^{3/2} e^{2s} \frac{\sqrt{2\pi}}{(2s)^{1/2}} \left(1 + O(s^{-1/2})\right).$$

Now since $s = \frac{\pi}{\sqrt{6}} (n - \frac{1}{24})^{1/2}$, $\frac{2\pi}{\sqrt{6}} = \pi\sqrt{\frac{2}{3}} = K$, and $\mu = \frac{\sqrt{6}}{12} (n - \frac{1}{24})^{-1/2}$, we obtain

$$p(n) = \frac{1}{4n\sqrt{3}} e^{K n^{1/2}} \left(1 + O\left(n^{-1/4}\right)\right),$$

and the first conclusion of the theorem is established.

To obtain the more exact conclusion (ii), we retrace our steps and use an additional device, which allows us to evaluate rather precisely the key integral. With $p_1(n)$ defined by (29), which is an integral taken over $\gamma' = \gamma'_n$, we write

$$p_1(n) = \frac{d}{dn} q(n) + e(n),$$

where

$$q(n) = \frac{1}{2\pi} \int_{\gamma'} (z/i)^{-1/2} e^{\frac{i\pi}{12z}} e^{\frac{i\pi z}{12}} e^{-2\pi i n z} \, dz,$$

and $e(n)$ is the term due to the variation of the contour $\gamma' = \gamma'_n$, when forming the derivative in n. By Cauchy's theorem this is easily seen to be dominated by $O(e^{2\pi n \delta})$, which we have seen is $O(e^{\frac{K}{4} n^{1/2}})$, and can be subsumed in the error term. To analyze $q(n)$, we proceed as before, first making the change of variables $z \mapsto \mu z$, and then replacing the resulting contour Γ by Γ^*. As a consequence, we have

(31) $$q(n) = \frac{\mu^{1/2}}{2\pi} \int_{\Gamma^*} e^{-sF(z)} (z/i)^{-1/2} \, dz,$$

with $F(z) = i(z - 1/z)$, $s = \frac{\pi}{\sqrt{6}} (n - \frac{1}{24})^{1/2}$, and $\mu = \frac{1}{2\sqrt{6}} (n - \frac{1}{24})^{-1/2}$.

Now the two segments $[-a_n, -1]$ and $[1, a_n]$ of the contour Γ^* make harmless contributions to $\frac{d}{dn} q(n)$, since F is purely imaginary on the real axis. Indeed, they yield terms which are $O(a_n^{1/2} \mu^{1/2}) = O(1)$.

The main part of (31) is the term arising from the integration on the semicircle. Thus setting $z = e^{i\theta}$, $dz = ie^{i\theta} \, d\theta$, and $i(z - 1/z) = -2\sin\theta$, it equals

$$-\frac{\mu^{1/2}}{2\pi} \int_0^\pi e^{2s\sin\theta} e^{i\theta/2} i^{3/2} \, d\theta = \frac{\mu^{1/2}}{2\pi} \int_{-\pi/2}^{\pi/2} e^{2s\cos\theta} (\cos(\theta/2) + i\sin(\theta/2)) d\theta$$

$$= \frac{\mu^{1/2}}{2\pi} \int_{-\pi/2}^{\pi/2} e^{2s\cos\theta} \cos(\theta/2) d\theta,$$

where we have used the fact that the integral $\int_{-\pi/2}^{\pi/2} e^{2s\cos\theta} \sin(\theta/2) \, d\theta$ vanishes since the integrand is odd.

Now $\cos\theta = 1 - 2(\sin(\theta/2))^2$, so setting $x = \sin(\theta/2)$ we see that the above integral becomes

$$\frac{\mu^{1/2}e^{2s}}{\pi} \int_{-\frac{\sqrt{2}}{2}}^{\frac{\sqrt{2}}{2}} e^{-4sx^2}\, dx.$$

However

$$\int_{-\frac{\sqrt{2}}{2}}^{\frac{\sqrt{2}}{2}} e^{-4sx^2}\, dx = \int_{-\infty}^{\infty} e^{-4sx^2}\, dx + O\left(\int_{\frac{\sqrt{2}}{2}}^{\infty} e^{-4sx^2}\, dx\right)$$

$$= \frac{\sqrt{\pi}}{2s^{1/2}} + O(e^{-2s}),$$

and also

$$\frac{d}{ds}\left(\int_{-\frac{\sqrt{2}}{2}}^{\frac{\sqrt{2}}{2}} e^{-4sx^2}\, dx\right) = \frac{d}{ds}\left(\frac{\sqrt{\pi}}{2s^{1/2}}\right) + O(e^{-2s}).$$

Gathering all the error terms together, we find

$$p(n) = \frac{d}{dn}\left(\mu^{1/2}\frac{e^{2s}}{\pi}\frac{\sqrt{\pi}}{2s^{1/2}}\right) + O(e^{\frac{K}{2}n^{1/2}}).$$

Since $s = \frac{\pi}{\sqrt{6}}(n - \frac{1}{24})^{1/2}$, $\mu = \frac{\sqrt{6}}{12}(n - \frac{1}{24})^{-1/2}$, and $K = \pi\sqrt{\frac{2}{3}}$, this is

$$p(n) = \frac{1}{2\pi\sqrt{2}}\frac{d}{dn}\left(\frac{e^{K(n-\frac{1}{24})^{1/2}}}{(n - \frac{1}{24})^{1/2}}\right) + O(e^{\frac{K}{2}n^{1/2}}),$$

and the theorem is proved.

5 Problems

1. Let η be an indefinitely differentiable function supported in a finite interval, so that $\eta(x) = 1$ for x near 0. Then, if $m > -1$ and $N > 0$,

$$\int_0^\infty e^{-sx}x^m\eta(x)\, dx = s^{-m-1}\Gamma(m+1) + O(s^{-N})$$

for $\operatorname{Re}(s) \geq 0$, $|s| \to \infty$.

(a) Consider first the case $-1 < m \leq 0$. It suffices to see that

$$\int_0^\infty e^{-sx}x^m(1 - \eta(x))\, dx = O(s^{-N}),$$

and this can be done by repeated integration by parts since $e^{-sx} = (-1)^N s^{-N} \left(\frac{d}{dx}\right)^N (e^{-sx})$.

(b) To extend this to all m, find an integer k so that $k - 1 < m \le k$, write

$$\int \left[\left(\frac{d}{dx} \right)^k (x^m) \right] e^{-sx} \eta(x) \, dx = c_{k,m} s^{-m+k-1} + O(s^{-N}),$$

and integrate by parts k times.

2. The following is a more precise version of Stirling's formula. There are real constant $a_1 = 1/12$, $a_2, \ldots, a_n, \ldots$, so that for every $N > 0$

$$\Gamma(s) = e^{s \log s} e^{-s} \frac{\sqrt{2\pi}}{s^{1/2}} \left(1 + \sum_{j=1}^{N} a_j s^{-j} + O(s^{-N}) \right) \qquad \text{when } s \in S_\delta.$$

This can be proved by using the results of Problem 1 in place of Proposition 1.3.

3. The Bessel functions and Airy function have the following power series expansions:

$$J_\nu(x) = \left(\frac{x}{2} \right)^\nu \sum_{m=0}^{\infty} \frac{(-1)^m \left(\frac{x^2}{4} \right)^m}{m! \Gamma(\nu + m + 1)},$$

$$\mathrm{Ai}(-x) = \frac{1}{\pi} \sum_{n=0}^{\infty} \frac{x^n}{n!} \sin(2\pi(n+1)/3) 3^{n/3 - 2/3} \Gamma(n/3 + 1/3).$$

(a) From this, verify that when $x > 0$,

$$\mathrm{Ai}(-x) = \frac{x^{1/2}}{3} \left(J_{1/3} \left(\frac{2}{3} x^{3/2} \right) + J_{-1/3} \left(\frac{2}{3} x^{3/2} \right) \right).$$

(b) The function $\mathrm{Ai}(x)$ extends to an entire function of order $3/2$.

[Hint: For (b), use (a), or alternatively, apply Problem 4 in Chapter 5 to the power series for Ai. Compare also with Problem 1, Chapter 4.]

4. Suppose $F(z) = \sum_{n=0}^{\infty} F_n w^n$ is meromorphic in a region containing the closed unit disc, and the only poles of F are on the unit circle at the points $\alpha_1, \ldots, \alpha_k$, and their orders are $r_1, \ldots, r_k$ respectively. Then for some $\epsilon > 0$

$$F_n = \sum_{j=1}^{k} P_j(n) + O(e^{-\epsilon n}) \qquad \text{as } n \to \infty.$$

Here

$$P_j(n) = \frac{1}{(r_j - 1)!} \left(\frac{d}{dw} \right)^{r_j - 1} \left[(w - \alpha_j)^{r_j} w^{-n-1} F(w) \right]_{w=\alpha_j}.$$

Note that each P_j is of the form $P_j(n) = A_j(\alpha_j^{-1}n)^{r_j-1} + O(n^{r_j-2})$.

To prove this, use the residue formula (Theorem 1.4, Chapter 3).

5.* The one shortcoming in our derivation of the asymptotic formula for $p(n)$ arose from the fact that while $f_1(z) = \sqrt{z/i}\; e^{\frac{i\pi}{12z}} e^{\frac{i\pi z}{12}}$ is a good approximation to the generating function $f(z)$ near $z = 0$, this fails near other points on the real axis, since f_1 is regular there, but f is not.

However, using the transformation law (24) and the identity $f(z+1) = f(z)$, one can derive the following generalization of (24): whenever p/q is a rational number in lowest form (so p and q are relatively prime) then

$$f\left(z - \frac{p}{q}\right) = \omega_{p/q}\sqrt{\frac{zq}{i}}\; e^{\frac{i\pi}{12zq^2}} e^{-\frac{i\pi z}{12}} f\left(-\frac{1}{zq^2} - \frac{p'}{q}\right),$$

where $pp' = 1 \mod q$. Here $\omega_{p/q}$ is an appropriate 24th root of unity. This formula leads to an analogous $f_{p/q}$, approximating f at $z = p/q$.

From this one can obtain for each p/q a contribution of the form

$$c_{p/q} \frac{1}{2\pi\sqrt{2}} \frac{d}{dn} \left(\frac{e^{\frac{K}{q}(n-\frac{1}{24})^{1/2}}}{(n - \frac{1}{24})^{1/2}} \right)$$

to the asymptotic formula for $p(n)$. When suitably modified, the resulting series, summed over all proper fractions p/q in $[0, 1)$, actually converges and gives an exact formula for $p(n)$.

Appendix B: Simple Connectivity and Jordan Curve Theorem

> Jordan was one of the precursors of the theory of functions of a real variable. He introduced in this part of analysis the capital notion of functions of bounded variation. Not less celebrated is his study of curves, universally called Jordan curves, which curves separate the plane in two distinct regions. We also owe him important propositions regarding the measure of sets that have led the way to numerous modern researches.
>
> *E. Picard*, 1922

The notion of simple connectivity is at the source of many basic and fundamental results in complex analysis. To clarify the meaning of this important concept, we have gathered in this appendix some further insights into the properties of simply connected sets. Closely tied to the idea of simple connectivity is the notion of the "interior" of a simple closed curve. The theorem of Jordan states that this interior is well-defined and is simply connected. We prove here the special case of this theorem for curves which are piecewise-smooth.

Recall the definition in Chapter 3, according to which a region Ω is simply connected if any two curves in Ω with the same end-points are homotopic. From this definition we deduced an important version of Cauchy's theorem which states that if Ω is simply connected and $\gamma \subset \Omega$ is any closed curve, then

$$(1) \qquad \int_\gamma f(\zeta)\, d\zeta = 0$$

whenever f is holomorphic in Ω. Here, we shall prove that a converse also holds, therefore:

(I) A region Ω is simply connected if and only if it is holomorphically simply connected; that is, whenever $\gamma \subset \Omega$ is closed and f holomorphic in Ω then (1) holds.

Besides this fundamental equivalence, which is analytic in nature, there are also topological conditions that can be used to describe simple connectivity. In fact, the definition in terms of homotopies suggests that a simply connected set has no "holes." In other words, one cannot find a closed curve in Ω that loops around points that do not belong to Ω. In the first part of this appendix we shall also turn these intuitive statements into tangible theorems:

(II) We show that a bounded region Ω is simply connected if and only if its complement is connected.

(III) We define the winding number of a curve around a point, and prove that Ω is simply connected if and only if no curve in Ω winds around points in the complement of Ω.

In the second part of this appendix we return to the problem of curves and their interior. The main question is the following: given a closed curve Γ that does not intersect itself (it is simple), can we make sense of the "region enclosed by Γ"? In other words, what is the "interior" of Γ? Naturally, we may expect the interior to be open, bounded, simply connected, and have Γ as its boundary. To solve this problem, at least when the curve is piecewise-smooth, we prove a theorem that guarantees the existence of a unique set which satisfies all the desired properties. This is a special case of the Jordan curve theorem, which is valid in the general case when the simple curve is assumed to be merely continuous. In particular, our result leads to a generalization of Cauchy's theorem in Chapter 2 which we formulated for toy contours.

We continue to follow the convention set in Chapter 1 by using the term "curve" synonymously with "piecewise-smooth curve," unless stated otherwise.

1 Equivalent formulations of simple connectivity

We first dispose of (I).

Theorem 1.1 *A region Ω is holomorphically simply connected if and only if Ω is simply connected.*

Proof. One direction is simply the version of Cauchy's theorem in Corollary 5.3, Chapter 3. Conversely, suppose that Ω is holomorphically simply connected. If $\Omega = \mathbb{C}$, then it is clearly simply connected. If Ω is not all of $\mathbb{C}$, recall that the Riemann mapping theorem still applies (see the remark following its proof in Chapter 8), hence Ω is conformally equivalent to the unit disc. Since the unit disc is simply connected, the same must be true of Ω.

Next, we turn to (II) and (III), which, as we mentioned, are both precise formulations of the fact that a simply connected region cannot have "holes."

Theorem 1.2 *If Ω is a bounded region in $\mathbb{C}$, then Ω is simply connected if and only if the complement of Ω is connected.*

Note that we assume that Ω is bounded. If this is not the case, then the theorem as stated does not hold, for example an infinite strip is simply connected yet its complement consists of two components. However, if the complement is taken with respect to the extended complex plane, that is, the Riemann sphere, then the conclusion of the theorem holds regardless of whether Ω is bounded or not.

Proof. We begin with the proof that if Ω^c is connected, then Ω is simply connected. This will be achieved by showing that Ω is holomorphically simply connected. Therefore, let γ be a closed curve in Ω and f a holomorphic function on Ω. Since Ω is bounded, the set[1]

$$K = \{z \in \Omega : d(z, \Omega^c) \geq \epsilon\}$$

is compact, and for sufficiently small ϵ, the set K contains γ. In an attempt to apply Runge's theorem (Theorem 5.7 in Chapter 2), we must first show that the complement K^c of K is connected.

If this is not the case, then K^c is the disjoint union of two non-empty open sets, say $K^c = \mathcal{O}_1 \cup \mathcal{O}_2$. Let

$$F_1 = \mathcal{O}_1 \cap \Omega^c \quad \text{and} \quad F_2 = \mathcal{O}_2 \cap \Omega^c.$$

Clearly, $\Omega^c = F_1 \cup F_2$, so if we can show that F_1 and F_2 are disjoint, closed, and non-empty, then we will conclude that Ω^c is not connected, thus contradicting the hypothesis in the theorem. Since $\mathcal{O}_1$ and $\mathcal{O}_2$ are disjoint, so are F_1 and F_2. To see why F_1 is closed, suppose $\{z_n\}$ is a sequence of points in F_1 that converges to z. Since Ω^c is closed we must have $z \in \Omega^c$, and since Ω^c is at a finite distance from K, we deduce that $z \in \mathcal{O}_1 \cup \mathcal{O}_2$. Now we observe that we cannot have $z \in \mathcal{O}_2$, for otherwise we would have $z_n \in \mathcal{O}_2$ for sufficiently large n because $\mathcal{O}_2$ is open, and this contradicts the fact that $z_n \in F_1$ and $\mathcal{O}_1 \cap \mathcal{O}_2 = \emptyset$. Hence $z \in \mathcal{O}_1$ and F_1 is closed, as desired. Finally, we claim that F_1 is non-empty. If otherwise, $\mathcal{O}_1$ is contained in Ω. Select any point $w \in \mathcal{O}_1$, and since $w \notin K$, there exists $z \in \Omega^c$ with $|w - z| < \epsilon$, and the entire line segment from w to z belongs to K^c. Since $z \in \mathcal{O}_2$ (because $\mathcal{O}_1 \subset \Omega$), some point

---

[1] Here, $d(z, \Omega^c) = \inf\{|z - w| : w \in \Omega^c\}$ denotes the distance from z to Ω^c.

on the line segment $[z, w]$ must belong to neither $\mathcal{O}_1$ nor $\mathcal{O}_2$, and this is a contradiction. More precisely, if we set

$$t^* = \sup\{0 \le t \le 1 : (1 - t)z + tw \in \mathcal{O}_2\},$$

then $0 < t^* < 1$, and the point $(1 - t^*)z + t^*w$, which is not in K, cannot belong to either $\mathcal{O}_1$ or $\mathcal{O}_2$ since these sets are open. Similar arguments imply the same conclusions for F_2, and we have reached the desired contradiction. Thus K^c is connected.

Therefore, Runge's theorem guarantees that f can be approximated uniformly on K, and hence on γ, by polynomials. However, $\int_\gamma P(z)\, dz = 0$ whenever P is a polynomial, so in the limit we conclude that $\int_\gamma f(z)\, dz = 0$, as desired.

The converse result, that Ω^c is connected whenever Ω is bounded and simply connected, will follow from the notion of winding numbers, which we discuss next.

Winding numbers

If γ is a closed curve in $\mathbb{C}$ and z a point not lying on γ, then we may calculate the number of times the curve γ winds around z by looking at the change of argument of the quantity $\zeta - z$ as ζ travels on γ. Every time γ loops around z, the quantity $(1/2\pi) \arg(\zeta - z)$ increases (or decreases) by 1. If we recall that $\log w = \log |w| + i \arg w$, and denote the beginning and ending points of γ by ζ_1 and ζ_2, then we may guess that the quantity

$$\frac{1}{2\pi i} \left[\log(\zeta_1 - z) - \log(\zeta_2 - z) \right], \quad \text{which "equals"} \quad \frac{1}{2\pi i} \int_\gamma \frac{d\zeta}{\zeta - z},$$

computes precisely the number of times γ loops around ζ.

These considerations lead to the following precise definition: the **winding number** of a closed curve γ around a point $z \notin \gamma$ is

$$W_\gamma(z) = \frac{1}{2\pi i} \int_\gamma \frac{d\zeta}{\zeta - z}.$$

Sometimes, $W_\gamma(z)$ is also called the index of z with respect to γ.

For example, if $\gamma(t) = e^{ikt}$, $0 \le t \le 2\pi$, is the unit circle traversed k times in the positive direction (with $k \in \mathbb{N}$), then $W_\gamma(0) = k$. In fact, one has

$$W_\gamma(z) = \begin{cases} k & \text{if } |z| < 1, \\ 0 & \text{if } |z| > 1. \end{cases}$$

Similarly, if $\gamma(t) = e^{-ikt}$, $0 \leq t \leq 2\pi$, is the unit circle traversed k times in the negative direction, then we find that $W_\gamma(z) = -k$ in the interior of the disc, and $W_\gamma(z) = 0$ in its exterior.

Note that, if γ denotes a positively oriented toy contour, then

$$W_\gamma(z) = \begin{cases} 1 & \text{if } z \in \text{interior of } \gamma, \\ 0 & \text{if } z \in \text{exterior of } \gamma. \end{cases}$$

In general we have the following natural facts about winding numbers.

Lemma 1.3 *Let γ be a closed curve in $\mathbb{C}$.*

(i) *If $z \notin \gamma$, then $W_\gamma(z) \in \mathbb{Z}$.*

(ii) *If z and w belong to the same open connected component in the complement of γ, then $W_\gamma(z) = W_\gamma(w)$.*

(iii) *If z belongs to the unbounded connected component in the complement of γ, then $W_\gamma(z) = 0$.*

Proof. To see why (i) is true, suppose that $\gamma : [0, 1] \to \mathbb{C}$ is a parametrization for the curve, and let

$$G(t) = \int_0^t \frac{\gamma'(s)}{\gamma(s) - z} \, ds.$$

Then G is continuous and, except possibly at finitely many points, it is differentiable with $G'(t) = \gamma'(t)/(\gamma(t) - z)$. This implies that, except possibly at finitely many points, the derivative of the continuous function $H(t) = (\gamma(t) - z)e^{-G(t)}$ is zero, and hence H must be constant. Putting $t = 0$ and recalling that γ is closed, so that $\gamma(0) = \gamma(1)$, we find

$$1 = e^{G(0)} = c(\gamma(0) - z) = c(\gamma(1) - z) = e^{G(1)}.$$

Therefore, $G(1)$ is an integral multiple of $2\pi i$, as desired.

For (ii), we simply note that $W_\gamma(z)$ is a continuous function of $z \notin \gamma$ that is integer-valued, so it must be constant in any open connected component in the complement of γ.

Finally, one observes that $\lim_{|z| \to \infty} W_\gamma(z) = 0$, and, combined with (ii), this establishes (iii).

We now show that the notion of a bounded simply connected set Ω may be understood in the following sense: no curve in Ω winds around points in Ω^c.

Theorem 1.4 *A bounded region Ω is simply connected if and only if $W_\gamma(z) = 0$ for any closed curve γ in Ω and any point z not in Ω.*

Proof. If Ω is simply connected and $z \notin \Omega$, then $f(\zeta) = 1/(\zeta - z)$ is holomorphic in Ω, and Cauchy's theorem gives $W_\gamma(z) = 0$.

For the converse, it suffices to prove that the complement of Ω is connected (Theorem 1.2). We argue by contradiction, and construct an explicit closed curve γ in Ω and find a point w so that $W_\gamma(w) \neq 0$.

If we suppose that Ω^c is not connected, then we may write $\Omega^c = F_1 \cup F_2$ where F_1, F_2 are disjoint, closed, and non-empty. Only one of these sets can be unbounded, so that we may assume that F_1 is bounded, thus compact. The curve γ will be constructed as part of the boundary of an appropriate union of squares.

Lemma 1.5 *Let w be any point in F_1. Under the above assumptions, there exists a finite collection of closed squares $\mathcal{Q} = \{Q_1, \ldots, Q_n\}$ that belong to a uniform grid $\mathcal{G}$ of the plane, and are such that:*

(i) *w belongs to the interior of Q_1.*

(ii) *The interiors of Q_j and Q_k are disjoint when $j \neq k$.*

(iii) *F_1 is contained in the interior of $\bigcup_{j=1}^{n} Q_j$.*

(iv) *$\bigcup_{j=1}^{n} Q_j$ is disjoint from F_2.*

(v) *The boundary of $\bigcup_{j=1}^{n} Q_j$ lies entirely in Ω, and consists of a finite union of disjoint simple closed polygonal curves.*

Assuming this lemma for now, we may easily finish the proof of the theorem. The boundary ∂Q_j of each square is equipped with the positive orientation. Since $w \in Q_1$, and $w \notin Q_j$ for all $j > 1$, we have

$$(2) \qquad \sum_{j=1}^{n} \frac{1}{2\pi i} \int_{\partial Q_j} \frac{d\zeta}{\zeta - w} = 1.$$

If $\gamma_1, \ldots, \gamma_M$ denotes the polygonal curves in (v) of the lemma, then, the cancellations arising from integrating over the same side but in opposite directions in (2) yield

$$\sum_{j=1}^{n} \frac{1}{2\pi i} \int_{\gamma_j} \frac{d\zeta}{\zeta - w} = 1,$$

and hence $W_{\gamma_{j_0}}(w) \neq 0$ for some j_0. The closed curve γ_{j_0} lies entirely in Ω, and this gives the desired contradiction.

Proof of the lemma. Since F_2 is closed, the sets F_1 and F_2 are at a finite non-zero distance d from one another. Now consider a uniform grid $\mathcal{G}_0$ of the plane consisting of closed squares of side length which is much smaller than d, say $< d/100$, and such that w lies at the center of a closed square R_1 in this grid. Let $\mathcal{R} = \{R_1, \ldots, R_m\}$ denote the finite collection of all closed squares in the grid that intersect F_1. Then, the collection $\mathcal{R}$ satisfies properties (i) through (iv) of the lemma. To guarantee (v), we argue as follows.

The boundary of each square in $\mathcal{R}$ is given the positive (counterclockwise) orientation. The boundary of $\bigcup_{j=1}^{m} R_j$ is then equal to the union of all boundary sides, that is, those sides that do not belong to two adjacent squares in the collection $\mathcal{R}$. Similarly, the boundary vertices are the end-points of all boundary sides. A boundary vertex is said to be "bad," if it is the end-point of more than two boundary sides. (See point P on Figure 1.)

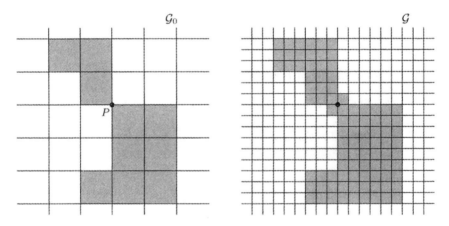

Figure 1. Eliminating bad boundary vertices

To eliminate the bad boundary vertices, we refine the grid $\mathcal{G}_0$ and possibly add some squares. More precisely, consider the grid $\mathcal{G}$ obtained as a refinement of the original grid, by dissecting all squares of $\mathcal{G}_0$ into nine equal subsquares. Then, let $Q_1, \ldots, Q_p$ denote all the squares in the grid $\mathcal{G}$ that are subsquares of squares in the collection $\mathcal{R}$ (so in particular, $p = 9n$), and where Q_1 is chosen so that $w \in Q_1$. Then, we may add finitely many squares from $\mathcal{G}$ near each bad boundary vertex, so that the resulting family $\mathcal{Q} = \{Q_1, \ldots, Q_n\}$ has no bad boundary vertices. (See Figure 1.)

Clearly, $\mathcal{Q}$ still satisfies (i) through (iv), and we claim this collection also satisfies (v). Indeed, let $[a_1, a_2]$ denote any boundary side of $\bigcup_{j=1}^{n} Q_j$ with its orientation from a_1 to a_2. By considering the three different possibilities, one sees that a_2 is the beginning point of another boundary side $[a_2, a_3]$. Continuing in this fashion, we obtain a sequence of boundary sides $[a_1, a_2], [a_2, a_3], \ldots, [a_n, a_{n+1}], \ldots$. Since there are only finitely many sides, we must have $a_n = a_m$ for some n and some $m > n$. We may choose the smallest m so that $a_n = a_m$, say $m = m'$. Then, we note that if $n > 1$, then $a_{m'}$ is an end-point of at least three boundary sides, namely $[a_{n-1}, a_n]$, $[a_n, a_{n+1}]$, and $[a_{m'-1}, a_{m'}]$, hence $a_{m'}$ is a bad boundary vertex. Since we arranged that $\mathcal{Q}$ had no such boundary vertices, we conclude that $n = 1$, and hence the polygon formed by $a_1, \ldots, a_{m'}$ is closed and simple. We may repeat this process and find that $\mathcal{Q}$ satisfies property (v), and the proof of Lemma 1.5 is complete.

Finally, we are now able to finish the proof of Theorem 1.2, namely, if Ω is bounded and simply connected we can conclude that Ω^c is connected. To see this, note that if Ω^c is not connected, then we have constructed a curve $\gamma \subset \Omega$ and found a point $w \notin \Omega$ so that $W_\gamma(w) \neq 0$, thus contradicting the fact that Ω is simply connected.

2 The Jordan curve theorem

Although we emphasize in the statement of the theorems which follow that the curves are piecewise-smooth, we note that the proofs involve the use of curves that may only be continuous, (the curves Γ_ϵ below).

The two main results in this section are the following.

Theorem 2.1 *Let Γ be curve in the plane that is simple and piecewise-smooth. Then, the complement of Γ is an open connected set whose boundary is precisely Γ.*

Theorem 2.2 *Let Γ be a curve in the plane which is simple, closed, and piecewise-smooth. Then, the complement of Γ consists of two disjoint connected open sets. Precisely one of these regions is bounded and simply connected; it is called the **interior** of Γ and denoted by Ω. The other component is unbounded, called the **exterior** of Γ, and denoted by $\mathcal{U}$.*

Moreover, with the appropriate orientation for Γ, we have

$$W_\Gamma(z) = \begin{cases} 1 & \text{if } z \in \Omega, \\ 0 & \text{if } z \in \mathcal{U}. \end{cases}$$

Remark. These two theorems continue to hold in the general case where we drop the assumption that the curves are piecewise-smooth. However, as it turns out, the proofs then are more difficult. Fortunately, the restricted setting of piecewise-smooth curves suffices for many applications.

As a consequence of the above propositions, we may state a version of Cauchy's theorem as follows:

Theorem 2.3 *Suppose f is a function that is holomorphic in the interior Ω of a simple closed curve Γ. Then*

$$\int_\eta f(\zeta)\, d\zeta = 0$$

whenever η is any closed curve contained in Ω.

The idea of the proof of Theorem 2.1 can be roughly summarized as follows. Since the complement of Γ is open, it is sufficient to show it is pathwise connected (Exercise 5, Chapter 1). Let z and w belong to the complement of Γ, and join these two points by a curve. If this curve intersects Γ, we first connect z to z' and w to w', where z' and w' are close to Γ, by curves that do not intersect Γ. Then, we join z' to w' by traveling "parallel" to the curve Γ and going around its end-points if necessary.

Therefore, the key is to construct a family of continuous curves that are "parallel" to Γ. This can be achieved because of the conditions imposed on the curve. Indeed, if γ is a parametrization for a smooth piece of Γ, then γ is continuously differentiable, and $\gamma'(t) \neq 0$. Moreover, the vector $\gamma'(t)$ is tangent to Γ. Consequently, $i\gamma'(t)$ is perpendicular to Γ, and if Γ is simple, considering $\gamma(t) + i\epsilon\gamma'(t)$ amounts to a new curve that is "parallel" to Γ. The details are as follows.

In the next three lemmas and two propositions, we emphasize that Γ_0 denotes a simple *smooth* curve. We recall that an arc-length parametrization γ for a smooth curve Γ_0 satisfies $|\gamma'(t)| = 1$ for all t. Every smooth curve has an arc-length parametrization.

Lemma 2.4 *Let Γ_0 be a simple smooth curve with an arc-length parametrization given by $\gamma : [0, L] \to \mathbb{C}$. For each real number ϵ, let Γ_ϵ be the continuous curve defined by the parametrization*

$$\gamma_\epsilon(t) = \gamma(t) + i\epsilon\gamma'(t), \quad \text{for } 0 \leq t \leq L.$$

Then, there exists $\kappa_1 > 0$ so that $\Gamma_0 \cap \Gamma_\epsilon = \emptyset$ whenever $0 < |\epsilon| < \kappa_1$.

Proof. We first prove the result locally. If s and t belong to $[0, L]$, then

$$\gamma_\epsilon(t) - \gamma(s) = \gamma(t) - \gamma(s) + i\epsilon\gamma'(t)$$

$$= \int_s^t \gamma'(u)\, du + i\epsilon\gamma'(t)$$

$$= \int_s^t [\gamma'(u) - \gamma'(t)]\, du + (t - s + i\epsilon)\gamma'(t).$$

Since γ' is uniformly continuous on $[0, L]$, there exists $\delta > 0$ so that $|\gamma'(x) - \gamma'(y)| < 1/2$ whenever $|x - y| < \delta$. In particular, if $|s - t| < \delta$ we find that

$$|\gamma_\epsilon(t) - \gamma(s)| > |t - s + i\epsilon|\,|\gamma'(t)| - \frac{|t - s|}{2}.$$

Since γ is an arc-length parametrization, we have $|\gamma'(t)| = 1$, and hence

$$|\gamma_\epsilon(t) - \gamma(s)| > |\epsilon|/2,$$

where we have used the simple fact that $2|a + ib| \geq |a| + |b|$ whenever a and b are real. This proves that $\gamma_\epsilon(t) \neq \gamma(s)$ whenever $|t - s| < \delta$ and $\epsilon \neq 0$.

To conclude the proof of the lemma, we argue as follows. (See Figure 2 for an illustration of the argument.)

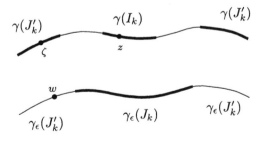

Figure 2. Situation in the proof of Lemma 2.4

Let $0 = t_0 < \cdots < t_n = L$ be a partition of $[0, L]$ with $|t_{k+1} - t_k| < \delta$ for all k, and consider

$$I_k = \{t : |t - t_k| \leq \delta/4\}, \qquad J_k = \{t : |t - t_k| \leq \delta/2\},$$

and

$$J'_k = \{t : |t - t_k| \geq \delta/2\}.$$

Then, we have just proved that

(3) $$\gamma(I_k) \cap \gamma_\epsilon(J_k) = \emptyset \quad \text{whenever } \epsilon \neq 0.$$

Since Γ_0 is simple, the distance d_k between the two compact sets $\gamma(I_k)$ and $\gamma(J'_k)$ is strictly positive. We now claim that

(4) $$\gamma(I_k) \cap \gamma_\epsilon(J'_k) = \emptyset \quad \text{whenever } |\epsilon| < d_k/2.$$

Indeed, if $z \in \gamma(I_k)$ and $w \in \gamma_\epsilon(J'_k)$, then we choose s in J'_k so that $w = \gamma_\epsilon(s)$ and let $\zeta = \gamma(s)$. The triangle inequality then implies

$$|z - w| \geq |z - \zeta| - |\zeta - w| \geq d_k - |\epsilon| \geq d_k/2,$$

and the claim is established. Finally, if we choose $\kappa_1 = \min_k d_k/2$, then (3) and (4) imply that $\Gamma_0 \cap \Gamma_\epsilon = \emptyset$ whenever $0 < |\epsilon| < \kappa_1$, as desired.

The next lemma shows that any point close to an interior point of the curve belongs to one of its parallel translates. By an interior point of the curve, we mean a point of the form $\gamma(t)$ with t in the open interval $(0, L)$. Such a point should not to be confused with an "interior" point of a curve, as in Theorem 2.2.

Lemma 2.5 *Suppose z is a point which does not belong to the smooth curve Γ_0, but that is closer to an interior point of the curve than to either of its end-points. Then z belongs to Γ_ϵ for some $\epsilon \neq 0$.*

More precisely, if $z_0 \in \Gamma_0$ is closest to z and $z_0 = \gamma(t_0)$ for some t_0 in the open interval $(0, L)$, then $z = \gamma(t_0) + i\epsilon\gamma'(t_0)$ for some $\epsilon \neq 0$.

Proof. For t in a neighborhood of t_0 the fact that γ is differentiable guarantees that

$$z - \gamma(t) = z - \gamma(t_0) - \gamma'(t_0)(t - t_0) + o(|t - t_0|).$$

Since $z_0 = \gamma(t_0)$ minimizes the distance from z to Γ_0, we find that

$$|z - z_0|^2 \leq |z - \gamma(t)|^2 \; = \; |z - z_0|^2 - 2(t - t_0)\mathrm{Re}\left([z - \gamma(t_0)]\overline{\gamma'(t_0)}\right) +$$
$$+ \; o\left(|t - t_0|\right).$$

Since $t - t_0$ can take on positive or negative values, we must have $\mathrm{Re}\left([z - \gamma(t_0)]\overline{\gamma'(t_0)}\right) = 0$, otherwise the above inequality can be violated for t close to t_0. As a result, there exists a real number ϵ with $[z - \gamma(t_0)]\overline{\gamma'(t_0)} = i\epsilon$. Since $|\gamma'(t_0)| = 1$ we have $\overline{\gamma'(t_0)} = 1/\gamma'(t_0)$, and therefore $z - \gamma(t_0) = i\epsilon\gamma'(t_0)$. The proof of the lemma is complete.

Suppose that z and w are close to interior points of Γ_0, so that $z \in \Gamma_\epsilon$ and $w \in \Gamma_\eta$ for some non-zero ϵ and η. If ϵ and η have the same sign, we say that the points z and w belong to the same side of Γ_0. Otherwise, z and w are said to be on opposite sides of Γ_0. We stress the fact that we do not attempt to define the "two sides of Γ_0," but only that given two points near Γ_0, we may infer if they are on the "same side" or on "opposite sides". Also, nothing we have done so far shows that these conditions are mutually exclusive.

Roughly speaking, points on the same side can be joined almost directly by a curve "parallel" to Γ_0, while for points on opposite sides, we also need to go around one of the end-points of Γ_0.

We first investigate the situation for points on the same side of Γ_0.

Proposition 2.6 *Let A and B denote the two end-points of a simple smooth curve Γ_0, and suppose that K is a compact set that satisfies either*

$$\Gamma_0 \cap K = \emptyset \quad or \quad \Gamma_0 \cap K = A \cup B.$$

If $z \notin \Gamma_0$ and $w \notin \Gamma_0$ lie on the same side of Γ_0, and are closer to interior points of Γ_0 than they are to K or to the end-points of Γ_0, then z and w can be joined by a continuous curve that lies entirely in the complement of $K \cup \Gamma_0$.

The unspecified compact set K will be chosen appropriately in the proof of the Jordan curve theorem.

Proof. By the previous lemma, consider $z_0 = \gamma(t_0)$ and $w_0 = \gamma(s_0)$ that are interior points of Γ_0 closest to z and w, respectively. Then

$$z = \gamma(t_0) + i\epsilon_0\gamma'(t_0) \quad and \quad w = \gamma(s_0) + i\eta_0\gamma'(s_0),$$

where ϵ_0 and η_0 have the same sign, which we may assume to be positive. We may also assume that $t_0 \leq s_0$.

The hypothesis of the lemma implies that the line segments joining z to z_0 and w to w_0 are entirely contained in the complement of K and Γ_0. Therefore, for all small $\epsilon > 0$, we may join z and w to the points

$$z_\epsilon = \gamma(t_0) + i\epsilon\gamma'(t_0) \quad \text{and} \quad w = \gamma(s_0) + i\epsilon\gamma'(s_0),$$

respectively. See Figure 3.

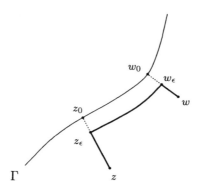

Figure 3. Situation in the proof of Proposition 2.6

Finally, if ϵ is chosen smaller than κ_1 in Lemma 2.4 and also smaller than the distance from K to the part of Γ_0 between z_0 and w_0, that is, $\{\gamma(t) : t_0 \leq t \leq s_0\}$, then the corresponding part of Γ_ϵ, namely $\{\gamma_\epsilon(t) : t_0 \leq t \leq s_0\}$, joins the point z_ϵ to w_ϵ. Moreover, this curve is contained in the complement of K and Γ_0. This proves the proposition.

To join points on opposite sides of Γ_0, we need the following preliminary result, which ensures that there is enough room necessary to travel around the end-points.

Lemma 2.7 *Let Γ_0 be a simple smooth curve. There exists $\kappa_2 > 0$ so that the set N, which consists of points of the form $z = \gamma(L) + \epsilon e^{i\theta}\gamma'(L)$ with $-\pi/2 \leq \theta \leq \pi/2$ and $0 < \epsilon < \kappa_2$, is disjoint from Γ_0.*

Proof. The argument is similar to the one given in the proof of Lemma 2.4. First, we note that

$$\gamma(L) + \epsilon e^{i\theta}\gamma'(L) - \gamma(t) = \int_t^L [\gamma'(u) - \gamma'(L)]\, du + (L - t + \epsilon e^{i\theta})\gamma'(L).$$

If we choose δ so that $|\gamma'(u) - \gamma'(L)| < 1/2$ when $|u - L| < \delta$, then $|t - L| < \delta$ implies

$$|\gamma(L) + \epsilon e^{i\theta}\gamma'(L) - \gamma(t)| \geq |\epsilon|/2.$$

Therefore $\gamma(t) \notin N$ whenever $L - \delta \leq t \leq L$. Finally, it suffices to choose κ_2 smaller than the distance from the end-point $\gamma(L)$ to the rest of the curve $\gamma(t)$ with $0 \leq t \leq L - \delta$, to conclude the proof.

Finally, we may state the result analogous to Proposition 2.6 for points that could lie on opposite sides of Γ_0.

Proposition 2.8 *Let A denote an end-point of the simple smooth curve Γ_0, and suppose that K is a compact set that satisfies either*

$$\Gamma_0 \cap K = \emptyset \quad or \quad \Gamma_0 \cap K = A.$$

If $z \notin \Gamma_0$ and $w \notin \Gamma_0$ are closer to interior points of Γ_0 than they are to K or to the end-points of Γ_0, then z and w can be joined by a continuous curve that lies entirely in the complement of $\Gamma_0 \cup K$.

We only provide an outline of the argument, which is similar to the proof of Proposition 2.6. It suffices to consider the case when z and w lie on opposite sides of Γ_0 and $A = \gamma(0)$. First, we may join

$$z_\epsilon = \gamma(t_0) + i\epsilon\gamma'(t_0) \quad \text{and} \quad w_\epsilon = \gamma(s_0) - i\epsilon\gamma'(s_0)$$

to the points

$$z'_\epsilon = \gamma(L) + i\epsilon\gamma'(L) \quad \text{and} \quad w'_\epsilon = \gamma(L) - i\epsilon\gamma'(L).$$

Then, z'_ϵ and w'_ϵ may be joined within the "half-neighborhood" N of Lemma 2.7. Here, if $t_0 \leq s_0$ we must select $|\epsilon|$ smaller than the distance from $\{\gamma(t) : t_0 \leq t \leq L\}$ to K, and also smaller than κ_1 and κ_2 of Lemmas 2.4 and 2.7.

Proof of Theorem 2.1

Let Γ be a simple piecewise-smooth curve.

First, we prove that the boundary of the set $\mathcal{O} = \Gamma^c$ is precisely Γ. Clearly, $\mathcal{O}$ is an open set whose boundary is contained in Γ. Moreover, any point where Γ is smooth also belongs to the boundary of $\mathcal{O}$ (by Lemma 2.4 for instance). Since the boundary of $\mathcal{O}$ must also be closed, we conclude it is equal to all of Γ.

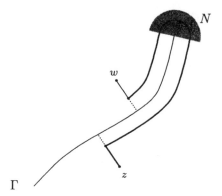

Figure 4. Situation in the proof of Proposition 2.8

The proof that $\mathcal{O}$ is connected is by induction on the number of smooth curves constituting Γ. Suppose first that Γ is simple and smooth, and let Z and W be any two points that do not lie on Γ. Let Λ be any smooth curve in $\mathbb{C}$ that joins Z and W, and which omits the two end-points of Γ. If Λ intersects Γ, it does so at interior points. Therefore, we may join Z by a piece of Λ that does not intersect Γ to a point z that is closer to the interior of Γ than to either of its end-points. Similarly, W can be joined in the complement of Γ to a point w also closer to the interior of Γ than to either of its end-points. Proposition 2.8 (with K empty) then shows that z and w can be joined by a continuous curve in the complement of Γ. Altogether, we may join any two points in the complement of Γ, and this proves the base step of the induction.

Suppose that the theorem is proved for all curves containing $n-1$ smooth curves, and let Γ consist of n smooth curves, so that we may write

$$\Gamma = K \cup \Gamma_0,$$

where K is the union of $n-1$ consecutive smooth curves, and Γ_0 is smooth. In particular, K is compact and intersects Γ_0 in a single one of its end-points. By the induction hypothesis, any two points Z and W in the complement of Γ can be joined by a curve that does not intersect K, and we may also assume that this curve omits both end-points of Γ_0. If this curve intersects Γ_0 in its interior, then we apply Proposition 2.8 to conclude the proof of the theorem.

Proof of Theorem 2.2

Let Γ denote a curve which is simple, closed, and piecewise-smooth. We first prove that the complement of Γ consists of *at most* two components.

Fix a point W that lies outside some large disc that contains Γ, and let $\mathcal{U}$ denote the set of all points that can be joined to W by a continuous curve that lies entirely in the complement of Γ. The set $\mathcal{U}$ is clearly open, and also connected since any two points can be joined by passing first through W. Now we define

$$\Omega = \Gamma^c - \mathcal{U}.$$

We must show that Ω is connected. To this end, let K denote the curve obtained by deleting a smooth piece Γ_0 of Γ. By the Jordan arc theorem, we may join any point $Z \in \Omega$ to W by a curve Λ_Z that does not intersect K. Since $Z \notin \mathcal{U}$, the curve Λ_Z must intersect Γ_0 at one of its interior points. We may therefore choose two points $z, w \in \Lambda_Z$ closer to interior points of Γ_0 than to either of its end-points, and so that the pieces of Λ_Z joining Z to z and W to w are entirely contained in the complement of Γ. Then, the points z and w are on opposite sides of Γ_0, for otherwise, we could apply Proposition 2.6 to find that Z can be joined to W by a curve lying in the complement of Γ, and this contradicts $Z \notin \mathcal{U}$. Finally, if Z_1 is another point in Ω, the two corresponding points z_1 and w_1 must also lie on opposite sides of Γ_0. Moreover, z and z_1 must lie on the same side of Γ_0, for otherwise z and w_1 do, and we can once again join Z to W without crossing Γ, thus contradicting $Z \notin \mathcal{U}$. Therefore, by Proposition 2.6 the points z and z_1 can be joined by a curve in the complement of Γ, and we conclude that Z and Z_1 belong to the same connected component.

The argument thus far proves that Γ^c contains at most two components, but nothing as yet guarantees that Ω is non-empty. To show that Γ^c has precisely two components, it suffices (by Lemma 1.3) to prove that there are points that have different winding numbers with respect to Γ. In fact, we claim that points that are on opposite sides of Γ have winding numbers that differ by 1. To see this, fix a point z_0 on a smooth part of Γ, say $z_0 = \gamma(t_0)$, let $\epsilon > 0$, and define

$$z_\epsilon = \gamma(t_0) + i\epsilon\gamma'(t_0) \quad \text{and} \quad w_\epsilon = \gamma(t_0) - i\epsilon\gamma'(t_0).$$

By our previous observations, points on the same side of Γ belong to the same connected component, and hence

$$\triangle = |W_\Gamma(z_\epsilon) - W_\Gamma(w_\epsilon)|$$

is constant for all small $\epsilon > 0$.

First, we may write

$$\left(\frac{\gamma'(t)}{\gamma(t) - z_\epsilon} - \frac{\gamma'(t)}{\gamma(t) - w_\epsilon} \right) = \frac{2i\epsilon\gamma'(t_0)\gamma'(t)}{[\gamma(t) - \gamma(t_0)]^2 + \epsilon^2\gamma'(t_0)^2}.$$

For the numerator, we use

$$\begin{aligned} \gamma'(t) &= \gamma'(t_0) + [\gamma'(t) - \gamma'(t_0)] \\ &= \gamma'(t_0) + \psi(t), \end{aligned}$$

where $\psi(t) \to 0$ as $t \to t_0$. For the denominator, we recall that $\gamma'(t_0) \neq 0$, so that

$$[\gamma(t) - \gamma(t_0)]^2 + \epsilon^2\gamma'(t_0)^2 = \gamma'(t_0)^2[(t - t_0)^2 + \epsilon^2] + o(|t - t_0|).$$

Putting these results together, we see that

$$\left(\frac{\gamma'(t)}{\gamma(t) - z_\epsilon} - \frac{\gamma'(t)}{\gamma(t) - w_\epsilon} \right) = \frac{2i\epsilon}{(t - t_0)^2 + \epsilon^2} + E(t),$$

where given $\eta > 0$, there exists $\delta > 0$ so that if $|t - t_0| \leq \delta$, the error term satisfies

$$|E(t)| \leq \eta \frac{\epsilon}{(t - t_0)^2 + \epsilon^2}.$$

We then write

$$\Delta = \frac{1}{2\pi i} \int_{|t-t_0| \geq \delta} \left(\frac{\gamma'(t)}{\gamma(t) - z_\epsilon} - \frac{\gamma'(t)}{\gamma(t) - w_\epsilon} \right) dt +$$

$$+ \frac{1}{2\pi i} \int_{|t-t_0| < \delta} \left(\frac{2i\epsilon}{(t - t_0)^2 + \epsilon^2} + E(t) \right) dt.$$

The first integral goes to 0 as $\epsilon \to 0$. In the second integral we make the change of variables $t - t_0 = \epsilon s$, and note that

$$\frac{1}{\pi} \int_{-\rho}^{\rho} \frac{ds}{s^2 + 1} = \frac{1}{\pi} [\arctan s]_{-\rho}^{\rho} \to 1 \qquad \text{as } \rho \to \infty.$$

We therefore see that letting $\epsilon \to 0$ gives

$$|\Delta - 1| < \eta.$$

We conclude that $\triangle = 1$, and hence Γ^c has precisely two components. Finally, only one of these components can be unbounded, namely $\mathcal{U}$, and the winding number of Γ in this component must therefore be zero. By our last result, we see that, after possibly reversing the orientation of the curve, the winding number of any point in the bounded component Ω is constant and equal to 1. Also, it is clear from what has been said that any smooth point on Γ can be approached by points in either component, and hence Γ is the boundary of both Ω and $\mathcal{U}$.

The final step in the proof is to show that the interior of the curve, that is, the bounded component Ω, is simply connected. By Theorem 1.2 it suffices to show that Ω^c is connected. If not, then

$$\Omega^c = F_1 \cup F_2,$$

where F_1 and F_2 are closed, disjoint, and non-empty. Let

$$\mathcal{O}_1 = \mathcal{U} \cap F_1 \quad \text{and} \quad \mathcal{O}_2 = \mathcal{U} \cap F_2.$$

Clearly, $\mathcal{O}_1$ and $\mathcal{O}_2$ are disjoint. If $z \in \mathcal{O}_1$, then $z \in \mathcal{U}$, and every small ball centered at z is contained in $\mathcal{U}$. If every such ball intersects F_2, then $z \in F_2$ since F_2 is closed. However, F_1 and F_2 are disjoint, so this cannot happen. Consequently, $\mathcal{O}_1$ is open, and by the same argument, so is $\mathcal{O}_2$. Finally, we claim that $\mathcal{O}_1$ is non-empty. If not, then F_1 is entirely contained in Γ and $\mathcal{U}$ is contained in F_2. Pick any point $z \in F_1$, which we know belongs to Γ. Now every ball centered at z intersects $\mathcal{U}$, hence F_2. But F_2 is closed and disjoint from F_1, so we get a contradiction. A similar argument for $\mathcal{O}_1$ proves that

$$\mathcal{U} = \mathcal{O}_1 \cup \mathcal{O}_2,$$

where $\mathcal{O}_1, \mathcal{O}_2$ are disjoint, open, and non-empty. This contradicts the fact that $\mathcal{U}$ is connected, and concludes the proof of the Jordan curve theorem for piecewise-smooth curves.

2.1 Proof of a general form of Cauchy's theorem

Theorem 2.9 *If a function f is holomorphic in an open set that contains a simple closed piecewise-smooth curve Γ and its interior, then*

$$\int_\Gamma f = 0.$$

Let $\mathcal{O}$ denote an open set on which f is holomorphic, and which contains Γ and its interior Ω. The idea is to construct a closed curve Λ

in Ω that is so close to Γ that $\int_\Gamma f = \int_\Lambda f$. Then, the integral on the right-hand side is 0, since f is holomorphic in the simply connected open set Ω. We build Λ as follows. Near the smooth parts of Γ, the curve Λ is essentially a curve like Γ_ϵ in Lemma 2.4. Near points where smooth parts of Γ join, we shall use for Λ an arc of a circle. This is illustrated in Figure 5.

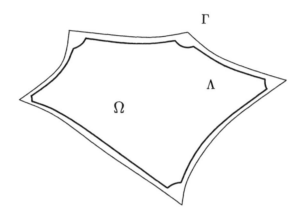

Figure 5. The curve Λ

To find the appropriate connecting arcs, we need the following preliminary result.

Lemma 2.10 *Let* $\gamma : [0,1] \to \mathbb{C}$ *be a simple smooth curve. Then, for all sufficiently small* $\delta > 0$ *the circle* C_δ *centered at* $\gamma(0)$ *and of radius* δ *intersects* γ *in precisely one point.*

Proof. We may assume that $\gamma(0) = 0$. Since $\gamma(0) \neq \gamma(1)$ it is clear that for each small $\delta > 0$, the circle C_δ intersects γ in at least one point. If the conclusion in the lemma is false, we can find a sequence of positive δ_j going to 0, and so that the equation $|\gamma(t)| = \delta_j$ has at least two distinct solutions. The mean value theorem applied to $h(t) = |\gamma(t)|^2$ provides a sequence of positive numbers t_j so that $t_j \to 0$ and $h'(t_j) = 0$. Thus

$$\gamma'(t_j) \cdot \gamma(t_j) = 0 \quad \text{for all } j.$$

However, the curve is smooth, so

$$\gamma(t) = \gamma(0) + \gamma'(0)t + t\varphi(t) \quad \text{and} \quad \gamma'(t) = \gamma'(0) + \psi(t),$$

where $|\varphi(t)| \to 0$ and $|\psi(t)| \to 0$ as t goes to 0. Then recalling that $\gamma(0) = 0$, we find $\gamma'(t) \cdot \gamma(t) = |\gamma'(0)|^2 t + o(|t|)$. The definition of a smooth curve also requires that $\gamma'(0) \neq 0$, so the above gives

$$\gamma'(t) \cdot \gamma(t) \neq 0 \quad \text{for all small } t.$$

This is the desired contradiction.

Returning to the proof of Cauchy's theorem, choose ϵ so small that the open set $\mathcal{U}$ of all points at a distance $< \epsilon$ of Γ is contained in $\mathcal{O}$.

Next, if $P_1, \ldots, P_n$ denote the consecutive points where smooth parts of Γ join, we may pick $\delta < \epsilon/10$ so small that each circle C_j centered at a point P_j and of radius δ intersects Γ in precisely two distinct points (this is possible by the previous lemma). These two points on C_j determine two arcs of circles, only one of which (denoted by C_j) has an interior entirely contained in Ω. To see this, it suffices to recall that if γ is a parametrization of a smooth part of Γ with end-point P_j, then for all small ϵ' the curves parametrized by $\gamma_{\epsilon'}$ and $\gamma_{-\epsilon'}$ of Lemma 2.4 lie on opposite sides of Γ and must intersect the circle C_j. By construction the disc D_j^* centered at P_j and of radius 2δ is also contained in $\mathcal{U}$, hence in $\mathcal{O}$.

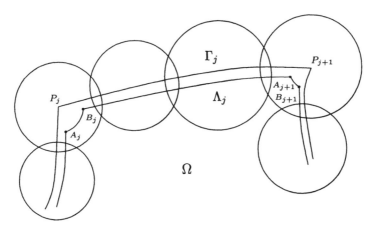

Figure 6. Construction of the curve Λ

We wish to construct Λ so that we may argue as in the proof of Theorem 5.1, Chapter 3 and establish $\int_\Gamma f = \int_\Lambda f$. To do so, we consider a chain of discs $\mathcal{D} = \{D_0, \ldots, D_K\}$ contained in $\mathcal{U}$, and so that Γ is contained in their union, with $D_k \cap D_{k+1} \neq \emptyset$, $D_0 = D_K$, and with the discs

D_j^* part of the chain $\mathcal{D}$. Suppose Γ_j is the smooth part of Γ that joins P_j to P_{j+1}. By Lemma 2.4 it is possible to construct a continuous curve Λ_j that is contained in Ω and in the union of the discs, and which connects a point on B_j on C_j to a point on A_{j+1} on C_{j+1} (see Figure 6). Since we only assumed that Γ has one continuous derivative, Λ_j need not be smooth, but by approximating this continuous curve by polygonal lines if necessary, we may actually assume that Λ_j is also smooth. Then, A_{j+1} is joined to B_{j+1} by a piece of C_{j+1}, and so on. This procedure provides a piecewise-smooth curve Λ that is closed and contained in Ω.

Since f has a primitive on each disc of the family $\mathcal{D}$, we may argue as in the proof of Theorem 5.1, Chapter 3 to find that $\int_\Gamma f = \int_\Lambda f$. Since Ω is simply connected, we have $\int_\Lambda f = 0$, and as a result

$$\int_\Gamma f = 0.$$

Notes and References

Useful references for many of the subjects treated here are Saks and Zygmund [34], Ahlfors [2], and Lang [23].

Introduction

The citation is from Riemann's dissertation [32].

Chapter 1

The citation is a free translation of a passage in Borel's book [6].

Chapter 2

The citation is a translation of an excerpt from Cauchy's memoir [7].

Results related to the natural boundaries of holomorphic functions in the unit disc can be found in Titchmarsh [36].

The construction of the universal functions in Problem 5 are due to G. D. Birkhoff and G.R. MacLane.

Chapter 3

The citation is a translation of a passage in Cauchy's memoir [8].

Problem 1 and other results related to injective holomorphic mappings (univalent functions) can be found in Duren [11].

Also, see Muskhelishvili [25] for more about the Cauchy integral introduced in Problem 5.

Chapter 4

The citation is from Wiener [40].

The argument in Exercise 1 was discovered by D. J. Newman; see [4].

The Paley-Wiener theorems appeared first in [28]; further generalizations can be found in Stein and Weiss [35].

Results related to the Borel transform (Problem 4) can be found in Boas [5].

Chapter 5

The citation is a translation from the German of a passage in a letter from K. Weierstrass to S. Kowalewskaja; see [38].

A classical reference for Nevanlinna theory is the book by R. Nevanlinna himself [27].

Chapter 6

A number of different proofs of the analytic continuation and functional equation for the zeta function can be found in Chapter 2 of Titchmarsch [37].

Chapter 7

The citation is from Hadamard [14]. Riemann's statement concerning the zeroes of the zeta function in the critical strip is a passage taken from his paper [33].

Further material related to the proof of the prime number theorem presented in the text is in Chapter 2 of Ingham [19], and Chapter 3 of Titchmarsch [37].

The "elementary" analysis of the distribution of primes (without using the analytic properties of the zeta function) was initiated by Tchebychev, and culminated in the Erdös-Selberg proof of the prime number theorem. See Chapter XXII in Hardy and Wright [17].

The results in Problems 2 and 3 can be found in Chapter 4 of Ingham [19].

For Problem 4, consult Estermann [13].

Chapter 8

The citation is from Christoffel [9].

A systematic treatment of conformal mappings is Nehari [26].

Some history related to the Riemann mapping theorem, as well as the details in Problem 7, can be found in Remmert [31].

Results related to the boundary behavior of holomorphic functions (Problem 6) are in Chapter XIV of Zygmund [41].

An introduction to the interplay between the Poincaré metric and complex analysis can be found in Ahlfors [1]. For further results on the Schwartz-Pick lemma and hyperbolicity, see Kobayashi [21].

For more on Bieberbach's conjecture, see Chapter 2 in Duren [11] and Chapter 8 in Hayman [18].

Chapter 9

The citation is taken from Poincaré [30].

Problems 2, 3, and 4 are in Saks and Zygmund [34].

Chapter 10

The citation is from Hardy, Chapter IX in [16].

A systematic account of the theory of theta functions and Jacobi's theory of elliptic functions is in Whittaker and Watson [39], Chapters 21 and 22.

Section 2. For more on the partition function, see Chapter XIX in Hardy and Wright [17].

Section 3. The more standard proofs of the theorems about the sum of two and four squares are in Hardy and Wright [17], Chapter XX. The approach we use was developed by Mordell and Hardy [15] to derive exact formulas for the number of representations as the sum of k squares, when $k \geq 5$. The special case $k = 8$ is in Problem 6. For $k \leq 4$ the method as given there breaks down because of the non-absolute convergence of the associated "Eisenstein series." In our presentation we get around this difficulty by using the "forbidden" Eisenstein series. When $k = 2$, an entirely different construction is needed, and the analysis centering around $C(\tau)$ is a further new aspect of this problem.

The theorem on the sum of three squares (Problem 1) is in Part I, Chapter 4 of Landau [22].

Appendix A

The citation is taken from the appendix in Airy's article [3].

For systematic accounts of Laplace's method, stationary phase, and the method of steepest descent, see Erdélyi [12] and Copson [10].

The more refined asymptotics of the partition function can be found in Chapter 8 of Hardy [16].

Appendix B

The citation is taken from Picard's address found in Jordan's collected works [20].

The proof of the Jordan curve theorem for piecewise-smooth curve due to Pederson [29] is an adaptation of the proof for polygonal curves which can be found in Saks and Zygmund [34].

For a proof of the Jordan theorem for continuous curves using notions of algebraic topology, see Munkres [24].

Bibliography

[1] L. V. Ahlfors. *Conformal Invariants*. McGraw-Hill, New York, 1973.

[2] L. V. Ahlfors. *Complex Analysis*. McGraw-Hill, New York, third edition, 1979.

[3] G. B. Airy. On the intensity of light in the neighbourhood of a caustic. *Transactions of the Cambridge Philosophical Society*, 6:379–402, 1838.

[4] J. Bak and D. J. Newman. *Complex Analysis*. Springer-Verlag, New York, second edition, 1997.

[5] R. P. Boas. *Entire Functions*. Academic Press, New York, 1954.

[6] E. Borel. *L'imaginaire et le réel en Mathématiques et en Physique*. Albin Michel, Paris, 1952.

[7] A. L. Cauchy. Mémoires sur les intégrales définies. *Oeuvres complètes d'Augustin Cauchy, Gauthier-Villars, Paris*, Iere Série(I), 1882.

[8] A. L. Cauchy. Sur un nouveau genre de calcul analogue au calcul infinitesimal. *Oeuvres complètes d'Augustin Cauchy, Gauthier-Villars, Paris*, IIeme Série(VI), 1887.

[9] E. B. Christoffel. Ueber die Abbildung einer Einblättrigen, Einfach Zusammenhägenden, Ebenen Fläche auf Einem Kreise. *Nachrichten von der Königl. Gesselschaft der Wissenchaft und der G. A. Universität zu Göttingen*, pages 283–298, 1870.

[10] E.T. Copson. *Asymptotic Expansions*, volume 55 of *Cambridge Tracts in Math. and Math Physics*. Cambridge University Press, 1965.

[11] P. L. Duren. *Univalent Functions*. Springer-Verlag, New York, 1983.

[12] A. Erdélyi. *Asymptotic Expansions*. Dover, New York, 1956.

[13] T. Estermann. *Introduction to Modern Prime Number Theory*. Cambridge University Press, 1952.

[14] J. Hadamard. *The Psychology of Invention in the Mathematical Field*. Princeton University Press, 1945.

[15] G. H. Hardy. On the representation of a number as the sum of any number of squares, and in particular five. *Trans. Amer. Math. Soc*, 21:255–284, 1920.

[16] G. H. Hardy. *Ramanujan*. Cambridge University Press, 1940.

[17] G. H. Hardy and E. M. Wright. *An introduction to the Theory of Numbers*. Oxford University Press, London, fifth edition, 1979.

[18] W. K. Hayman. *Multivalent Functions*. Cambridge University Press, second edition, 1994.

[19] A. E. Ingham. *The Distribution of Prime Numbers*. Cambridge University Press, 1990.

[20] C. Jordan. *Oeuvres de Camille Jordan*, volume IV. Gauthier-Villars, Paris, 1964.

[21] S. Kobayashi. *Hyperbolic Manifolds and Holomorphic Mappings*. M. Dekker, New York, 1970.

[22] E. Landau. *Vorlesungen über Zahlentheorie*, volume 1. S. Hirzel, Leipzig, 1927.

[23] S. Lang. *Complex Analysis*. Springer-Verlag, New York, fourth edition, 1999.

[24] J. R. Munkres. *Elements of Algebraic Topology*. Addison-Wesley, Reading, MA, 1984.

[25] N. I. Muskhelishvili. *Singular Integral Equations*. Noordhott International Publishing, Leyden, 1977.

[26] Z. Nehari. *Comformal Mapping*. McGraw-Hill, New York, 1952.

[27] R. Nevanlinna. *Analytic functions*. Die Grundlehren der mathematischen Wissenschaften in Einzeldarstellung. Springer-Verlag, New York, 1970.

[28] R. Paley and N. Wiener. *Fourier Transforms in the Complex Domain*, volume XIX of *Colloquium publications*. American Mathematical Society, Providence, RI, 1934.

[29] R. N. Pederson. The Jordan curve theorem for piecewise smooth curves. *Amer. Math. Monthly*, 76:605–610, 1969.

[30] H. Poincaré. L'Oeuvre mathématiques de Weierstrass. *Acta Mathematica*, 22, 1899.

[31] R. Remmert. *Classical Topics in Complex Function Theory.* Springer-Verlag, New York, 1998.

[32] B. Riemann. Grundlagen für eine Allgemeine Theorie der Functionen einer Veränderlichen Complexen Grösse. *Inauguraldissertation, Göttingen, 1851,* Collected Works, Springer-Verlag, 1990.

[33] B. Riemann. Ueber die Anzahl der Primzahlen unter einer gegebenen Grösse. *Monat. Preuss. Akad. Wissen., 1859,* Collected Works, Springer-Verlag, 1990.

[34] S. Saks and Z. Zygmund. *Analytic Functions.* Elsevier, PWN-Polish Scientific, third edition, 1971.

[35] E. M. Stein and G. Weiss. *Introduction to Fourier Analysis on Euclidean Spaces.* Princeton University Press, 1971.

[36] E. C. Titchmarsh. *The Theory of Functions.* Oxford University Press, London, second edition, 1939.

[37] E. C. Titchmarsh. *The Theory of the Riemann Zeta-Function.* Oxford University Press, 1951.

[38] K. Weierstrass. *Briefe von Karl Weierstrass an Sofie Kowalewskaja 1871-1891.* Moskva, Nauka, 1973.

[39] E.T. Whittaker and G.N. Watson. *A Course in Modern Analysis.* Cambridge University Press, 1927.

[40] N. Wiener. "R. E. A. C. Paley - in Memoriam". *Bull. Amer. Math. Soc.,* 39:476, 1933.

[41] A. Zygmund. *Trigonometric Series,* volume I and II. Cambridge University Press, second edition, 1959. Reprinted 1993.

Symbol Glossary

The page numbers on the right indicate the first time the symbol or notation is defined or used. As usual, $\mathbb{Z}$, $\mathbb{Q}$, $\mathbb{R}$, and $\mathbb{C}$ denote the integers, the rationals, the reals, and the complex numbers respectively.

$\Gamma(s)$	Gamma function	160
$\zeta(s)$	Riemann zeta function	168
$\vartheta,\ \Theta(z\vert\tau),\ \theta(\tau)$	Theta function	169, 284, 284
$\xi(s)$	Xi function	169
J_ν	Bessel functions	176
B_m	Bernoulli number	179
$\pi(x)$	Number of primes $\leq x$	182
$f(x) \approx g(x)$	Asymptotic relation	182
$\psi(x),\ \Lambda(n),\ \psi_1(x)$	Functions of Tchebychev	188, 189, 190
$d(n)$	Number of divisors of n	200
$\sigma_a(n)$	Sum of the ath powers of divisors of n	200
$\mu(n)$	Möbius function	200
$\text{Li}(x)$	Approximation to $\pi(x)$	202
$\mathbb{H}$	Upper half-plane	208
$\text{Aut}(\Omega)$	Automorphism group of Ω	219
$\text{SL}_2(\mathbb{R})$	Special linear group	222
$\text{PSL}_2(\mathbb{R})$	Projective special linear group	223
$\text{SU}(1,1)$	Group of fractional linear transformations	257
$\Lambda,\ \Lambda^*$	Lattice and lattice minus the origin	262, 267
$\wp$	Weierstrass elliptic function	269
$E_k(\tau),\ E_2^*(\tau)$	Eisenstein series	273, 305
$F(\tau),\ \tilde{F}(\tau)$	Forbidden Eisenstein series and its reverse	278, 305
$\Pi(z\vert\tau)$	Triple product	286
$\eta(\tau)$	Dedekind eta function	292
$p(n)$	Partition function	293
$r_2(n)$	Number of ways n is a sum of two squares	296
$r_4(n)$	Number of ways n is a sum of four squares	297
$d_1(n),\ d_3(n),\ \sigma_1^*(n)$	Divisor functions	297, 304
$\text{Ai}(s)$	Airy function	328
$W_\gamma(z)$	Winding number	347

Index

Relevant items that also arise in Book I are listed in this index, preceeded by the numeral I.